AUCTORES BRITANNICI MEDII AEVI · II
New edition [v.22]

JOHN BLUND

TREATISE ON THE SOUL

Edited by

D. A. CALLUS & R. W. HUNT

With a new Introduction and English translation by

MICHAEL W. DUNNE

Published for THE BRITISH ACADEMY
by OXFORD UNIVERSITY PRESS

Oxford University Press, Great Clarendon Street, Oxford OX2 6DP

© The British Academy 2013

Database right The British Academy (maker)

First edition published in 2013

British Library Cataloguing in Publication Data

Data available

Library of Congress Cataloging in Publication Data

Data available

*Typeset by John Waś, Oxford
Printed in Great Britain by
TJ International Ltd,
Padstow, Cornwall*

ISBN 978–0–19–726514–7

CONTENTS

PREFACE

THE idea of making a translation of Blund's text was first suggested to me by my colleague Professor James McEvoy, who knew of the importance of the work from his own studies on Oxford's first chancellor, Robert Grosseteste. The project was warmly welcomed by Professor David Luscombe, then the General Editor of the Academy's Medieval Texts Editorial Committee, who has remained a constant help and supporter of my work in the years it has taken for the project to come to fruition. My thanks are also due to the support and encouragement given by his successors, Professor Richard Sharpe and Professor John Marenbon. In the preparation of the volume I was assisted by Ann Gleeson, who typed the manuscript, by Michael Regan, who helped in preparing a summary of the text, and by Denise Ryan, who compiled a list of translated Latin terms. Conleth Loonan assisted in the preparation of the appendix. Much of the work of translation was done in the tranquil setting of the lovely town of Auzances (Creuse) in central France over a number of summer holidays, so my thanks for their patience are due to my wife Mary Rose Walker (who also compiled the subject index) and to our four children, Kate, Killian, Erika, and Carla.

MICHAEL W. DUNNE
National University of Ireland, Maynooth

PREFACE TO THE
LATIN EDITION OF 1970

FATHER DANIEL CALLUS was the first modern scholar to call attention to John Blund's *Tractatus de anima*. He gave a brief preliminary account of the work and printed the list of chapters in his paper on 'The Introduction of Aristotelian Learning to Oxford', which was published in the *Proceedings of the British Academy*, XXIX (1943), 241–52, 280–1. He prepared an edition of the text from the Cambridge manuscript which was then the only one known, and he did much work on the investigation of sources. In 1955 he contributed a further paper on John Blund to the volume in honour of Mgr A. Mansion (*Autour d'Aristote*, Louvain, pp. 471–95). In an addendum to this paper he announced the discovery of a second manuscript in the Vatican Library. Shortly afterwards Père B.-G. Guyot O.P. discovered a third manuscript in Prague. The discovery of these two new manuscripts was very welcome, but it made necessary a reconstitution of the text. At the time of his death in 1965 Fr. Callus had completed the text and *apparatus criticus*, but the *apparatus fontium* was in its final form only down to no. 337. It was the wish of Fr. Callus that I should complete the edition. The introduction is my work: the notes which Fr. Callus left were not sufficiently full for publication. I am also responsible for the indexes.

I found no note of the acknowledgements which Fr. Callus wished to make, and I must ask the forgiveness of those whose help is not acknowledged. I know that he would have wished to thank Mlle Marthe Dulong and Mr. Walter Mitchell for their help. Acknowledgement of the help of Dr. L. Minio-Paluello is found in the *apparatus*, and I have to thank him on my own behalf. Mrs. S. P. Hall helped Fr. Callus in various ways, and I have to thank her for indispensable help in checking references and in striving for consistency in the citations in the *apparatus fontium*. Dom G. D. Schlogel helped me with the unfinished part of the *apparatus fontium*. Mlle M.-T. d'Alverny, Fr. Leonard Boyle O.P., Dom David Knowles, Monsieur l'Abbé P. Michaud-Quantin, Miss B. Smalley, and Professor R. W. Southern, all friends of Fr. Callus, have been good enough to answer my questions. Mlle d'Alverny supplied me with the reference to the 'dator formarum' in Avicenna (see no. 344). Fr. Callus was in the habit of

viiiPREFACE TO THE LATIN EDITION OF 1970

consulting Dr. A. B. Emden on questions relating to the early history of Oxford, and he has kindly given me the benefit of his advice on the career of John Blund. My considerable debt to Dr. Brian Lawn is acknowledged in the introduction. Miss D. W. Pearson, the assistant secretary of the Academy, has been an unfailing source of strength. Finally I should like to thank the Printer and his learned readers for the skill with which they have handled a difficult task.

Fr. Daniel Callus was a great and much-loved man, and those who did not know him may like to be referred to the notices of him by Professor R. W. Southern in *Oxford Studies Presented to Daniel Callus* (Oxford Historical Society, N.S. XVI, 1964), by M. D. Knowles in *New Blackfriars* (1965), 609–12, and by A. Vella O.P. in the *Journal of the Faculty of Arts* of the University of Malta III (1965), 66–72.

R. W. HUNT

INTRODUCTION TO
THE NEW EDITION

SINCE the publication of the edition of John Blund's *Tractatus de anima* by the British Academy in 1970[1] there has been widespread acceptance of the importance of this text for the history of thought. As Dag Hasse has noted: 'John Blund is the first master of arts we know of who wrote a treatise on the soul.'[2] Moreover, Blund was probably one of the first commentators on the *libri naturales* at Paris before the prohibition of 1210, and later introduced them to Oxford.[3] As James McEvoy has remarked: 'Blund is a significant figure in the development of philosophy in England, for his is the first writing we have from a master who was teaching the liberal arts . . . at Oxford.'[4]

Indeed, apart from the prohibitions of 1210 and 1215, the *De anima* of Blund is the one text which sheds light on the first reception of Aristotle at Paris.[5] The text was probably composed at Paris, before 1204 since Alexander Nequam drew upon it in his *Speculum speculationum* (*c*.1213). When Nequam quotes Aristotle's definition of the soul as 'the soul is the perfection of an organized body potentially having life', he did so in the form found in the *Tractatus de anima* (ch. II. i, 14), and Nequam's source for his teaching on sight and vision in the *De naturis rerum* was also Blund's text.[6]

John Blund (*c*.1175–1248) taught arts at Paris *c*.1200–5, then at Oxford around 1207–9. He returned to study theology at Paris during the interdict (1208–14) and the contemporaneous suspension

[1] Iohannes Blund, *Tractatus de anima*, edd. D. A. Callus O.P. and R. W. Hunt (Auctores Britannici Medii Aevi II), London: published for the British Academy by the Oxford University Press, 1970.

[2] D. N. Hasse, *Avicenna's* De anima *in the West: The Formation of a Peripatetic Philosophy of the Soul 1160–1300* (Warburg Institute Studies and Texts 1), London and Turin, 2000, p. 18.

[3] For details of Blund's life see C. H. Lawrence, 'Blund, John (*c*.1175–1248)', in *The Oxford Dictionary of National Biography*, Oxford, 2004.

[4] J. McEvoy, 'Liberal Arts, Science, Philosophy, Theology and Wisdom at Oxford, 1200–1250', in Jan A. Aertsen and Andreas Speer (eds.), *Was ist Philosophie im Mittelalter? Akten des X. Internationalen Kongresses für mittelalterliche Philosophie der Société Internationale pour l'Étude de la Philosophie Médiévale, 25. bis 30. August 1997 in Erfurt* (Miscellanea Mediaevalia, 26), Berlin and New York, 1998, pp. 560–70 at p. 563.

[5] See F. Van Steenberghen, *La Philosophie au XIIIe siècle*, 2nd rev. edn., Paris, 1991, p. 90: 'Quant aux premières années du siècle, à part le *De Anima* de Jean Blund, seules les interdictions de 1210 et de 1215 jettent un certain jour sur la période immédiatement antérieure, qui a été témoin de la première pénétration de l'aristotélisme à Paris.'

[6] McEvoy, p. 563.

of the schools at Oxford (1209–14).[7] He was regent in theology at Paris for twelve years, and taught theology at Oxford after 1229. The *Tractatus* is the only known writing of John to have survived, and even it is ascribed to him in only one of the three extant manuscripts of the work.

With the *Tractatus* a whole area of philosophical speculation, namely Greek and Arabic psychology, arrived at Oxford, where it would continue to grow and be debated throughout the century. Although the work of a master of arts, Blund's text set the agenda for psychology, something which would be fully taken up by theologians within a generation.[8] Blund did not know Averroës but he reflects the state of Latin Aristotelianism during the first third of the thirteenth century. Like his contemporaries, he regarded the *De anima* of Avicenna as a commentary on Aristotle: indeed, they found it clearer than the text of Aristotle and were guided by it. Blund is faithful to Aristotle and to Avicenna, rejecting, for example, the *binarium famosissimum* drawn from the *Fons vitae* of Avicebron.[9] In expounding the doctrine of Aristotle and following the plan laid out by Avicenna, he considers the arguments for and against before offering his own reasoned position in the *solutio*. He defends the role of the philosopher as considering the nature of the soul and as distinct from theological considerations.

As Hasse has observed,[10] few have taken Avicenna seriously as a philosopher because in the history of medieval Western philosophy he loses out in comparison with Aristotle and Averroës. Certainly, the time in which his influence dominates is merely a few decades, as against the centuries of influence of Aristotle and Averroës. There is therefore the danger of misrepresenting not only the work of Avicenna as a commentator but also that of Blund as an Aristotelian *manqué*. Again, as Hasse argues,[11] Blund was not trying to use Avicenna as a guide to understanding Aristotle—his aim was to understand what the soul was, and 'He had two new sources at his disposal, Aristotle's *Peri psychēs* and Avicenna's *De anima*, and of these he very much preferred the latter.'

---

[7] Ibid.

[8] Ibid.

[9] As Van Steenberghen comments (*La Philosophie au XIIIᵉ siècle*, p. 160 n. 142), the fact that Blund rejected the theories of Avicebron, namely universal hylomorphism and the plurality of forms, shows that they were not as 'traditional' as Kilwardby and Peckham later insisted.

[10] Hasse, *Avicenna's* De anima, preface, p. v.

[11] Ibid., p. 23.

Blund's sources[12]

Blund, it seems, regarded the *De anima* of Avicenna as a 'commentary' on the homonymous treatise of Aristotle.[13] He had access to both but, like Gundissalinus before him, made a conscious choice in favour of Avicenna over the relative obscurity of Aristotle's text. It might also have been the case that the text of Avicenna fitted in better with what his expectations of a 'Peripatetic' commentary might have been. There was the 'national' tradition in scientific enquiry going back to Adelard of Bath, and he would have been aware of the new 'medical' texts, translated, for instance, by Constantine the African.[14] The *De anima* was written as part of the *Kitāb aš-šifā* (*Book of Healing*, *Sufficientia* in Latin) some time between 1021 and 1024 at Hamadān. In his introduction Avicenna makes it clear that he had neither the time nor the inclination for close textual analysis, which is what later came to be expected from a 'commentary' in the Latin West. Instead he wrote a work which deals comprehensively with the topics that one would discuss when writing within the Peripatetic tradition.

Therefore, it might be suggested that when he came to write a treatise on the topic of the soul, Blund was primarily motivated by the aim of giving a clear answer to the kinds of topic that had become part of the canon of psychology. Thus, Blund also refers to al-Ġhazālī, Ps. Alexander of Aphrodisias, and to many of the physical and logical works of Aristotle. Among Latin authors he cites above all Augustine, Boethius, and Anselm, but also the more recent *Sentences* of Peter Lombard. This, again, might explain why he adds a section dealing with free will, since it would have been judged to be a major omission by Avicenna of a central topic as the discussion had developed in the Christian West. Indeed, it would seem that Blund's framework

---

[12] The best treatment of Blund's sources is to be found in John Blund, *Traktat über die Seele*, ed. and trans. Dorothée Werner, Freiburg i.Br., 2005, pp. 17–31, 'Zu Blunds Quellen'. Indeed, the most extended and comprehensive treatment of Blund's text is that given by Werner in her introduction, pp. 9–93.

[13] As Hasse notes (*Avicenna's De anima*, p. 20), Blund's use of the term *commentator* for Avicenna and *commentarium* for the text is quite exceptional, since the usual term was the *Liber de anima* or the *Liber sextus de naturalibus*. This might have been his own idiosyncrasy, since he also refers to the *Metaphysica* of al-Ġhazālī as a *commentum primae philosophiae*.

[14] As Werner comments (p. 18), 'Das 12. Jahrhundert war geprägt von einem Aufschwung in der Kosmologie, ermöglicht vor allem durch naturkundliches und medizinisches Wissen, vermittelt durch Werke wie die *Pantegni* des Constantinus Africanus, die *Naturales quaestiones* Senecas, den Traktat *De natura hominis* des Nemesius von Emesa und nicht zuletzt Platons *Timaios* in der Übersetzung und Kommentierung des Chalcidius. Die Psychologie spielte dabei allerdings eine eher untergeordnete Rolle. Ihre dominante Einscheinungsform war die Einbindung in umfassendere Sammlungen von *Quaestiones naturales*, wie zum Beispiel diejenigen des Adelard von Bath, die *Quaestiones Salernitanae* oder Wilhelms von Conches zwei Schriften *Philosophia mundi* und *Dragmaticon*.'

is not Aristotelian or indeed Avicennian, but that through integrat-
ing these with the Christian tradition, he produces something new.[15]
We should also avoid considering Blund only from the perspective of
what would be achieved later on in the century. If we look at what
came before him, then what he succeeded in doing with the material
at his disposal can be seen to be ground-breaking.

The text of Avicenna which Blund uses is that which was translated
between 1152 and 1166 by Avendauth Israelita and Dominicus Gundissalinus. There were two versions of this translation, identified by
S. Van Riet as versions A and B. Van Riet has shown that someone
in the late twelfth or early thirteenth century reworked the translation from the original Arabic, especially at I. 5. However, it was the
possibly earlier version A that was used by Blund. As Hasse points
out,[16] version A was used by Gundissalinus in his *Liber de anima*
(c.1170), by Blund around 1200, and then by Jean de la Rochelle in
his *Tractatus de anima* (c.1233–5). However, version B was used by
Jean de la Rochelle in his *Summa de anima* (c.1235–6) and by Albert
the Great in his *Summa de homine* (1242–3). Thus, B seems to have
been a later version which entered into circulation some time in the
mid-1230s.

In any case, Avicenna's work proved to be very popular, and since
it was somewhat consonant with the already existing Western speculation on the soul, rather than being seen as running counter to it
(as Aristotle and Averroës came to be seen), it gave rise to increased
reflection on and renewal of traditional sources. Whatever the criti-
cisms of Latin Avicennism (or Avicennized Augustinianism), it aimed
at reconciliation rather than rupture. Again, Christian treatises on the
soul had a long tradition and had shown themselves capable of adapt-
ing to new ideas. The new medical theories coming from the School
of Salerno regarding the four elements and humours, the theory of
the three spirits (vital, animal, and spiritual) in the body and their
respective organs (liver, heart, and brain), and the explanation of the
functions of the external senses had, in a sense, prepared the way for
the reception of a work such as that of Avicenna. From a religious
point of view, of course, much of this medical and philosophical in-
formation is neutral; the crisis when it comes will centre on the per-

---

[15] Werner, p. 27: 'so scheint Blund in einem durch Aristoteles legitimierten avicenni-
schen Rahmen Material nicht nur aus Avicennas und Aristoteles' Schriften, sondern auch
aus denen der klassischen Autoren der christlichen Tradition und der frühmittelalterliche-
europäischen naturwissenschaftlichen Forschung zu integrieren, obwohl weder Aristoteles
noch die frühmittelalterliche naturwissenschaftliche Abweichungen von Avicenna verant-
wortlich gemacht werden kann.'   [16] Hasse, *Avicenna's* De anima, p. 8.

sonal nature of the human intellect, and the associated issues of free will, personal immortality, reward and punishment. Here again, in trying to understand the nature of the intellect from the sources at his disposal, Blund's achievement is quite considerable. In particular, on the relationship between the *intelligentia* and the *intellectus*, Blund had to make sense of five different new terminologies—those of Aristotle, Alexander of Aphrodisias, al-Fārābī, Avicenna, and al-Ġhazālī—and then to synthesize them. Again, whereas he had the notions of the agent and potential intellect from Aristotle, he had to integrate Avicenna's four intellects[17] with this, and in view of later developments, he does quite a good job.

Nature and readership of the text

According to Hasse,[18] 'Blund's work is an early example of a fully fledged *quaestio* scheme in psychological literature.' To be sure, this is confirmed by the structure of the text itself, since it presents a question to be answered, arguments for and against, the authoritative determination by the master (*dicendum quod*), and then the replies to the objections. Olga Weijers, however, voiced some doubts as to whether Blund's text was written on the basis of his teaching at Paris.[19] While conceding that it does have many of the characteristics of *quaestiones*, she holds that it is still far from being a commentary, or consisting of disputed questions. I think it will readily be agreed that Blund's text is not directly a commentary *per modum sententiarum et quaestionum* but is a treatise on a topic using the form of the *disputatio*. In this respect one could compare it to the *Summa theologiae* of Aquinas, where the form of the *disputatio* is followed without its being the record of actual teaching. Incidentally, it was the question form that would emerge triumphant in the fourteenth century, when 'commentaries' are only loosely bound to a text.[20] The structure of Blund's treatise, however, is somewhat uneven and may never have been finalized by the author. As Weijers notes, the text is essentially made up of questions, some of which have features of disputed ques-

[17] See D. N. Hasse, 'Das Lehrstück von den vier Intellekten in der Scholastik: Von den arabischen Quellen bis zu Albertus Magnus', *Recherches de théologie et philosophie médié-vales*, LXVI (1999), 21–77.   [18] Hasse, *Avicenna's De anima*, p. 20.

[19] O. Weijers, *La 'disputatio' à la Faculté des arts de Paris (1200–1350 environ): esquisse d'une typologie* (Studia artistarum 2), Turnhout, 1995, p. 40 n. 38.

[20] I have already noted this progression in 13th- and 14th-cent. commentaries on the *De longitudine et brevitate vitae*: see M. Dunne, 'Thirteenth- and Fourteenth-Century Commentaries on Aristotle's *De longitudine et brevitate vitae*', *Early Science and Medicine*, VIII (2003), 320–35.

tions, though this is by no means true in all cases.[21] Still, there seems
to be clear evidence in some cases at least that the text did begin as part
of a *disputatio*, especially in ch. XXII, no. 312: 'Moreover, the sensitive
soul is a corporeal substance, as the Respondent says, and is extended
throughout the human body.'[22] It may be the case that the text as it
has come down to us began as a record of actual teaching in the form
of disputed questions, that this text was then incompletely edited or
partially restructured either by the author or by someone else, with
a view to publication. In any case, nothing in the above observations
detracts from the importance of the text for the history of both the
University of Paris and the University of Oxford.

If the *Treatise* originated in the classroom, then the textbook was
the *De anima* of Avicenna and not that of Aristotle. Indeed, Blund
was mostly interested (see *Treatise*, chs. 5–20) in the arrangement of
the faculties of the soul as given by Avicenna in his *De anima*, part I,
ch. 5 (something which is to be found in later authors).[23] However—
and this may have been incidental to the aims of the author—Blund's
Treatise marks the entry of Aristotle's psychological theories into the
Latin West and into the universities. As Henry of Avranches's poem
of 1232 noted, Blund was 'the first man to investigate deeply the books
of Aristotle, when the Arabs had recently handed them over to the
Latins, and the man who lectured on Aristotle first and with the most
renown in both Oxford and Paris'.[24]

Summary of the text

Broadly speaking, Blund follows the plan of Avicenna but adds a
chapter on free will (necessary, it would seem, to counteract the de-

[21] Weijers, p. 40 n. 38: 'Ce traité est composé essentiellement de questions, et certaines
d'entre elles sont semblables aux questions disputées. Mais c'est loin d'être le cas de toutes
les questions: parfois l'auteur cite simplement plusieurs opinions, avec des arguments à
l'appui, et puis sa propre opinion (*solutio*), suivie de quelques autres questions (voir par ex.
la question IV); dans d'autres cas, il cite des opinions accompagnées d'objections pour les
refuter, suivies de sa propre opinion (par ex. qu. III); quelquefois aussi la question consiste
en une objection et la réponse (cf. par ex. la qu. II. i). Bref, ce texte paraît dans l'ensemble
assez différent des commentaires sous forme de questions disputées.'

[22] Already noted in D. A. Callus, 'The Treatise of John Blund "On the Soul"', in *Re-
cueil d'études de philosophie ancienne et médiévale offert à Msgr. A. Mansion*, Leuven, 1955,
pp. 471–95 at p. 483.

[23] Although there is at present no English version of Avicenna's *De anima*, there is a
translation of the *Kitāb al-najāt*, II, c. vi, in F. Rahman, *Avicenna's Psychology*, Oxford,
1952, where much of the same material is treated.

[24] Quoted in Callus, 'The Treatise of John Blund', p. 471. Translated by C. Burnett,
'The Introduction of Aristotle's Natural Philosophy in Great Britain: A Preliminary Sur-
vey of the Manuscript Evidence', in *Aristotle in Britain in the Middle Ages*, ed. J. Marenbon,
Turnhout, 1996, pp. 21–50 at p. 37.

terministic trend of Arabic thought). D. A. Callus has already shown how the plan of the *Treatise* (apart from the final chapter, where the influence is the *De libero arbitrio* of Anselm) follows the *De anima* of Avicenna by placing the chapter headings in parallel columns.[25]

The present descriptive arrangement of the edited text is based upon chapter headings given in two of the manuscripts.[26] A more analytical approach is found in the layout proposed by Werner[27] and her description of the contents based upon the *quaestiones* and objections.

1. *Introduction (chapters I–IV)*

Blund begins with an enquiry into the existence of the soul (*anima*) and associated questions regarding the distinction between voluntary movement and natural movement. Following Aristotle, he espouses the principle that all grounded and justified knowledge is derived from sense data. Voluntary movement either pertains to the nature of the body in itself, or not; if it were from the nature of the body, then every body would be moved voluntarily (which does not happen), so voluntary movement arises from soul. Taking the example of the movement of the heavenly bodies (and that of the sun in particular), Blund examines (3–13) whether their movement is natural or voluntary. Again, he proposes that nothing in itself, consistent with its nature, desires and flees from one and the same thing. Applying this principle to circular motion, Blund concludes that a body such as the sun would not naturally return to a point from which it had moved unless a will is involved. Therefore, the circularity of solar motion emanates from soul, as does the orbital motion of other heavenly bodies. Ultimately, however, all movement is due to the First Intelligence, which contemplates the First Cause as the Highest Good and, moving because of desire, communicates motion to the rest of the universe.

Blund now moves on to examine the essence of the soul (14–16), quoting the famous definition of Aristotle that it is what completes an already organized body which has the potential for life. An interesting demarcation of subject areas now follows (17–22) when Blund

[25] Callus, 'The Treatise of John Blund', pp. 484–6.

[26] The status of the chapter headings is problematic. As Hunt pointed out (Introduction, p. xl below), the list of chapters at the beginning may have been a plan for the work which was not revised after the work was finished. This would explain why some of the chapter headings (e.g. xxv.iii and xxv.iv) do not correspond to the contents of the text. Again, the last chapter (xxvii), on divine providence, is not announced in the list of chapters. This might suggest that the text was left in an unfinished state by the author.

[27] See Werner, pp. 12–17, 'Textrekonstruktion'.

asks whether the topic under consideration, namely the soul, should be examined by the natural philosopher or the metaphysician. It is acknowledged that both study the soul but from different aspects and considerations. Blund, however, seems particularly keen to restrict the scope of the theologian in this area. The theologian is not to investigate what the soul is or how it comes to be in the body, but only how the soul gains or loses merit and what leads to its salvation.

Various opinions concerning the soul are next surveyed by Blund (23–34): some have claimed that the soul is blood, nature, or air. It is interesting to note that many of the arguments employed by Blund to dismiss these theories are drawn from experience or natural philosophy. He concludes, however, that the soul is both a substance and incorruptible.

None the less, as Blund expounds (35–45), each soul is not devoid of corruption, such as the vegetative soul (found in trees) and the sensitive soul (located in brute animals). The case of the rational soul is different. How, then, is the word 'soul' used in these three cases? Surely not in a univocal manner? The solution offered by Blund also allows him to reject the notion that there are three souls in the human person and to assert that there is one only.

2. *Vegetative soul* (*chapter* v)

Investigating the powers of the vegetative soul, Blund classifies its tripartite structure (46–8): (i) the nutritive faculty, which converts food from a material state into a semblance of the body and so preserves the body; (ii) the augmentative faculty, which increases the size of a body by means of the assimilation of another body; (iii) the generative faculty, which takes a part of the body (potentially similar to the body) and ultimately engenders an entity which is akin to the original body. Taken abstractly, the word 'power' denotes solely the capacity of the soul only in relation to some or other of its effects; concretely, it signifies both the soul itself and the power of the soul in general at the same time (51). For Blund, the aforementioned faculties are not subsumed within an individual power: the separation process of the generative faculty is at variance with the nutritive faculty, while the replenishment process of the nutritive faculty may indeed occur exclusive of bodily growth (as observed in the elderly and the decrepit).

3. *Sensitive soul*

3 (*a*) Its powers (chapters VI–VIII)

Surveying the sensitive soul, a division of two general powers is formulated by Blund (55–61): (1) the motive power—with the subdivisions (*a*) motive and directing power (concupiscible/desire and irascible/rejection), and (*b*) motive and motion-making power (nerves and muscles); (2) the perceptive power, with the subdivisions (*a*) external perceptive power (five senses), and (*b*) internal perceptive power (central sense, imaging, estimative, memorizing, and recollective). Pertaining to the five categories of external perceptive power, Blund presents a distinction between (i) active understanding, (ii) the material understanding, (iii) the formal/obtaining understanding, (iv) the understanding which radiates, as it were, into an effect. For instance, the *active* sight is a faculty of the soul which perceives entities endowed with colour, the *material* sight receives modifications from a bright entity acting upon it, the *formal* sight is the modification received, and the *radiated* sight is active sight's perception of formal sight's similitude in the effect. Thus all senses, in so far as they are *active*, are most specifically of the same species. Nevertheless, the divergent bodily organs and their associative modifications refer to differing kinds.

Blund next turns to the concupiscible power and the irascible power (62–82). Concretely, 'power' connotes a definite understanding of the soul and a correlative accident by way of the verb 'to be vigorous' (71). Taken as such, the concupiscible and irascible powers are one and the same in the subject. Taken abstractly, however, they are different powers in so far as they signify diverse acts, the soul desiring that which is pleasurable and rejecting that which is detrimental. Against the notion of these faculties being perceptive powers, Blund (73) argues that the appetitive power does not perceive what the soul desires by means of it, but rather the soul perceives either through the power of sense or through the imagination. Again (75), against the notion of these powers being parts of prudence, he clarifies that prudence derives from reason and from the comparison of good with evil, while the concupiscible power is devoid of a discriminative capacity. Against the notion that desire and the will stem from the same source, he distinguishes (78) between (i) the sensual will derived from the concupiscible power and (ii) the rational will derived from the rational power. Moreover, justice, he states (80), dwells in the soul following temperance (concupiscible power), fortitude (irascible power), and prudence (rational power).

With regard to the perceptive powers of the sensitive soul (83–8), Blund holds that only the particular is seen. Following Avicenna, he holds that being is the first impression on the soul and so in itself comes under the understanding. However, being also comes under the power of sensation. Blund argues that being as such does not comes under sensation but only by way of sensible accidents (87). Again, following Boethius, Blund states that the particular and the universal refer to the same thing, particular when grasped by sense, universal when grasped by thought. Therefore, the universal is sensed, yet not the universal *per se*.

3(*b*) External senses (chapters IX–XVI)
Concerning sight, Blund begins (89) by explaining that the vitreous humour in the eye is aqueous and so can receive, yet not retain, any impression or semblance of the form acting upon it. That which is seen is perceived within an angle, the lines of which converge at the eye and are not extended equidistantly, thereby permitting perception of entities larger than the eye itself (91–3). Al-Fārābī points out (96) that an active bright object forms an impression upon the eye through the medium of air; however, the air is not subject to the same impression. Again, while colour is a proper object of sensation, Aristotle holds that size, number, and movement are not but are grasped by the central sense. Further problems concerning vision are discussed *more geometrico* (102–7). Examples of the refraction of light are discussed and the chapter finishes with problems associated with the contemporary beliefs regarding diseases of the eye (114–17).

When discussing light (118–36), Blund holds that while both light and colour comprise distinct accidents (translucence and colour, respectively), both are subsumed under one subject—light being *essentially* the perfection of translucence, and colour being light, solely inasmuch as it yields brightness or colours its subject. Additionally, he clarifies that the light of the sun and the light of fire are not of the same most specific kind, yet 'light' in a *universal* abstraction is a genus in respect of both (125–7). Taking darkness next (128–32), Blund notes that it can be said in two ways: (i) as opposed to light in a privative manner, where both light and darkness can arise with respect to a single subject via the reversible order—that is, the same thing is now bright, now dark; and (ii) as opposed in a possessive manner, where darkness and light cannot exist in a single subject where light is an accident which is inseparable from its subject, as in the case of fire and the sun. Can darkness be seen (133–6)? When it is in darkness, the eye is not subjected to any impression, likeness, or modification, and thus darkness cannot be seen by way of the organ of sight. Following Aristotle, who states that sensing can be said in many ways, according to the body and according to the soul, Blund states (136) that darkness may be 'sensed' in that we can be aware of the absence of light.

Geometrical propositions concerning darkness are next depicted by Blund (137–44; see also the appendix below): (a) if a bright body and a dark body are dimensionally equivalent, the former projects an infinitely long cylindrical shadow; (b) if the dark body is smaller than the bright body, a finite conical shadow will be cast; (c) if the dark body is greater than the bright body, a 'basket-shaped' shadow is cast. In the case of (b), the nearer the dark body approaches the bright body, the smaller the shadow it casts, while the more it recedes, the greater the shadow. In the case of (c), the nearer the dark body approaches the bright body, the greater the shadow it casts, whereas the more it recedes, the smaller the shadow. In the case of (a), the equidistant sides in question give rise to an infinitely long shadow.

Citing Avicenna (145), Blund portrays hearing as a faculty residing in an extended nerve so as to perceive the form of an external disturbance of air that emanates from the compression of bodies. Furthermore, as Blund states, while sound is the proper sensible of hearing, and thus is in the air, air is heard in an accidental manner, whereas sound is heard essentially (154). He argues that the pleasure stemming from sound dwells in the soul (164), for there is on the one hand the movement and the undergoing of action in the body and, on the other hand, in the soul there is a species attained by means of the motion which gives rise to pleasure. All sounds which share a common measure are concordant, but discordant when not commensurate with a comparable aliquot part (165). One and the same sound may appear harmonious to some and dissonant to others as a consequence of the variety of dispositions in the hearers. Blund elaborates (170), stating that, for instance, if the hearer enjoys good health and constitution, with a proportionate symmetry of humours, he will grasp the sound as concordant and agreeable. For, in order that the being of the soul may be preserved in a good constitution, the soul encounters pleasure when the being of the body is preserved.

Blund next examines the reverberation of sound or the echo (176–92). It is asserted that the air acted upon through a collision communicates a movement which is successively formed as proportionate to the initial formation. When this air encounters resistance in a solid concave body, this gives rise to a rebound of the initial noise. In the case where sound is near to a solid object, the hearing of the rebound and the first sound formed are not grasped as distinct occurrences owing to the rapidity of movement throughout the air (180). Blund highlights the fact that while one sound is received in two ears, yet two sounds do not appear to the soul (184). Modifications are like- nesses of corresponding sounds, and the eye of the mind intuits them by correlating them with one and the same sound. In addition, given that the sense of hearing is a composite organ, diverse representations of sounds can be found in different parts of the ear at the same time (187). Thus, since the likenesses are distinct, the soul assesses that the sounds are different also. Moreover, the soul's preference regarding sound may be due to chance and fortune; at times, however, there is a habitual inclination. Furthermore, that the organ of hearing is more influenced by sound than by any other modification is owing to its predominantly airy quality (192).

Blund next examines the sense of smell (193–207), which, following Avicenna, he locates in the front part of the brain. What, however, is a smell? Various writers have said that it is a 'fume' emitted from an odorous body, mainly composed of water and air. When it receives the modification, the soul directs itself towards the impression and per- ceives the odour. A distinction is made between a fume and an odour (201). Does the organ of smell itself have a smell (202–7)? The solution is found in the words of Aristotle: like cannot be acted upon by like, and thus, if the organ of smell were of a certain odour, it would not receive an impression from something else. Yet it does receive such an impression. Thus, the organ of smell is devoid of a certain smell.

Blund next portrays taste as a power in an extended nerve upon which an impression is collected by means of flavour (208–16). The will focuses the eye of the mind towards the modification (in the sense of taste) and the soul perceives the external flavour through a correla- tion with the said modification. Thus, it is the eye of the mind, rather than the nerve or power of the nerve, that perceives. Again, according to the words of Aristotle (216), saliva is the medium which transports flavour to taste, and the tongue's flexibility, mobility, and malleability are attributable to its moist nature.

Touch (217–31) is more prevalent in the body than the aforemen- tioned senses, for its role is to safeguard its subject and evade in- jurious contact. Such prevalence, Blund clarifies, is exclusive of the brain, bones, and lungs, as the nerves of touch do not extend to these parts (221). In addition, those properties perceived by touch are either primary qualities (hot, cold, dry, moist) or derived from primary qualities (viscosity from humidity, aridity from dryness, heaviness from coldness, lightness from heat and dryness). Furthermore, that which is received in the organ of touch is obtained by the flesh or skin which presses upon the nerve (223). Finally, in accordance with Aristotle's *On Generation and Corruption*, Blund echoes the principle that the lesser is destroyed by the greater; when something is destroyed it is acted upon, as when skin is burnt by fire; therefore pure elements such as fire can be sensed.

3(*c*) Internal senses (chapters XVII–XXI)
Concerning the central sense (232–49), Blund first of all points out that sensation is not abstractive as it solely treats of particulars, whereas the intellect alone is abstractive, dealing as it does with universals by way of abstraction from particulars and their accidents. As maintained by Avicenna and Augustine (237), the central sense is a 'common' faculty residing in the frontal lobe of the brain, which receives impressions derived from the five external senses. Furthermore, the will of the soul directs the eye of the mind towards the likeness of the body, which is situated in the central sense, and joins it with what gives rise to it, which is situated in the external sense (248).

The imagination or imaging power is located in the furthest extremity of the front part of the brain, the first ventricle, which receives impressions from the central sense (250). The imagination treats of absent entities, so how does it differ from memory and recollection (252)? Blund states that in imagination only the image of the thing perceived is preserved but no apprehension of a 'before' is found as in the case of memory, nor any linking or separating of any property as occurs in recollection.

The estimative faculty is positioned in the middle ventricle of the brain to facilitate the perception of non-sensed intentions which dwell in individual and sensed entities (254). It judges whether something should be shunned because of a harmful intention or else desired because of a useful intention. Estimation, Blund argues, perceives only individuals. Thus, it does not perceive that wolves should be fled from, but more accurately that *this* wolf should be fled from. Moreover, while there is composition or division in the estimative faculty, it is not through the estimative power that truth or falsity is perceived, but more precisely through the intellect and reason, and so it follows that this perception is deficient in brute animals.

Memory, positioned in the rear ventricle of the brain, receives impressions from estimation and retains them (262). When the will focuses the eye of the mind towards memory, it detects images there by means of which it intuits entities formerly perceived. Certainly, the image of 'previousness' arises in memory through that of 'presentness'—that image which is a semblance of the present will be a semblance of the past, after the impression occurs. Furthermore, Blund notes, numerous images are deposited in memory (274), yet a knowledge of the associated entities may be absent, for knowledge of something is not attainable through a single perception but rather by means of various perceptions. Blund highlights the fact that if something is wholly forgotten (279), memory is incapable of recollecting that it has forgotten the thing in question. On the other hand, if memory preserves a component of the thing, it recalls the knowledge of it.

4. *Rational soul* (*chapters XXII–XXVI*)

The remainder of the text is devoted to the examination of the rational soul. The rational soul, an incorporeal substance, has its being in the body in order to attain its perfection through the acquisition of virtue and knowledge (304). However, although a soul is in a body and can intuit images formed in the body itself, it is not to be regarded as, essentially, one with the images themselves (308). Thus, if I see a wall I am not in the wall, and if I feel a pain in my finger I am not my finger. Again, although the soul gives life to all of the body, the soul does not have to be extended throughout all of the body. Just as the bright centre of a circle can illuminate all of the circle while being at the centre and without having extension, so life and growth can emanate from the soul, go forth into all of the body, and be diffused throughout, yet the soul itself is devoid of extension and is incorporeal (309). Perhaps, then, a human being has three souls, a mortal vegetative soul, a mortal sensitive soul, and an immortal rational soul (310). Blund argues against this, relying on arguments drawn from natural philosophy against the impossibility of a vegetated body and a sensitive body occupying the same space. A final argument centres on the nature of the nerve impulses running throughout the body, asking whether they are corporeal or incorporeal (316).

Blund next examines the question as to whether the rational soul is mortal or immortal (317–28). In giving his solution (324) he proposes that the soul has a twofold being: (i) from its first perfection, its initial creation, it has an immortality of which it *cannot* be deprived; (ii) from the acquisition of knowledge and virtue it has a being of which it *can* be deprived through turning away from the truth and giving way to falsity. A human being is made up of two components, a body and a soul. At death, the human body returns to its first material elements, the soul to its source in God. Furthermore, Blund argues that the soul is simple (329–34) rather than composed. The First Cause, thus, is the most simple and so those entities immediately derived from it (the soul and the Intelligences) have a simple being (333).

In examining the powers of the rational soul (335–6), Blund lists two faculties: (i) the power to act and (ii) the power to know. The former entails the individual actions of man, which are chosen in accordance with the judgement and intention of reason. This power can be construed in three ways: (*a*) as regards its correlation with the vital and appetitive power and in proportion to the temperament with which one is affected; (*b*) as regards its correlation with itself in so far as it produces skill and understanding in itself; (*c*) as regards the connection of the active power to the imaginative and estimative power, directed in line with the judgement of the contemplative power. The power of knowing is a contemplative faculty by which wisdom is generated (and virtues perfected) in man though the intuition of those things which are above him (336). Following Avicenna, Blund states that the human soul comprises, as it were, two faces, one directed below (towards the production of morals) and the other directed above (towards the production of wisdom).

As regards the understanding, Blund gives its fourfold division as derived from Avicenna (337–41): (i) the material intellect, the soul without acquired dispositions, also called the intellect in potency as it is devoid of form; (ii) acquired or formal understanding, which is a modification produced in the soul that is a likeness of an external thing; as such it completes the material intellect by giving it being in effect; (iii) intellect in effect, a correlation between (i) and (ii); (iv) the agent intellect which grasps universals through abstraction. Blund concludes this section by stating (361) that there is from nature a preparation in an organized body so that it will be better disposed towards receiving the *rational soul* than any other. The soul is, subsequently, infused into the body, a potentially vivified organized body, by the first Giver of Forms. This leads on to the subsequent topic: is every soul which is infused into a body a new one? In other words, is reincarnation possible? It would seem good that no soul would in the end be damned and that each would be given further opportunities to

acquire merit and turn away from evil (363). Blund's answer (365) is
that a soul which has already freely spurned the good and submitted
itself to evil should not be given another chance to turn away from
good. Hence, the infusion into the body of a new soul is argued for
so that it may, through utilizing reason and understanding, conserve
the uprightness of its will with the choice of reason for the sake of
that uprightness. It follows, thus, that the will shall be preserved in
uprightness solely when one wishes that which God desires one to
wish. Additionally, it is argued that the soul, when separated from the
body, is capable of using the intellect (366–375bis). The rather amus-
ing question as to whether the separated soul of a philosopher knows
more than the separated soul of an idiot is also debated (370).

A penultimate question deals with the location of the soul (376–
81), which seems to mirror the discussion above (23–4), in which the
soul was identified with a material thing. Two locations are proposed,
the heart or the brain—or, towards the end, blood (381). One possible
solution is that the soul is complete in every part of the body (379).
No clear conclusion is reached, and the argument in this section is far
from satisfactory.

Finally, Blund examines the difference between reason and the freedom of the will (382–400). He propounds that there is solely an accidental difference between the former and the latter. In the will there
is reason, which is a power of judging by way of reasoning what is
good and evil, what is true and what is false (385); and again, in the
will there is freedom because, given judgement or discernment, it is
able to pursue truth and preserve itself in goodness. Furthermore, the
power to sin is not derived from free will, for free will cannot be indifferent with respect to the choice of good or evil (392). If the will
opts for the good with the intention of love, it is good, while if it opts
for evil, it is evil. Through no necessity does it sin, but rather solely
by means of the will. The ability to sin is, in fact, an inability, for it is
from the weakness of the mind that the choice of evil occurs.

The final chapter[28] deals with whether there is free will in God and
the angels, since presumably they cannot and will not sin (401–15).
Anselm is quoted to the effect that it would be absurd to say that God
cannot choose to do what he wishes, and since freedom of the will is
related to reason, it must be found in angels, who are higher in reasoning than men. The problem is solved by stating that the ability to sin
does arise from weakness and imperfection rather than from freedom

---

[28] A promised chapter on divine providence given in the list as chapter XXVII is missing,
either because it has not survived or because it was never written.

itself. The ability to sin is not a power but a lack of power (405), since it is better never to fail in preserving a good will. Finally, Blund asks whether animals have free will (411–13), since they appear to have desires and to make choices from among various objects. They also clearly distinguish between what is good and what is harmful. Blund acknowledges that on the basis of sense knowledge they distinguish between various things, but denies that they do so on the basis of reason, which is the true mark of freedom. Discernment and judgement derive from reason, and although human beings may be affected by desire or flight, reason can overrule emotion (415).

Blund's use of Avicenna

It has been observed that in the *solutiones* and descriptive passages which are not part of a *quaestio* Blund is far more reliant upon Avicenna than Aristotle for his doctrine.[29] Indeed, Hasse notes that 'all of the theories of Aristotle mentioned can also be found in Avicenna but that most of Avicenna's theses go beyond Aristotle's philosophy. This is especially true of the Avicennian definitions and descriptions of the external and internal senses.'[30] Thus, rather than supposing, as some have, that Blund represents an intermediate stage in the reception of Aristotle when the thought of the Stagirite was not fully appreciated because not fully understood, we should conclude, on the contrary, that Blund consciously chose Avicenna over Aristotle on these topics because of the superiority of the former over the latter. In doing so he was followed by later authors such as Roland of Cremona, William of Auvergne, and Jean de la Rochelle, who also seem to have preferred Avicenna. As Hasse writes:

Avicenna and his psychology dominated the structure and much of the content of psychological writings in the West for over half a century, from John Blund to Albertus Magnus and Petrus Hispanus. Avicenna had developed a theory of the soul which combined Peripatetic philosophical argumentation with an elaborate system of faculties, based on a great deal of physiological material. Avicenna's *De Anima* offered the latest and best in philosophical subtlety and comprehensiveness and in terms of scientific discoveries. It was for these reasons that he was preferred even to Aristotle by many writers of the period examined. A factor which contributed greatly to Avicenna's success was the compatibility of his theory with the teachings of the medical tradition, which had begun to influence even theological discus-

---

[29] Hasse, *Avicenna's* De anima, pp. 21–2.   [30] Ibid., p. 22.

sions of the soul. Aristotle's treatise could not compete with Avicenna's on the level of physiology.[31]

Nor should it be surmised that Blund merely repeats what he finds. He occasionally chooses a compromise between Aristotle and Avicenna, as in his views on the organ of touch.[32] Moreover, he has an excellent understanding of Avicenna's theory of *aestimatio* and *intentiones*, which he carefully paraphrases for the sake of the reader, and he transforms the doctrine of the four intellects.[33] Again, evidence of independence of thought is to be found in the fact that although Blund is indebted to Avicenna for his theory of the vegetative and animal faculties, he is less so where the intellect is concerned. Indeed, he cannot be counted among the Avicennized Augustinianists since he does not identify the active intellect with God.

Perhaps something of the quality of Blund's achievement can be seen in his explanation of Avicenna's teaching on estimation.

The Commentator (Avicenna) calls an intention an individual quality which is not picked up by sensation, which is either harmful or useful to a thing. Harmful, such as that property which is in a wolf and because of which the sheep flees from it; useful, such as that property which is in the sheep and because of which the lamb approaches it. (254)

Avicenna's theory of estimation was quite successful among Western authors, since it seemed to offer an explanation for the seemingly rational behaviour of animals, crediting them not only with sensation but with discernment as well. Blund has, it seems, a very accurate understanding of Avicenna's theory of *intentiones* as being the qualities and attributes of perceived objects.[34] This allows Blund to puts his finger on a central issue within Avicenna's theory. If the *intentiones* reach the power of estimation after first passing through sense perception and the imagination, why are they not picked up by the latter? Blund puts forward an objection which sees *intentiones* as an equivalent of sense data:

Moreover, since the wolf is a separate thing from the sheep, in what way is a likeness of the intention which exists in the wolf constituted in the estimation unless there was first of all an impression formed in the sensation of

[31] Ibid., p. 225.

[32] Ibid., pp. 22, 101, 103. As Hasse points out, Avicenna does not speak of a medium where the sense of touch is concerned, but believes instead that touch is located in the nerves and the flesh, and is affected there directly. Aristotle would oppose the view that there is no medium, and locates the organ of touch near the heart. Blund (whether intentionally or not) compromised by accepting from Avicenna that the organ of touch is to be found in the nerves and from Aristotle the view that the flesh or skin is the medium.

[33] Ibid., p. 22.   [34] Ibid., p. 145.

the sheep by the intention which exists in the wolf, since sensation is a medium between the thing sensed and the estimation? For in what way can a fire
which is far off from a man heat the man, unless the air in between receives
heat from the heat of the fire? (256)

Avicenna's text does not provide an answer, but Blund himself does,
by stating that estimation directly perceives the *intentiones* without
any intermediate perception by the senses:

However, because this might seem to someone to be difficult to understand, it can be said that a likeness of the intention occurs in sensation and
in imagination, but the soul does not perceive according to them, because
sensation and imagination are not by nature in agreement with the proper
subject of the intention. Yet the organ of estimation is of a similar nature to
that which is in itself and properly the subject of an intention, and thus according to the estimative power the perception of an intention occurs. (257)

As Hasse clarifies (p. 146), by 'the proper subject of the intention'
Blund means the wolf, and its likeness or image is that which reaches
the power of estimation. The intentions are in the object only and
their likenesses are to be found in the mind. Blund then concludes
that the estimative power grasps only particulars and not universals.
Hasse concludes: 'The quality of John Blund's discussion is exceptional. Most writers of the first half of the thirteenth century simply
adopt the doctrine.' It may well be the case that Blund contributed in
no small way to the widespread adoption of the doctrine of the *virtus
existimativa*, whereas other famous Avicennian doctrines such as the
separate agent intellect were known but not accepted.

Avicenna's idea that the Agent Intelligence plays an intermediary
role in abstraction was also taken up by Blund. Here again he was one
of the first to explain the idea:

With regard to this it should be said, as has been shown above, that the soul
has to turn itself towards the body which it has to direct, and towards its dispositions, and towards the likeness of images found in memory. The formal
intellect is impressed on the soul by means of the First Giver of Forms; or, as
many authors seem to wish, that form is an impression from an Intelligence
as its servant, and from the First Giver of Forms as by its authority. That
Intelligence is said by many authors to be an angel, who is a servant of the
soul of man. For a man has two angels, one good and one bad, each of whom
is an attendant of the human soul. (344)

While this is taken from Avicenna's *De anima* v. 5, it seems to have
been Blund's idea to introduce the notion of the Dator Formarum
in an epistemological context, and the term *intellectus formalis* may

well be Blund's own invention to parallel the notion of the *intellectus materialis*.[35] The section on angels will find an echo in Roland of Cremona and may have been suggested by Gundissalinus or even Peter Lombard.[36]

Note on the translation

It goes without saying that the decline in the teaching of Latin in schools in the English-speaking world means that the vast majority of undergraduate students are unable to read medieval philosophical texts in the original language and must rely upon translations. The question remains, however, as to what kind of translation would be best for such students: a free or a literal one. I have chosen to follow the latter course, rendering complex philosophical terms into their long-established English equivalents, reasoning that it gives a flavour of the original and that it allows one to follow the argumentation closely. A literal translation also has the advantage that it enables the reader to compare the translation and the original more closely, so that it will be of use to the student of medieval philosophy who is beginning to learn how to read such university texts in their original Latin.

Even the decision to follow the model of a literal translation is not without its problems. Despite the fact that we possess a centuries-old scholastic vocabulary in English, sometimes there is no meaningful English equivalent for a Latin word, or if there is, sometimes that word has also changed its meaning with time. To take one example, the Latin *passio* cannot now be translated by 'passion', and one struggles to find the right word for such a central technical term, which means 'the state of being acted upon'. Perhaps one can take some consolation from the fact that the original translators of Avicenna into Latin, Avendauth and Gundissalinus, also had their difficulties, as is clear from their efforts to express Avicenna's teaching on light. As Hasse points out, they translated the terms which Avicenna uses for two kinds of light, *dau'* and *nūr*, as *lux* and *lumen*. Both words in Arabic and Latin mean 'light', and the distinction which Avicenna makes is between the 'natural' light of a source such as the sun or a flame, which is seen but has no colour, and the reflected light from an object which does have colour. The problem, as Hasse notes, is that the Latin translators did not use the two terms consistently but interchangeably, and also had a liking for *splendor*,

---

[35] See Hasse, *Avicenna's* De anima, p. 201 n. 684.   [36] Ibid., p. 23 n. 58.

which they sometimes used for either *lux* or *lumen*. When Blund came to discuss the matter, the addition of the term *splendor* did not help. He wrote:

> In the commentary ⟨of Avicenna⟩ a distinction is made between light, brightness, and brilliance [*inter lucem et lumen et splendorem*]. The Commentator (Avicenna) calls light the perfection of the translucent; brightness, however, he calls the modification produced in the translucent, such as in air; brilliance, however, he states to be an effect produced by any colour in a translucent thing, such as by redness or something similar. (123)

Hasse comments (p. 115):

> There are several remarkable features about this passage. Firstly, the definition of *lux* is quoted in an abbreviated version which omits the core concept of *lux* as a natural quality of certain bodies. What remains is part of Avicenna's definition which sounds most Aristotelian. Secondly, *lumen* is defined as an effect created in the translucent—instead of, correctly, the non-translucent—which turns Avicenna into an Aristotelian. Thirdly . . . what appears as a definition of *splendor* is in fact Avicenna's definition of *lumen*.

It was hardly Blund's fault that he sought to make sense of an unreliable text. He was not alone, since most of the writers who followed (with the exception of the careful reading of Albert) adopted the same consistent (mistaken) interpretation.

* * *

Blund's philosophical achievement was considerable. As McEvoy points out,[37] it is a pity that we have none of his theological works that would allow us to examine how his thought developed. However, it is enough to know that with his arrival and the circulation of his *Tractatus* at Oxford, Greek and Arabic psychology came to a place where it was to maintain its presence through considerable developments in the centuries that followed.

Note on the Latin text

All of the original preliminary material from the 1970 edition has been reset to take into account, among other things, the changes in the cross-references necessitated by the new pagination. However, the pages of the edition proper are reproduced from the 1970 publication, as are also the Index auctorum and the Index nominum et verborum potiorum.

[37] McEvoy, 'Liberal Arts', p. 564.

INTRODUCTION TO THE
LATIN EDITION OF 1970

THE fixed dates in the career of John Blund[1] are that he was a clerk
in the service of Henry III, king of England, from 1227, that he was
elected archbishop of Canterbury on 26 August 1232, but never con-
secrated, that he was made chancellor of York Cathedral in 1234, and
that he held this office until his death in 1248. Our only source of in-
formation about the earlier part of his life is the poem which Master
Henry of Avranches addressed to Pope Gregory IX in 1232 in support of John's election to the see of Canterbury. It is a poem of nearly
250 lines, but it is only necessary to quote the passage on his studies
and on his career as teacher:[2]

> Adde quod a puero studiis electus inhesit,
> primus Aristotilis satagens perquirere libros,
> quando recenter eos Arabes misere Latinis,
> quos numquam fertur legisse celebrius alter
> aut prius, ut perhibent Oxonia Parisiusque.
> Non tamen est contentus eo quasi fine, nec artis
> illi mundane suffecit adeptio, donec
> humanos regeret divina sciencia sensus,
> ad quam translatus lustris duobus et annis
> insudans totidem rexit dominanter in ipsa.

From these verses we learn that he had lectured on Aristotle both at
Oxford and at Paris, and that he had gone on to the study of theology
in which he was regent (*rexit*) for twelve years.

It is not possible to fix with certainty the periods at which he studied at Oxford and Paris. The most probable hypothesis is that he was
born about 1175, that he studied first at Paris, and that he had incep-
ted as Master of Arts by about 1200. A period of teaching the arts in
Oxford followed. It was perhaps brought to an end by the dispersal of
masters and scholars during the interdict laid upon England by Pope
Innocent III, which lasted from 1209 to 1214.[3] He then returned to

---

[1] The details were discussed by Callus, *Aristotelian Learning*, pp. 241–7, and *The Treatise*, pp. 473–5. See also A. B. Emden, *Biographical Register of the University of Oxford to
1500*, i, Oxford, 1957, 206–7.

[2] *The Shorter Latin Poems of Master Henry of Avranches Relating to England*, edd. J. C.
Russell and J. P. Heironimus, Cambridge (Mass.), 1935, p. 131, ll. 77–86.

[3] Dr. Emden (*Biog. Reg. of the University of Cambridge*, 1963, p. xii) suggested the possi-
bility that John Blund was one of the Oxford masters who migrated to Cambridge as a result

Paris and studied theology. According to Matthew Paris he was one of the Englishmen who left Paris at the great dispersal of masters and students in 1229, but he had probably left somewhat earlier, since he was receiving instructions for a mission to the Roman curia in May 1227 on behalf of King Henry III.

Whatever may be thought about the details of this reconstruction, it is reasonably certain that the period of John's regency in arts fell within the first decade of the thirteenth century. This is important, because the *Tractatus de anima* is the work of an artist, not of a theologian.[4] Its composition, therefore, must fall within the same period. Among contemporary writers the closest connection so far established is with the Englishman, Alexander Nequam (1157–1217). In his unfinished *summa*, the *Speculum speculationum*, which was written about 1213, Alexander appears to be drawing on the *Tractatus* when he quotes the Aristotelian definition of the soul in the same wording as that used by John Blund (no. 14), which differs from other versions then current: 'Gaudeat ergo philosophus sua consideratione animam sic describens: Anima est corporis organici perfectio vitam habentis in potentia.'[5] This impression is strengthened by the words which immediately follow in the *Speculum speculationum*:

> Cum enim omnis perfectio sit ex forma, videtur secundum hoc quod anima sit forma. Quid? Immo omnis anima substantia est.[6]

These words should be compared with the section of John Blund which immediately follows his enunciation of the definition (no. 15). There is another passage in the *Speculum speculationum* which was not

of the dispersal, since a master of that name is found witnessing charters of Eustace, bishop of Ely (1198–1220). The question is complicated by the fact that two men with the name 'John Blund' and with the title 'Magister' are found as witnesses to episcopal charters in the early thirteenth century. They were very close contemporaries, since our John Blund died in 1248, and the other in 1239, and the separation of the career of the one from the other is very difficult; see the notices in Emden's Oxford and Cambridge registers. Fr. Benedict Hackett in his forthcoming edition of the earliest Cambridge statutes shows with some persuasiveness that the move from Oxford to 'that distant marsh town' which Rashdall found so inexplicable was due to the presence at Oxford of several natives of Cambridge who are among the first group of Masters at the new university. He shows that, for example, both Grim (Emden, op. cit. ii. 826 and p. 272) and Blund were well-known Cambridge names. The John Blund who migrated was therefore most probably not our John Blund. [M. B. Hackett, *The Original Statutes of Cambridge University: The Text and its History*, Cambridge, 1970, 44–6.]

[4] Callus, *The Treatise*, pp. 480–2; see the text below, especially nos. 17–21. The numbers here and elsewhere refer to the sections into which Fr. Callus divided the text; see p. 21.

[5] Callus, ibid., pp. 490–1. The passage from the *Speculum speculationum* is in London, British Museum, MS. Royal 7 F. I, fol. 71[vb]. [*Speculum speculationum*, III. 89, ed. R. M. Thomson, *Alexander Nequam: Speculum speculationum* (Auctores Britannici Medii Aevi, XI), Oxford, 1988, 359.]   [6] Callus, op. cit., p. 491; ed. Thomson, loc. cit.

considered by Fr. Callus in this connection. It is part of the chapter 'De sensualitate':

Virium ergo sensitivarum quedam est vis motiva, quedam vis est appre-hensiva. Vis motiva continet appetitivam et animositatem et vim effectivam motus, que operatur in nervis et musculis. Vis apprehensiva dividitur, ut aiunt, in illam que est apprehensiva deforis et eam que est apprehensiva dein-tus. Vis apprehensiva deforis est sensus qui dividitur in quinque species. Vis apprehensiva deintus dividitur in sensum interiorem et ymaginationem et estimationem et vim memorialem.[7]

The corresponding passage in John Blund is nos. 55 and 61 (below, pp. 31, 35). The main source is Avicenna *De anima* I. v, but detailed comparison makes it probable that Alexander derived his knowledge of Avicenna through John Blund. Compare in the passage just cited

Avicenna: Vis motiva secundum quod est efficiens est vis infusa nervis et musculis.
John Blund: Vis autem motiva et efficiens motum est illa vis que operatur in nervis et musculis.

The probability is turned to certainty by the appearance of 'animosi-tas' in Alexander Nequam, which is not found in Avicenna, but which appears in John Blund. He had stated that two parts of the 'vis motiva' were 'vis concupiscibilis' and 'vis irascibilis', and, he continues, 'this "vis" is sometimes called by Aristotle in the *Topics* "animositas"'.

Father Callus had begun by supposing that John Blund had bor-rowed from Alexander Nequam, a very natural first view, since John was the younger man. On further consideration he became convinced, surely rightly, that the borrowing was the other way. He had noted in the apparatus to the text the close resemblance between a passage in the *Tractatus* and a part of the chapter on sight in the *De naturis re-rum* of Alexander Nequam,[8] but he had not considered its possible chronological implications. First it is necessary to set out the texts:

John Blund	Alexander Nequam
nos. 102–3	

Item. Quod sub maiori angulo vi-detur, maius videtur, et quod sub minori minus, et quanto res visa ma-gis elongatur ab oculo, tanto minor

Generale est quoniam quanto res remotior est, tanto videtur minor . . . In geometricis etiam speculationibus accidit quiddam admiratione di-

---

[7] III. 94, fol. 74^{rb-va}.

[8] Alexander Neckam, *De naturis rerum*, ed. T. Wright (Rolls Series), London, 1863, II. 153, pp. 234–5.

videtur. Unde quanto aliquid a re-motiori videtur tanto minus videtur, quanto a propinquiori tanto maius. Ergo quanto periferia a remotiori videtur, tanto minor videtur. Sed quanto periferia est minor, tanto angulus contingentie est maior. Ergo quanto periferia a remotiori videtur, tanto angulus contingentie maior videtur: non ergo quanto aliquid a remotiori videtur tanto minus videtur.

gnum. Est enim aliquid quod quanto remotius est tanto videtur maius. Angulus enim contingentiae quanto propinquior est, tanto videtur mi-nor. Cuius rei haec est causa. Quanto circulus maior est tanto angulus contingentiae est minor. Et quanto circulus minor, tanto angulus con-tingentiae maior est, ut est videre in subiecta figura. Sed quanto circulus est remotior, tanto videtur minor. Quanto namque est remotior, tanto sub minore angulo videtur. Quanto autem circulus videtur minor, tanto angulus contingentiae videtur maior. Igitur quanto angulus contingentiae remotior est, tanto videtur maior . . .

Solutio. Hec argumentatio non valet: quanto periferia est minor tanto an-gulus contingentie est maior; ergo quanto periferia a remotiori videtur tanto angulus contingentie minor videtur. Quia sicut decrescit an-gulus sub quo videtur circulus per elongationem ipsius ab oculo, ita decrescit angulus sub quo videtur angulus contingentie per elongatio-nem anguli contingentie ab oculo, quia quanto plus et plus recedit an-gulus ab oculo tanto magis minoratur angulus linearum concurrentium in oculo, ut ipse recessus sit proportio-nalis diminutioni ipsius oculi.

Videtur tamen quibusdam quod haec argumentatio non valeat. Quanto cir-cumferentia est minor, tanto angulus contingentiae est maior, ergo quanto circumferentia videtur minor, tanto angulus contingentiae videtur maior. Quia sicut decrescit angulus sub quo videtur circulus per elongationem ipsius ab oculo, ita decrescit angulus sub quo videtur angulus per remo-tionem. Sed quid? Sub quo angulo videtur angulus contingentiae?

These passages are clearly related, and no common source has been found. Alexander quite often borrows passages of some length from other writers without naming them. An example is found on the next page of the *De naturis rerum*, where thirty lines are copied word for word from Urso of Calabria without any indication that they are a borrowing.[9] John Blund does not proceed in this way, as the reader of the text may see. Further, the run of the passage makes it probable that Alexander is drawing on John Blund. Compare

[9] pp. 236–7: 'Vide quod quaedam animalia . . . nimia disgregatione' from Urso, *Apho-rismi*, Comm. on Aph. 38, ed. R. Creutz, *Quellen and Studien zur Gesch. der Naturwissen-schaften and der Medizin*, v. i, Berlin, 1936.

John Blund	Alexander Nequam
Hec argumentatio non valet.	Videtur tamen quibusdam quod haec argumentatio non valeat.

In the elaboration of this argument there is close verbal correspondence. If this is a correct reading of the evidence, the *Tractatus* must have been written before the *De naturis rerum*. Now the *De naturis rerum* can be dated within fairly narrow limits, *c.*1197–1204.[10] The *Tractatus*, therefore, was written not later than 1204, and probably not later than about 1200.

An early date gains further support from comparison with the surviving fragments of the *Quaternuli* of David of Dinant,[11] which must be earlier than 1210, the year in which they were condemned to be burnt. If John Blund had been teaching in the Faculty of Arts at Paris at a time when David's views became known there, he could hardly have failed to take notice of them, and of such notice there is no trace in the *Tractatus*.

Even if we did not know the name of the author, there would be little doubt that the *Tractatus* was written by an Englishman. Its peculiar mark is a fondness for dwelling on topics of natural science,[12] which is a special characteristic of English scholars at this period. Where it was written is more difficult to determine. All the evidence points to Paris or Oxford, but we have no firm ground for deciding between these two possibilities. It is not, however, a question of great moment, since Oxford at this time was dependent on Paris. It is almost certain that John Blund first became acquainted with the *libri naturales* and the *Commenta* on them in Paris. For him, as for the authors of the condemnation of 1210, the *De anima* of Avicenna was a *commentum* on the *De anima* of Aristotle.[13] The great interest of the *Tractatus* lies in the fact that we can study in it the ways in which a master of the beginning of the thirteenth century used the new philosophical and scientific texts.

[10] The limiting dates are fixed because Alexander was already an Augustinian canon at Cirencester when he wrote the work, and the date of his entry to the abbey is *c.*1197. The lower limit is given by a reference to the 'invictissimi cives Rotomagenses' (I. 27, pp. 80 f.), which must have been written before the news of the fall of Rouen on 24 June 1204 reached him.

[11] Ed. M. Kurdzialek, *Davidis de Dinanto Quaternulorum Fragmenta* (Studia mediewistyczne 3), Warsaw, 1963.   [12] Callus, *The Treatise*, p. 487.

[13] The words of the condemnation of 1210 are: 'Nec libri Aristotelis de naturali philosophia nec commenta legantur Parisius publice vel secreta' (H. Denifle and Ae. Chatelain, *Chartularium Universitatis Parisiensis*, I (1889), p. 70). For John Blund see especially no. 64 below: 'Dicit Aristoteles in libro de Anima . . . Et Avicenna idem dicit in commento super librum de Anima.'

No other work by John Blund is known, but in the library catalogue of St. Augustine's Canterbury there is a volume of *Collectiones* of Abbot Roger II (1253–73) which contained 'Quidam sermones Johannis Blond'.[14] Whether they were the work of our John Blund is, of course, uncertain, but no other John Blund is known to have left written work.

Three manuscripts of the *Tractatus de anima* are at present known. They are:

C. Cambridge, St. John's College MS. 120,[15] saec. XIII med.

The manuscript is made up of several parts originally distinct. Only the fourth part (fols. 123–34,141–202) concerns us here.

> fol. 123[ra] List of Chapters.[16]
> fols. 123[rb]–134[v], 141–55 Text headed in red 'Tractatus de anima secundum Iohannem Blondum.'

The text is followed (fol. 155) by extracts from Seneca, Ps. Boethius *De disciplina scolarium*, Boethius *De consolatione philosophiae*, Cicero, Isaac *De diffinitionibus*, a short treatise on the cardinal virtues, and Aristotle's *Nicomachean Ethics* I–III, VI–X. The intervening leaves (fols. 135–40) were misplaced in binding, and belong to the third part of the manuscript. They should follow fol. 122. The spaces for the initials were never filled in, but the guide letters were written in the margin.

The medieval provenance of the manuscript is unknown. It was given to the college by 'master Gent', either John, rector of Birdbrook, co. Essex, 1632–51, or his brother, Nicholas, rector of the same place, 1670–7. Three other manuscripts were given to the college by the same donor (37, 123, 190).

P. Prague, University Library IV D. 13 (667). saec. XIII med.

This manuscript was discovered by Père B.-G. Guyot OP, who published a full description of it in *Archivum fratrum praedicatorum* XXXII (1962), 5–125. The part which concerns us here is the beginning of the manuscript, A.I in Guyot's description, but originally eleven

---

[14] M. R. James, *Ancient Libraries of Canterbury and Dover*, Cambridge, 1903, p. 382, no. 1576; A. B. Emden, *Donors of Books to S. Augustine's Canterbury* (Oxford Bibliographical Society Occasional Publication, no. 4, 1968), p. 3.

[15] M. R. James, *Descriptive Catalogue of the Manuscripts in the Library of St. John's College, Cambridge*, Cambridge, 1913, pp. 153–5, corrected by Callus, *The Treatise*, pp. 475–7.

[16] James gives 'De diversis opinionibus circa anime essentiam' before the capitula. It is the title of c. III, which was accidentally omitted by the scribe in its place, and was added in the top margin by another hand.

quires preceded it, since the two surviving quires are numbered XI and XII. It contains:

1. fol. $1^{ra–vb}$. Anon. *Quaestio de substantia corporea et incorporea.*
2. fol. 2ra. Blank. It should have contained the capitula for the *Tractatus* of John Blund.
    fols. 2rb–14rb. Text of the *Tractatus* without heading or author's name.
3. fols. 14rb–15va. Anon. *Quaestio de synderesi.*
4–6. fols. 15va–17vb. Three anon. *Quaestiones de conscientia.*
7. fol. $18^{ra–va}$. Anon. *Quaestio de lege naturali* (perhaps by Guiard).
8–10. fols. 19ra–22rb. Anon. *Quaestio de libero arbitrio* (by Alexander of Hales 'antequam esset frater').
11. fols. 23va–24vb. Short anonymous treatise on the soul.

The bulk of the manuscript is taken up with extracts from the *quaestiones* of Alexander of Hales, from the *quaestiones* and *quodlibeta* of Guerric of St. Quentin and of other masters at Paris, chiefly of the years 1240–50. There are also extracts from the Commentary of Richard Fishacre on the Sentences. The manuscript was written by a number of different copyists, who appear to have been set to work by a master who wrote a few short stretches of text and added annotations throughout. Guyot calls him 'la main D'. This hand wrote three notes in the margins at the beginning of the text of the *Tractatus*:

fol. 2rb. Definitions of the soul.
fol. 2^{v}. Schema on the composition of the soul.
fol. 2vb. Views of Greek philosophers on the essence of the soul.

The arrangement of the *quaestiones* that follow this part of the manuscript appears to be haphazard, but from fols. 98 to 248 the figure 'IIII' is written in the top margin, and the subject-matter is arranged in accordance with that treated in Book IV of the Sentences.

The conclusion of Guyot is that the manuscript contains the collections of a master or bachelor made for purposes of teaching. The number of copyists suggests that he was working in a house of regulars where he could command the help of *socii*. Some of the scripts, including the part that concerns us, appear to be English, and the decoration is English in style. Three of the writers excerpted or copied, John Blund, Richard Fishacre, Alexander of Hales, are Englishmen. The manuscript may probably be regarded as the product of some Franciscan or Dominican convent in England, despite the fact that the majority of the *quaestiones* are by Parisian masters. There is no indication of the way the manuscript reached Czechoslovakia, but there is on fol. 164ᵛ a draft of a letter written with a plummet. It is not continuously legible, but a few words can be read, including *fr. S. de ordine sancti Benedicti* and *Moraviam.*

V. Vatican City, MS. Vat. lat. 833.[17] saec. XIV.

This manuscript contains:

fols. 1–80ᵛ. Aegidius Romanus *In rhetoricam Aristotelis.*
fols. 81–88ᵛ. Albertus Magnus *De mineralibus*, incomplete at the end owing to the loss of two gatherings.
fol. 89. Table of chapters of the *Tractatus.*
fols. 89–102ᵛ. Text of the *Tractatus* without heading or author's name.

The manuscript belonged to Coluccio Salutati, Chancellor of Florence.[18] On fol. 80ᵛ is *liber est Coluccii pyeri* and on fol. 102ᵛ *liber Colucci de Salutatis cancellarii Florentie.* It was bought from his heirs through the Florentine bookseller, Vespasiano da Bisticci, by Tommaso Parentucelli of Sarzana, who in 1447 became Pope Nicholas V. On fol. 102ᵛ is *Liber Thome de Sarzana emptus Florentie ab heredibus ser Colucii per manum Vespasiani.*

The three manuscripts are not closely related to each other, an indication that the text had a wider circulation than would appear in the present state of our knowledge. They all share one common fault, that of frequent omission by homoeoteleuton. It is a fault to which a text is particularly liable where the argument is cast in syllogistic form. In most instances each manuscript stands alone in making the omission, but in a few either P or V singly contains the omitted passage, e.g. p. 138, ll. 13–15 P against CV, p. 194, ll. 9–10 V against CP. The two earlier manuscripts are the work of scribes who were often very careless. There are stretches in C where the carelessness is especially noticeable, e.g. pp. 128–32, 148–52, 160–6, 192–4. The singular merit of C is that it gives us the name of the author and the title of the work. The peculiarity of V is that it contains a number of short additional passages which are not found in CP. To give one example: after a discussion of the reasons why some sounds are consonant and others dissonant, the conclusion is reached (p. 86, ll. 22–3):

Unde commensuratio est causa consonantie, incommensurabilitas est causa inconsonantie.

V adds:

[17] A. Pelzer *Bibl. apostolicae Vaticanae codices manuscripti recensiti. Codd. Vat. lat.* Pars prior, t. II, Vatican City, 1931, p. 200.
[18] B. L. Ullman, *The Humanism of Coluccio Salutati*, Padua, 1963, pp. 182, 279.

> Et tamen non quelibet superparticularis proportio in sonis facit consonan-
> tiam.

This and other additions[19] appear to be glosses of some later student
which have crept into the text. They have accordingly been relegated
to the *apparatus criticus*.

There are a small number of instances where V alone is right against
CP, usually in supplying omissions of single words.[20] At p. 26, l. 17 V
is certainly right in reading 'augens' against 'agens' CP, and at p. 48,
l. 6 in reading 'humore' against 'humido' C 'humano' P. There are,
however, a certain number of instances of agreement between PV
against C where ordinary scribal carelessness does not appear to be
the explanation. Consider the following passages:

> p. 38, l. 11 Ea sunt contraria opposita que sunt sub contrariis oppositis
> potentiis C
> Preterea. Contrarii actus a contrariis potentiis sunt PV
> p. 80, ll. 28–9 quoniam similiter contingit frigiditatem videri si albedo sit
> effectus frigiditatis C
> quoniam similiter posset concedi (credi V) frigiditatem videri si albedo
> esset in frigiditate PV
> p. 112, l. 1 Sequitur de gustu C[21]
> Dicto de olfacto dicendum est de gustu P
> Consequenter dicendum est de gustu V
> p. 138, l. 25 quia nec exigitur C
> ita quod non exigatur PV
> p. 138, l. 30–p. 140, l. 1 sicut superius dictum est de visu, fit ymago rei
> vise C
> sicut superius dictum est de visu, quod ymago rei vise sit in visu PV

In these passages C offers a text which makes sense, but is more
clumsy in expression than that offered by PV. It is tempting to suppose that PV contain a slightly revised text, but as so often in matters
of textual criticism, the question is not so simple as this. There are
instances where V agrees with C against P, where the differences are
just those which we have been illustrating, for example,

> p. 72, l. 9 Si corpus tenebrosum fuerit equale corpori luminoso CV
> Si corpus luminosum et corpus tenebrosum sunt equalia P.

[19] See the *apparatus* at pp. 8, l. 9; 20, ll. 10–11; 22, l. 6; 32, l. 11; 70, l. 29; 82, l. 3; 102,
l. 16; 112, ll. 11–14; 122, l. 9; 134, l. 14; 180, l. 29; 204, ll. 28–9. There is one place (p. 130,
ll. 28–9) where an addition of V has been admitted into the text.

[20] At pp. 46, ll. 10, 22; 52, l. 5; 56, l. 16; 174, l. 20; 206, l. 23 V supplies single words
omitted by CP. At pp. 26, ll. 6, 8; 58, l. 27; 102, l. 20; 122, l. 7; 136, l. 4; 150, l. 29; 156,
ll. 11, 17 P supplies single words omitted by CV.

[21] Cf. p. 116, l. 14, where all three manuscripts read 'Sequitur de tactu'.

The truth may be that we have one of those cases where scribes copied what they thought to be the sense without too nice a regard for the actual words of the author. It is a situation that came to be all too common in university texts of the thirteenth century.[22] Fr. Callus had in these passages adopted the readings of P, and I have thought it best not to make alterations in the text on which he had lavished such care and attention. Only in one place have I ventured to make a change. At p. 144, l. 16 (no. 268) PV read 'Zenonis', where C has the unintelligible 'mentionis'. The context shows that the correct reading is 'Menonis', which it seemed best to put in the text.

In working over a text prepared by another hand it is inevitable that one should occasionally question the choice of reading. Nearly always it is a choice which does not affect the sense. For instance, at p. 220, l. 4 the text reads 'servandi' with PV (and the source, Anselm), where C gives 'conservandi', but at pp. 212, l. 23 and 220, l. 17 the text reads 'conservandi' with CV, where P had 'servandi'. There are, however, one or two places where the sense is affected, and where the readings of CV seem preferable to those of P which are in the text:

> p. 16, l. 23 Et ideo ponebant aera esse id quod corpus vivificat CV
>   Et ideo ponebant aera esse animam, quia illud quod corpus vivificat P
> p. 80, l. 20 coloris CV   corporis P
> p. 86, l. 25 inconcinnus CV   concinnus P
>   l. 26 alia CV   illa P.

The chapter headings call for a word of comment. As we have seen, there is a list of chapters at the beginning in CV, and space left for one in P; but the headings in the text are only found in C, except for c. XIV and c. XVI which are added in the margin in V. Both in P and in V there are breaks between the chapters, but not space enough for the headings. The wording of the headings in the text often differs from that given in the list of chapters at the beginning, sometimes only slightly, as in c. III:

> *List*: De diversis opinionibus circa anime essentiam.
> *Text*: De diversorum opinione de anima.

sometimes more widely as in c. XVII:

> *List*: Ostensio quod sensus communis sit, et ad quid sit sensus communis.
> *Text*: De sensu communi. (C, V *marg.*)

---

[22] Cf. A. Dondaine, *Secrétaires de Saint Thomas*, Rome, 1956, especially his conclusion, pp. 205–7.

Further there are subordinate headings in the text in C in cc. II, XIV, XXV, XXVI. In c. XIV they show a certain correspondence with the list:

> *List*: De olfactu et odore, et quare odor potius immutet olfactum quam aliquem alium sensum.
> *Text*: XIV. i De olfactu.   XIV. ii De odore, et quare potius inmutat olfactum quam alium sensum.   XIV. iii Utrum instrumentum odoratus sit alicuius odoris.

Sometimes there is no subordinate heading where one would expect to find it, as in c. XXV:

> *List*: De vi contemplativa, et ratione et intellectu; et utrum mundus habeat animam sicut corpus humanum habet animam.

But there is no subordinate heading at no. 351, where the discussion of the *anima mundi* begins, and the text is so articulated that one could not conveniently be inserted. In fact we find:

> *Text*: XXV. i De viribus anime rationalis.   XXV. ii De intellectu. XXV. iii Inquirit utrum nova anima infundatur corpori vel antiqua. XXV. iv In qua parte corporis sit anima.

There is nothing in the list to correspond with the last two headings. It looks as if the list of chapters at the beginning was a draft for the work which was not revised after it had been written. Nothing corresponding to the last chapter (XXVII) in the list of headings 'De divina providentia' is found in any of the manuscripts.

I regret that I had not studied these discrepancies until the text was set up in type, and the *apparatus criticus* is not always consistent. In cc. III, VI, XII–XV, XVII to the end the *apparatus* is silent. It should be noted that the headings in the text in cc. I. i, ii, and II. i are editorial additions.

All the numbering is editorial, both in the list of chapters at the beginning, and in the text. The chapters of the *Tractatus* are too long for convenience of citation, and Fr. Callus divided the text into short numbered sections, which I have used in this introduction and in making the indexes.

There is no general agreement about the orthography to be adopted for editions of Medieval Latin texts. Fr. Callus, rightly it seems to me, decided to keep the spelling of the manuscripts, and in particular of C. Variant spellings of note are included in the index. It is sometimes difficult to distinguish between a genuine medieval form and a scribal mistake. For example, at p. 194, l. 11 'decontinuationem', which is written out in C, has not been changed to 'discontinuationem'. It is a possible form, although I have not been able to find another instance of its use.

The most difficult part of the task of an editor of a text like the *Tractatus* is the investigation of the sources. Fr. Callus has shown in his papers the indebtedness of John Blund to Avicenna *De anima*. 'There can be no doubt', he says, 'that Blund's main inspiration is Avicenna.' The student of the text will find in the *apparatus* not only references and parallels to Avicenna but also to Algazel and Ps. Alexander of Aphrodisias and to many works of Aristotle. But John Blund was grafting this newer learning onto older stocks. To quote Fr. Callus again: 'There are other influences at work . . . particularly William of Conches in the chapter "De umbra". Nor is Adelard of Bath to be ruled out; certain *quaestiunculae* closely resemble the *Quaestiones Naturales*, though it is possible that both are based on a common source.'[23] Fr. Callus had made a very wide search for relevant material, as will be plain from the *apparatus;* but he did not live to gather up all the threads, as he had intended. For me to have attempted this task would have further delayed his long-awaited edition; and I have judged it better to confine myself to the verification of the passages he had cited in that part of the text which he had completed. Apart from the passage already mentioned (no. 268), I have only once altered his *apparatus* (no. 227), where the source that certainly lies behind John Blund's text has been discovered by Dr. Brian Lawn. In nos. 337 to the end I have tried to follow the method which Fr. Callus had used.

It is now nearly forty years since Alexander Birkenmajer[24] drew attention to the important part played by medical writers and naturalists in the spread of new philosophical and scientific ideas in the twelfth century. Very little has been done since then to make their works available in print, and without editions it is not easy to study the texts. Recently Dr. Lawn, to whose edition of the Salernitan *Quaestiones phisicales*[25] Fr. Callus referred in the *apparatus* to no. 111, has been exploring this field, and he has very generously spent much time on the *Tractatus*. I have to thank him for supplying valuable references, which I have incorporated in the *apparatus* from no. 337 onwards. Those on the earlier part of the text I print here, together with other

---

[23] *The Treatise*, p. 486.

[24] *Le rôle joué par les médecins et les naturalistes dans la réception d'Aristote au XIIe et XIIIe siècles* (La Pologne au VIe Congrès international des sciences historiques), Warsaw, 1930.

[25] *The Salernitan Questions*, Oxford, 1963.

addenda. I regret that it has not been possible to insert a reference to more than a few of these passages in the *apparatus*, but all the additional references are included in the index of authors cited, where they are marked with an asterisk.

The edition of Avicenna *De anima*, books iv–v, by Simone van Riet, did not come into my hands until the text was printed, but I have been able to rectify the quotations from these two books, and I have given the page references to her edition in the index of authors cited.

ADDENDA TO TEXT
AND APPARATUS (1970)

No. 22, ll. 16–18. Cf. Callus *The Treatise*, p. 482 and n. 29. To the references there given Fr. Callus added in his own copy a reference to *Sententie Atrebatenses*, ed. Lottin, *Psychol. et morale* v (1959), 414, nos. 54–62.

No. 46, ll. 16–17. Cf. Constantin. *Pantegn. theor.* IV. i (xiv^{vb}): 'Nutrimentum est in membris restauratio dissolute rei ut animal permanere queat'.

No. 54, p. 32, ll. 3–5. Cf. Maurus *Super Isagoge Ioannitii*: 'Perfecta igitur celebrata digestione inferior porta stomachi aperitur, et residuum predicte sucositatis pure ad portonarium mandatur, in quo puri a non puro fit separatio . . . In eo igitur purior pars illius sucositatis ab impura sequestratur, et pura pars per venas meseriacas ad epar transmittitur, impura vero ad duodenum mandatur' (cod. Paris. lat. 18499, fol. 10). A similar passage is found in *Anatomia magistri Nicolai phisici*, ed. Franz Redeker, Leipzig Diss. 1917, pp. 50–1, which is probably based on Maurus. Cf. also no. 352 below.

No. 84, ll. 1–3. Cf. Guill. de Conchis *Glosae super Platonem* (100); Maurus op. cit.: 'Dum ergo aliquid calens vel frigens vel cuiuslibet proprietatis propinquat instrumento sensus, spiritus existens in illo instrumento immutatur secundum proprietatem rei sense. Spiritus immutatus ibidem representat anime illas immutationes, et sic in membro sensus perficitur' (fol. 17^{v}). Avic. *De An.* II. ii (7ra B).

Nos. 111–17. This is a small group of *quaestiones naturales*. The same questions are found in the English collection of Salernitan questions (Oxford, Bodl. Lib. cod. Auct. F. 3. 10), for which see Lawn, *Salernitan Questions*, pp. 34–6,[1] but the answers given to nos. 111–13 are not the same. The questions posed in nos. 111 and 114 are found in the *Problemata* of Ps. Alexander of Aphrodisias (I. 68, II. 42 Ideler), and those posed in nos. 112 and 113 in Ps. Aristotle *Problemata* XI. 49, 58 (904^{b}, 905^{a}). Neither of these collections was translated into Latin until later, Ps. Aristotle by Bartholomew of Messina in the middle of the thirteenth century, Ps. Alexander of Aphrodisias by Peter of Abano in 1302 (Lawn, op. cit., pp. 79–80). David of Dinant had access to a manuscript of the Aristotelian *Problemata* in Greek.

No. 113. Cf. Guill. de Conchis *Dragm.* VI (294).

Nos. 114, 117. The question comes up in the Salernitan questions mentioned above in answer to the question: 'Quare quidam habent oculos fascinantes?': 'Aer ab animali spiritu inficitur, quod potest videri in lippo, quia si lippus ali-

---

[1] Dr. Lawn is preparing an edition of these *quaestiones*, which will form the next volume in this series [ABMA V, 1979].

quem sanos oculos habentem recto intuitu respiciat, lippus efficitur et ille.'
Cf. Guill. de Conchis *Glosae super Platonem* (237–8).

No. 212, ll. 24–5. Cf. Constantin. *Pantegn. theor.* I. vii (ii^v): 'Homo enim omnibus temperatior est animalis speciebus'.

No. 221, ll. 19–21. Cf. Macrob. *Sat.* VII. 9. 11–15.

No. 226. Cf. Isaac *Liber dietarum universalium*, x (xxxii$^{rb–va}$): 'Omne autem
cibarium duobus modis dividitur, aut enim sensui est passivum aut nichil
sensui illativum. Nichil enim inferens dupliciter dividitur, aut enim corpus
compositum, sicut albugo ovi et similia, aut simplex, sicut aqua et cetera elementa . . . Passionem inferentes qualitates sunt diverse . . . Nullam vero passionem gustui inferentes si sunt simplices nil erga sensus sunt operantes cum
sint simplices. Nam res sensus operantes sunt composite, sensus enim compositi. (Nam res . . . compositi *om. ed. Lugd. restitui secundum codd. Bodl.
Auct. F. 5. 30, fol. 51^v et Bodl. 355, fol. 56vb*). Singularia vero non faciunt in
compositis quod composita in compositis . . . Non ergo elementa sunt sensui
passionem inferentes, cum sensus sint compositi, et ipsa simplicia.'

No. 231, l. 2. For the phrase 'celebrata digestione' cf. the passage of Maurus
cited above (no. 54).

No. 273, l. 29. Read 'effectu' with V.

No. 325, ll. 19–21. Cf. Constantin. *Pantegn. theor.* I. iv (ii^{ra}): 'Quecumque destruuntur necesse est ad hec quatuor (*scil.* elementa) revertantur . . . Elementa
ergo ex toto non destruuntur, quia hec destructa que ex ipsis construuntur ad
suum principium particulariter revertuntur', and the passages cited in Lawn
op. cit., p. 23, n. 6.

No. 336, p. 184, l. 3. mores: Avic. *De An.* v. i (p. 81 van Riet). Cf. 351.

No. 363, ll. 23–4. inquisitionem: The sense requires 'adquisitionem', which
is the word used in the parallel phrase in no. 324.

No. 366, p. 202, l. 14. habeatur intellectum: The text is corrupt. Dr. L.
Minio-Paluello suggests 'habet uti intellectu'. For this use of 'habet' cf.
no. 308 'habet convertere suas intentiones'.

No. 375, l. 23. et per hoc: A new section should start here, since it is the
solution of the question propounded in no. 374.

SIGLA CODICUM

C   Cantabrigiae, Bibl. Collegii S. Iohannis, cod. 120

P   Pragae, Bibl. Universitatis, cod. IV. D. 13 (667)

V   Civitatis Vaticanae, cod. Vat. lat. 833

Tituli operum in apparatu fontium adhibitorum plenius
in indice auctorum (pp. 241–6) reperientur

INDEX CAPITULORUM

Index capitulorum deest in P.   3 firmamentum: et quid sit anima *add.* C
4 capitulum *om.* C   6 oppinionibus V   8 et² *om.* V   8–9 incorruptibilis:
corruptibilis V   13 vi *om.* C   16 aliqua: aliquam V   ut *om.* V   23 eorum:
eorundem V   24 soni: sonitu V   ut *om.* V   echo: hecco C   26 aliquem alium
om. C

CHAPTER INDEX

2 vel: et V   11 xxvii. De divina providentia *om*. V

JOHN BLUND
Treatise on the Soul

TEXT AND TRANSLATION

TRACTATUS DE ANIMA SECUNDUM IOHANNEM BLONDUM

1. Ut habetur ab Aristotele, omnis scientia est ab anima et in anima principaliter fundata. Et ideo cum verecundum sit ignorare illud a quo est scientia, dicendum est de anima et de dispositionibus eius. Unde 5 cum prima questio de re sit, an sit, primo loco inquirendum est an anima sit.

I. i. *An anima sit*

2. Quod anima sit sic ostenditur. Nos videmus quod quedam sunt que moventur voluntarie, et omnis doctrina et omnis disciplina sumit 10 ortum a cognitione sensitiva; unde ab hoc quod a sensu habemus possumus ostendere quod anima est. Illud enim quod movetur voluntarie, aut movetur secundum naturam corporis per se et proprie in quantum ipsum est, aut non. Non ex natura corporis in quantum ipsum est, quia secundum hoc omne corpus moveretur voluntarie; ergo movetur per aliquid quod est aliud a natura corporis. Quicquid illud sit, illud est 15 anima; et ita habetur quod anima est.

ii. *An motus orbicularis sit naturalis vel voluntarius*

3. Sed inter ea que moventur, dubium est de pluribus an motus dans motum ipsi mobili sit anima vel natura, ut in eis que moventur 20

1-2 *Inscriptione carent* PV   3 ab: de C   et *om.* P   4 principaliter *om.* P
5 de² *om.* C   eius: anime PV   7 anima sit: et postea utrum anima sit corporea an incorporea; utrum corruptibilis an incorruptibilis; utrum etiam sit composita ex materia et forma vel non; et cetera dubitabilia adnectenda sunt *add.* P   8 *Titulum om.* codd.   9 anima: que P   15 moveretur: movetur PV   16 corporis *om.* P   18 *Titulum om.* codd.

3-4 Cf. Arist. *De An.* I. 1 (402ᵃ4-7): 'Videtur autem ad veritatem omnem cognitio ipsius etiam maxime proficere; maxime autem ad naturam: est enim tanquam principium animalium' (cod. Coll. Corporis Christi Oxon. 111, fol. 252ʳᵃ). Gundiss. *De Divisione Philosophiae* (44-45): 'Scientia (dicitur) cum iam in anima retinetur; nam, dicente Aristotele, omnis scientia in anima est'; Abaelard. *Glossae super Praedicamenta* (212): 'Occurrit quaestio, quare dicat perempto animali scientiam permanere non posse, cum videlicet scientia in animali non fundatur, sed in anima tantum'.
4-5 Cf. *Prologus translatoris Avicennae de Anima*; Gundiss. *Prologus in Tractatu De Anima* (31); Ps. Guill. de Conchis *Philosophia* (19).
9-17 Cf. Avic. *De An.* I. i (1ʳᵇA); Gundiss. *De An.* i (32).
10-11 Cf. Arist. *Anal. Post.* I. i (71ᵃ1); II. xix (99ᵇ35 seq.). Johan. Saresberiensis *Metalogicon* iv. 20 (186); Petrus Pictaviensis *Sententiae* I. xv (i. 153): 'Ex sensu a quo omnis doctrina initium sumit'.
19-p. 2, l. 1 Cf. Daniel de Morley *Philosophia* ii (26-27).

TREATISE ON THE SOUL
BY JOHN BLUND

1. As we know from Aristotle, all knowledge is from the soul and is rooted in the soul as its source. And so, since it would be shameful not to know that from which knowledge is, we should speak about the soul and how it is arranged. Whence, since the first question about anything is whether it is, in the first place we should enquire as to whether the soul exists.[1]

I. i. *Does the soul exist?*

2. That the soul exists can be shown as follows. We can see that there are some things that are moved voluntarily and every learned subject and every branch of knowledge takes its origin from sense knowledge. Thus, from what we have from the senses we can demonstrate that the soul exists. For that which is moved voluntarily is either moved according to the nature of the body in itself and precisely in so far as the body is, or it is not. Not from the nature of the body in so far as it is, because then every body would be moved voluntarily. Therefore, it is moved by something which is other than the nature of the body. Whatever this is, it is the soul. Thus, we know that the soul exists.

I. ii. *Whether circular motion is natural or voluntary*

3. However, among those things that are moved, a doubt arises regarding many of them as to whether a movement giving motion to something moveable is from the soul or from nature, as in those which

---

[1] P adds: And after that whether the soul is corporeal or incorporeal; whether it is corruptible or incorruptible; and further whether or not it is composed of matter or form; and a number of other matters for discussion [*dubitabilia*] are appended.

orbiculariter. Et ideo videndum est utrum motus orbicularis sit naturalis vel voluntarius.

4. Quod non sit a natura sic ostenditur. Nihil movetur naturaliter nisi quando ipsum participat dispositionem sibi non naturalem. Hec propositio satis nota est in principio intellectus: quoniam existens 5 in dispositione sibi naturali in ea quiescit, ut ponderosum existens in centro terre. Si autem sursum esset, ipsum haberet inclinationem tendentem deorsum; unde si relinqueretur sue nature, transmutaretur a loco sursum usque ad locum deorsum. Supponatur ergo hec propositio, cum per se sit nota in intellectu: nihil movetur naturaliter nisi 10 quia ipsum existit in dispositione sibi non naturali. Sol ergo existens in aliquo puncto non movetur ab illo nisi quia ipse non est in suo proprio 'ubi', et movetur ad 'ubi' quod est ei proprium, in quo, cum ipsum sit ei naturale, intentionem habet quiescendi in illo. Si ergo sol existens in oriente moveatur ad occidens naturaliter, perveniens ad occidens, 15 oportet quod tunc sit in dispositione sibi naturali, et tunc habeat in ea quiescere de sui natura. Si ergo ab illo 'ubi' exeat sol, non exibit ab eo per sui naturam, sed per violentiam; et ita motus solis, qui erit ab occidente in oriens, non erit naturalis, immo potius violentus. Ut si ignis descendat deorsum, ille motus est violentus et non naturalis, quoniam 20 proprium 'ubi' ignis est sursum. Sed qua ratione motus ab occidente in oriens non est naturalis, pari ratione motus ab oriente in occidens non erit naturalis. Sed ille motus non est naturalis; ergo nec iste. Consimili modo potest ostendi quod alii motus orbiculares non sunt a natura.   25

5. Item. Nihil secundum se et secundum sui naturam appetit et fugit unum et idem. Hec propositio per se nota est. Ergo si aliquid fugit aliud naturaliter, non redit ad illud idem naturaliter. Ergo si sol fugit aliquem punctum naturaliter, non redit ad eundem punctum

3 non *om.* V   4 sibi non naturalem: non naturalem sibi P   5 satis *om.* P   in principio intellectus: a sensu P   6 in ea *om.* P   11 quia: vel quando *add.* V   existit: existat P   12 ipse: sol *add.* V   13 ipsum: ipse V   14 naturale: naturalem P   in illo *om.* PV   sol *om.* P   16 habeat: habet CV   ea: eo CV   20 violentus et *om.* P   non naturalis: innaturalis P 21–22 ab occidente in oriens: ab oriente in occidens V   22 est: erit V   oriente in occidens: occidente in oriens V   23–24 Consimili: Simili P   24 modo *om.* V 26 secundum² *om.* C   27 unum et idem: et idem unum P   28 aliud: aliquid V   28 naturaliter² *om.* P

p. 1, l. 20–p. 2, l. 1 moventur orbiculariter: 'orbiculata', 'orbiculata circuitione' occurrunt apud Platonem, *Timaeus* 37A in versione Calcidii (29, 153).
1 motus orbicularis: cf. Adelard. Bath. *Quaest. Nat.* lxxvi (68).
3–11 Cf. Arist. *De Caelo* I. ii; I. ix (279^b2); *Physic.* II. i (192^b11–23).
3 Cf. Avic. *Metaphys.* IX. ii (102rb–va^A).
11–25 Cf. Algazel *Metaphys.* I. iv. 1 (92).
26–p. 3, l. 3 Cf. Avic. loc. cit.; Algazel *Metaphys.* I. iv. 3 (105).

are moved in an orbital fashion. Thus, we have to examine whether orbital motion is natural or voluntary.

4. That it is not natural can be shown in the following way. Nothing is moved naturally except when it finds itself in a place which is not natural to it. This proposition is sufficiently known at the beginning of understanding: for something which exists in a place which is natural to it will remain at rest there, such as a heavy body existing at the centre of the earth. If, however, the heavy body were up high it would have a tendency to go downwards; thus if it were left to its nature, it would be moved from the place up high to the place below. Therefore, if we accept this proposition, since it is known in itself in the understanding, nothing is moved naturally except if it exists in a place which is not natural to it. Therefore, when the sun is at any point it will not be moved from there unless it is not in its own proper place, and it is moved to the place which is proper to it, in which, since it is the place natural to it, it has the tendency to be at rest. If, then, when the sun is in the east, it is moved naturally towards the west, when it reaches the west it is the case that then it is in a place which is natural to it, and will be at rest there according to its nature. If, therefore, the sun moves from that position, it would not move from there because of its nature, but because of violence; and thus the motion of the sun from west to east would not be natural—on the contrary, it would be violent. Just as, if fire were to move downwards, that is a violent and unnatural motion, because the proper place of fire is up. However, the reason why the movement from west to east is not natural is the same reason why the movement from east to west is not natural. But the former motion is not natural, therefore neither is the latter. In like manner it can be shown that other orbital movements do not arise from nature.

5. Again, nothing in itself and according to its nature desires and flees from one and the same thing. This proposition is self-evident. Therefore, if something naturally flees from something else, it will not naturally return to the same. Therefore, if the sun naturally moves away from a certain point, it will not return to the same point natur-

naturaliter. Sed cum eius motus sit orbicularis, exit ab uno puncto et redit ad illud idem. Relinquitur ergo quod eius motus non sit a natura. Et est a natura vel ab anima. Relinquitur ergo quod ab anima.

6. Similiter potest ostendi de qualibet parte firmamenti, cum ipsa moveatur orbiculariter. Hac ergo ratione videtur posse haberi quod 5 firmamentum et alia superiora motorem habeant animam.

7. Amplius. In libro *de Consolatione*, loquens de anima in illo capitulo, *O qui perpetua* etc., Boetius dicit:

In semet reditura meat, mentemque profundam
Circuit, et simili convertit ymagine celum.   10

Et sic vult Boetius quod celum moveatur ab anima.

8. Item. In omogeniis totum et quelibet eius pars eiusdem nature sunt; et ideo dicit Aristoteles in *Physica* quod eadem natura et est totius terre et unius glebe, quia tota terra et quelibet eius pars appetit esse circa centrum. Firmamentum autem est res omogenia, quia non 15 componitur ex rebus diversarum naturarum. Ergo quelibet eius pars est eiusdem nature cum suo toto. Sed firmamentum de sui natura habet specialiter moveri circa centrum terre; ergo et eius quelibet pars; quod falsum est. Ut si intelligatur semispera firmamenti versus alterum polorum dividi in duas partes equales, et altera earum partium maneat 20 et reliqua abstrahatur quantum ad intellectum, non habebit centrum terre pro suo centro. Sed si a natura moveretur, haberet illud centrum. Sed non habet; ergo a natura non movetur. Et movetur a natura vel ab anima: non a natura; ergo ab anima.

9. Et est corpus. Ergo est corpus animatum et mobile voluntarie; 25 ergo est animal. Non enim movetur ab anima vegetabili, quoniam

---

1 orbicularis: naturalis C   2 sit: est V   3 Relinquitur *om.* P   quod
om. P   4 ostendi: obici CP   5 haberi: probari P   6 et alia . . .
habeant: a natura superiora motione habent P   7–11 Amplius . . . ab anima
om. P   7 Amplius: Boetius *add.* V   8 Boetius *om.* V   dicit: dixit V
12 Item: Amplius V   14 quia: quod C   15 autem: omnino C   19 si
om. V   semispera: semisperii C; semper P   firmamenti: firmamentum P
alterum: altitudinem P   20 maneat: maneatur P   21 reliqua abstrahatur: relique abstrahantur PV   intellectum: si pars removens moveatur *add.* V
habebit: habet C   24 non a natura *om.* P   26 Non: hoc C

7–10 Boeth. *De Consol. Phil.* III metr. ix. 16–17.
13–14 Arist. in *Physica*: i.e. *De Caelo* II. xiv (297^{b}8–9) (iuxta versionem arabicam):
'Et hic quidem sermo conveniens est in gleba una et una partium terre, sicut in
totalitate terre' (cod. Coll. Corporis Christi Oxon. III, fol. 179vb). Vid. Daniel de
Morley *Philosophia* ii (27).
25–p. 4, l. 4, Cf. Arist. *De Caelo* II. ii; Algazel ibid. (106).
26 animal: Avic. *Metaphys.* IX. i ad fin. (102^{ra}A): 'celum est animal obediens deo';
Algazel loc. cit.; Adelard. Bath. *De Eodem et Diverso* (32): 'Superiora quippe illa
divinaque animalia inferiorum naturarum et principium et causae sunt'; *Quaest. Nat.*
lxxii (64). Cf. Bernardus Silvestris *De Mundi Universitate* II. vii (48).

ally. However, since its motion is circular, it moves away from one point and returns to that very same point. It follows, therefore, that its movement does not arise from nature. Now its movement arises either from nature or from a soul; therefore, it can only follow that it arises from a soul.

6. The same can be shown with regard to any part of the heavens if it moves in a circular way. Thus, by means of this argument it seems that we can know that the firmament and the other heavenly bodies have a soul as a mover.

7. Furthermore, in the book *On the Consolation of Philosophy*, when speaking about the soul in the chapter which begins '*You who rule the universe*', etc., Boethius says: 'The soul in going round returns to itself, it encircles Mind, and changes the heavens into an image of itself.' So in this way Boethius holds that the heavens are moved by soul.

8. Again, in homogeneous things, the whole and any of its parts are of the same nature. For this reason Aristotle states in his natural philosophy that the same nature is to be found in all of the earth and in a piece of soil, since the whole earth and any part of it seeks to be round the centre. Now the firmament is a homogeneous thing since it is not composed of things with varied natures; therefore any of its parts is of the same nature as its whole. However, the firmament of its nature tends to move specifically towards the centre of the earth; therefore any part of it as well; which is false. Now, if we take the hemisphere of the firmament to be divided into two equal parts from one pole to the other, and if one part stays and we take the other part to be removed, it will not have the centre of the earth as its centre. However, if it is moved by nature, it will have that centre. Yet it does not have that centre, therefore it is not moved by nature. Now, either it is moved by nature or by a soul; not by nature, therefore by a soul.

9. Moreover, the firmament is a body; therefore it is an animated body and can be moveable voluntarily; therefore it has awareness. For it is not moved by the vegetative soul, because the vegetative soul is

anima vegetabilis illud quod sua vegetabilitate movet in diversas partes movet, sursum, deorsum et in latus, ut patet in plantis et in arboribus. Et etiam, in quo est anima vegetabilis est vis generativa et nutritiva et augmentativa, et sex species motus. Sed huiusmodi motus non sunt in superioribus. Si ergo dicatur quod sit animal: ergo participat quinque 5 sensus animalis, videlicet gustum, olfactum, visum, tactum, auditum; quod est falsum.

10. Solutio. Dicimus quod firmamentum movetur a natura, non ab anima, et alia supercelestia; et ille motus naturalis est propter perfectionem habendam in inferioribus. Per motus enim superiorum 10 reguntur inferiora, et adquisita perfectione omnino inferiorum, cessabunt superiora a motu, et erit in eis quies naturalis. Unde contingit quod superiora que moventur ibi participant sibi dispositiones innaturales et moventur ad quietem sibi naturalem.

11. Ad hoc quod obicitur, quod nihil naturaliter fugit et appetit 15 idem, dicendum est quod hoc verum est, ita quod locutio intelligatur per se sumpta. Si autem per accidens intelligatur, falsa est locutio; ut si lapis existens sursum tendat ad centrum, omne 'ubi' intermedium appetitur per accidens, quoniam propter aliud.

12. Similiter: illa pars firmamenti que modo est in oriente iam erit 20 in occidente, et fugit oriens per accidens et redit ad illud idem per accidens, hoc est propter aliud, quoniam propter quietem habendam. Nec est omnis motus naturalis secundum rectum, immo quidam secundum obliquum, ut motus orbicularis.

---

1 diversas: tres V   2 in³ *om.* P   3–4 et etiam . . . motus *om.* P   3 quo: quocumque V   3–4 et augmentativa *om.* C   4 et sex species motus: harum virium V   4 sunt: est P   5 igitur: ergo V   6 videlicet: scilicet P   8 movetur: motu *add.* P   9 supercelestia: et hoc testatur Albumasar dixisse philosophos, dicens: dixerunt philosophi quod ipsi planete sunt corpora rotunda, lucida et mobilia motu naturali, *add.* V   11 omnino *om.* V   13 ibi *om.* V   14 naturalem: materialem V   15 quod² *om.* C   et appetit: ad oppositum P   16 est¹ *om.* C   ita *om.* V   17 intelligatur *om.* P   locutio *om.* P   18 tendat: tendit P   23 quidam: est *add.* V

---

4 sex species motus: cf. Arist. *Cat.* xiv (15ª13–14) (77).
5 Iohan. Damasc. *De Fide Orthodoxa* (iuxta versionem Burgundionis) xx. 11 (83): 'Nullus autem animatos caelos vel luminaria existimet: inanimati enim sunt et insensibiles'; Alex. Nequam *De Nat. Rer.* 1. ix (45): 'Absit ut planetas esse animalia censeamus'; cf. Daniel de Morley *Philosophia* ii (23, 27).
8–9 Guill. de Conchis *Phil. Mundi* 11. vii (59D): 'Alii dicunt eas (stellas) etiam proprio motu moveri, quia igneae sunt naturae, nec aliquid in aethere vel in aere sine motu possit sustineri, sed semper in eodem loco et circum se moveri'; cf. xvi (61D); *Dragm.* 111 (87).
10–11 Cf. Adelard. Bath. *Quaest. Nat.* lxxi–lxxiv (63–65); Guill. de Conchis *Phil. Mundi* 11. xxvi–xxvii (67–70).
15 Ad hoc quod obicitur: cf. supra, n. 5.
20 Similiter: cf. supra, n. 4.
23–24 Cf. Guill. de Conchis ibid. 11. xvi (61D); xxvi (67D); xxvii (70D).

that which, through its vegetative activity, moves different parts upwards, downwards, and sideways, as is clear in plants and trees. Again, inasmuch as it is a vegetative soul, it is a generative, nutritive, and augmentative power and together with six kinds of movement. However, these kinds of movement are not to be found in the heavenly bodies. Therefore, if it is said that the firmament has awareness, it follows that it would share in the five animal senses, namely, taste, smell, sight, touch, hearing, which is false.

10. Solution: We state that the firmament, as well as the other heavenly bodies, is moved by nature, not by a soul,[2] and this motion is natural owing to the perfection which is to be found in lower bodies. For it is through the motion of the superior bodies that the lower bodies are regulated, and when the lower bodies have acquired perfection in every way, the superior bodies would cease to move, and would achieve a natural state of rest. Thus, it must be the case that the superior bodies which are moved partake in dispositions there which are unnatural for them and are moved to a state of rest which is natural for them.

11. Regarding that which was objected,[3] that nothing naturally flees from and desires the same thing, it should be said that this is true on condition that the statement is understood to be taken without qualification. If, however, it is understood in an accidental manner, the statement is false. For example, if a stone which is up high seeks to be at the centre of the earth, each intermediate position would be desired in an accidental manner, because it is desired because of another position.

12. In a similar way, that part of the firmament which is now in the east will be in the west later, and it moves from the east in an accidental manner and returns to the same accidentally. This is due to another factor, namely, in order to achieve a state of rest. Nor is all natural motion in a straight line, rather some motion is curvilinear, as in the case of the motion of the planets.

---

[2] V adds: and this is what Abū Ma'shar affirms the philosophers to have stated: the philosophers said that the planets themselves are round luminous bodies moved by a natural motion.

[3] See above, no. 5.

13. Multi autem sunt qui dicunt superiora moveri ab anima, dicentes se illud habere ab Avicenna et ab aliis auctoribus. Dicunt enim quod ab intelligentia intelligente seipsam, in quantum ipsa habet esse possibile, fluit in eam anima celi et perfectio celi, que anima per ymaginationem apprehendit causam primam, que summum bonum est, et motu 5 desiderii vehementis accensa, movetur ut sibi adquirat id quod summum bonum est; et ita movetur a prima causa sicut amator a suo amato. Ut enim dicunt Algazel et Avicenna, sicut amatum movet suum amatorem, et sicut scientia questionis desiderata movet studentem et tendentem in illam adquirendam, ita pura bonitas, summe desiderata 10 et ab anima celi ymaginata, movet animam celi ut ipsa adquirat sibi puram bonitatem ut ea fruatur et ei assimiletur.

II. i. *De essentia anime*

14. Habito nunc quod anima sit, sequitur videre quid sit anima. Ab Aristotele habetur quod anima est corporis organici perfectio vitam 15 habentis in potentia.

15. Sed obicitur. Forma dat esse, et materia in se est imperfecta: unde omnis perfectio est a forma. Ergo cum perfectio corporis organici habentis vitam in potentia sit anima, anima est forma. Sed nulla forma est res per se existens separata a substantia. Ergo cum anima sit forma, 20 anima non habet dici res per se existens separata a substantia. Ergo anima non potest separari a corpore, sed perit cum corpore.

16. Ad hoc dicendum quod hoc nomen 'anima' designat rem suam in concretione. Significat enim substantiam sub quodam accidente in

---

1 autem *om.* P   4 in eam *om.* P; in esse V   6 id *om.* P   11 ipsa *om.* V
13 II i. De essentia anime *om.* codd.   14 nunc: vero V; *om.* P   videre: ut
videamus V   20 res *om.* P   21 habet dici: potest esse P   separata:
separatam P   22 potest separari: est separabilis P   sed: ergo P

1-2 Avic. *Metaphys.* IX. ii (102^{rb}A): 'Quod propinquus motor celestium non est natura, nec intelligentia, sed anima; et quod principium longinquum est intelligentia'; *De An.* I. i (2^{ra}Q); Ps. Avic. *De Celo et Mundo* vii (39vb): 'Quod motus circuli stellarum fixarum ab oriente in occidentem non potest esse a corpore'; Algazel *Metaphys.* I. iv. 3 (104-6): 'Celestia corpora sunt mobilia per animam et per voluntatem'. Vid. Arist. *De Caelo* II. xii (292^{a}20; cf. I. ix (279^{a}28); II. ii (285^{a}29-30). Boeth. *In Isagogen* II. ed.: 'Solem ceteraque caelestia corpora quae animata esse cum Plato tum plurimus doctorum chorus arbitratus est' (209; cf. 184; 257; 259). Calcidius *In Platonis Timaeum* cxxx-cxxxi (172-3). Isaac Israelita *De Elementis* iii (x^{rb}). Adelard. Bath. *Quaest. Nat.* lxxiv-lxxvi (65-69).
3-12 Algazel ibid. (112-15); Avic. *Metaphys.* IX. ii (103^{r}D). Cf. Arist. *Metaphys.* Λ (XII). vii (1072^{a}19-35).
14-16 Arist. *De An.* II. i (412^{a}27-28; 412^{b}5-6).
17 Cf. Boeth. *De Trin.* ii (1250B): 'Omne esse ex forma est'.
17-18 Cf. Avic. *De An.* I. i (1^{va}F).
23-p. 6, l. 3 Cf. Avic. ibid. (1^{vb}MN): 'Hoc enim nomen "anima" non est indictum ei ex substantia sua, sed ex hoc quod regit corpora et refertur ad illa, et idcirco recipitur

13. There are, however, many people who say that the heavenly bodies are moved by soul, saying that they know this from Avicenna and from other authors. Indeed, they say that from an Intelligence contemplating itself, inasmuch as this has a possible being, the soul of the sphere and perfection of the sphere flows into it, and that the soul through imagination apprehends the First Cause which is the Highest Good. Now the movement of intense desire having been ignited, it moves so that it can obtain for itself that which is the Highest Good; and thus it is moved by the First Cause just as a lover by its beloved. As al-Ġhazālī and Avicenna say, just as the lover moves the beloved and the desired solution to a problem gives rise to the studying and striving to acquire that knowledge, so it is that Pure Goodness, the highest thing desired and imagined by the soul of the heavens, moves the soul of the heavens so that it acquires Pure Goodness for itself so that it is enjoyed and becomes like it.

II. i. *On the essence of the soul*

14. Now that we know that the soul exists, we must next see what the soul is. From Aristotle, we have that the soul is the perfection of an organized body potentially having life.

15. However, it is objected: It is the form which gives existence, and matter in itself is incomplete, so all perfection is from form. Therefore, since the perfection of an organized body potentially having life is from the soul, the soul is the form. But no form is a thing existing in itself separate from a substance. Therefore, since the soul is a form, the soul should not be said to be a thing existing in itself separate from a substance. Therefore, the soul cannot be separated from the body, but perishes with the body.

16. With regard to this, it should be said that this word 'soul' refers to its object in a concrete manner. For it connotes a substance sub-

relatione ad corpus organicum in quantum ipsum animatur et vivificatur per ipsam, et gratia illius accidentis dicitur esse perfectio ipsius, eo
scilicet quod ipsa ipsum animat.

II. ii. *Cuius artificis sit speculatio de anima*

Sequitur videre cuius artificis sit speculatio de anima, an meta- 5
physici an physici.

17. Quod metaphysici sic ostenditur. Esse anime est esse creationis;
ergo eius esse non est a natura. Sed physicus habet tantum agere de
rebus naturalibus; ergo non habet agere de anima. Et esse quod est
a creatione, est supra esse naturale, et metaphysicus habet considerare 10
ea que sunt supra naturalia; ergo eius speculationi subiacet consideratio
de anima; et ita metaphysicus habet agere de anima.

18. Preterea. Ab Aristotele et ab Avicenna habemus quod subiectum
physici est corpus mobile in quantum ipsum est subiectum motus et
quietis. Sed nullius facultatis consideratio est de eis que sunt supra 15
suum subiectum, immo quascumque construit passiones habet con-
struere de suo subiecto. Sed esse anime non est sub corpore mobili, nec
est eius passio; ergo non subiacet considerationi physici. Et subiacet
considerationi physici vel metaphysici; ergo subiacet considerationi
metaphysici.   20

19. Contra. Ab Aristotele habemus quod anima est perfectio corporis organici vitam habentis in potentia, et corpus organicum pertinet
ad inquisitionem physici; ergo et eius perfectio. Sed anima est eius
perfectio; ergo anima subiacet considerationi physici.

20. Preterea. Infusio anime est terminus motus naturalis, sicut 25
igneitas; et anima est id ad quod tendit motus nature. Ex quo ergo
terminus motus naturalis et eius perfectio est de speculatione physici,
anima subiacet considerationi physici.

3 ipsa *om.* P   4 *Titulum om.* PV   5 Sequitur videre: Queritur P; ut
videamus V   7 sic ostenditur: videtur P   9 non *om.* V   10 est *om.* V
11 speculationi: speculum P   14 corpus: cuius P   17 de: cum P   est *om.* V
18 eius *om.* P   18–19 Et. . . vel *om.* V   23 inquisitionem: dispositionem P
25–28 Preterea. Infusio . . . physici *om. per homoeoteleuton* CP   28 considerationi:
consideratione V   physici: Forte . . . essentia *add.* V (*habentur infra*, n. 22)

corpus in sui diffinitione'; vid. etiam 2rb; Gundiss. *De An.* i (33); Ps. Aug. *De Spiritu
et Anima* ix (784): 'Anima vero ex eo dicta est quod animet corpus ad vivendum, hoc
est, vivificet'; cf. infra, Solutio n. 21.
    5–6 Cf. Arist. *De An.* I. i (403^{a}27–^{b}19); *Metaphys. E* (VI). i (1026^{a}5–6). Avic.
De An. I. i (I^{vb}N); v. v (25^{va}B). Alfredus Anglicus *De Motu Cordis* Prologus (1–4).
    7 esse creationis: cf. Avic. ibid. v. iii (24^{rb}B).
    13–15 Arist. *Physic.* I. ii (184^{b}18; 185^{a}12); II. i (192^{b}11–23); etc. Avic. *Metaphys.*
I. i (70^{ra}A); I. ii (70^{va}A).
    17–18 Avic. *De An.* v. v (25^{va}B): 'doctrine naturalis non est proprium speculari nisi
quod est rerum naturalium, et que habent comparationem ad materiam et motum'.
    21 Arist. *De An.* II. i (cf. supra, n. 14).

sumed under a certain accident in relation to an organized body inasmuch as the latter is animated and vivified by the soul, and thanks to
this accident it is said to be the perfection of the body, namely because
the soul animates the body.

II. ii. *Whose task it is to investigate the soul*

It now follows that we should examine whose task it is to investigate the soul, whether it is that of the metaphysician or the natural
philosopher.

17. That it is the task of the metaphysician is shown in the following manner. The being of the soul is a being which comes from creation; therefore its being does not arise from nature. However, the task
of the natural philosopher is to deal only with natural things; therefore
he does not treat of the soul. Furthermore, a being which arises from
creation surpasses natural things. Now the task of the metaphysician
is to consider those things which surpass natural things; therefore the
investigation of the soul falls under his consideration; and thus it is
the task of the metaphysician to treat of the soul.

18. Moreover, from Aristotle and from Avicenna we know that the
area of competence of the natural philosopher is the moveable body
inasmuch as it is subject to movement and rest. However, no branch of
study investigates those things which surpass its area of competence;
rather whatever attributes it examines, it must do so within its area of
competence. But the being of the soul cannot be classified as a moveable body, nor as an attribute of one; therefore it does not come under
the consideration of the natural philosopher. Now it comes under
the consideration of either the natural philosopher or metaphysician;
therefore it comes under the consideration of the metaphysician.

19. Against this: We know from Aristotle that the soul is the perfection of an organized body potentially having life, and an organized
body is something which pertains to the investigation of the natural
philosopher; therefore, the perfection of the organized body does as
well. But the soul is its perfection; therefore the soul comes under the
consideration of the natural philosopher.

20. Moreover, the infusion of the soul is the end of a natural motion, just like fieriness; and the soul is that towards which a process/
activity of nature aims. Since the end of a natural motion and its perfection is part of the consideration of the natural philosopher, the soul
comes under the consideration of the natural philosopher.

21. Solutio. Dicimus quod hoc nomen 'anima' est nomen concretum in concretione dans intelligere substantiam sub accidente quod copulatur per hoc verbum 'animo', 'animas'. Quantum autem ad illud accidens, dicimus quod dicit Aristoteles, animam esse perfectionem corporis organici habentis etc., et in hac comparatione anime ad corpus 5 est anima perfectio ipsius, scilicet in quantum ipsa vivificat corpus, et sub hac comparatione subiacet anima physici speculationi; preter autem illud accidens considerata, subiacet speculationi metaphysici. Ita hoc ipsum anima secundum suum esse concretum partim est de physica et partim est de metaphysica. Sicut triangulus secundum quid 10 est quantitas, et secundum hoc subiacet geometrice facultati; quantum autem ad formam a qua fit impositio ipsius nominis, triangulus est qualitas et non est de geometrica facultate.

22. Forte dicet aliquis quod theologi est tractare de anima.

Contra. Theologus habet inquirere qua via contingat animam mereri 15 et demereri, et quid sit ad salutem, quid ad penam. Quid autem anima sit, et in quo predicamento sit, et qualiter infundatur corpori, non habet ipse inquirere. Ex quo ista scire magis pertinent ad alium artificem. Ex quo ergo theologus solum habet docere qualiter sit merendum et demerendum, non habet ipse proprie docere quid sit anima nec quid 20 sit eius essentia.

III. *De diversorum opinione de anima*

23. Consequenter sciendum est quod diversi diversa sentiebant de anima. Quidam enim dicebant animam esse sanguinem quidam autem naturam, quidam vero aera.   25

24. Quod sanguis erat anima hac habebant industria. Per extractionem sanguinis contingit animam educi a corpore. Preterea, ex calore

---

1 concretum: concretivum C   5 hac: hoc, scilicet inquantum vivificat corpus *add. in marg.* C   comparatione anime: comparando animam C   6 scilicet inquantum ipsa vivificat corpus *om.* CP   7 sub: in V   anima: anime P   physici: physice C   10 et *om.* P   est *om.* P   Sicut: ipsum *add.* P   13 geometrica: geometrice P   14–21 Forte ... essentia: *om.* P; *habentur ante solutionem* (n. 20, l. 28) V   15 qua via: quam viam V   16 sit *om.* V   Quid: quod V   18 scire: inquirem V   23 sentiebant: sentiunt V   25 quidam vero aera *om.* CV   26 anima *om.* P   habebant: dicebant V

1–10 Cf. supra, n. 16.

16–17 Guill. de S. Theod. *De Natura Corporis et Animae* II (717D): 'Data autem anima quaerendum est quid sit, quare sit, quomodo sit. Cum quid sit quaerimus spectat ad physicam'.

24–25 sanguinem: cf. Arist. *De An.* I. ii (405[b]5); Avic. *De An.* I. ii (2[va]A); Gundiss *De An.* ii (37).

26–27 Cf. Nemesius *De Natura Hominis* ii (541); Avic. *De An.* I. ii (2[va]A; 3[rb]F).

21. Solution: We state that this word 'soul' [*anima*] is a concrete term, which conveys the understanding in a concrete situation of the substance which underlies an accident, which is linked to the substance by the verb 'to animate' ['*animo*', '*animas*']. However, with regard to that accident, we state what Aristotle said, that the soul is the perfection of an organized body, etc., and in this relation of the soul to the body, the soul is the perfection of the latter, namely inasmuch as the soul vivifies the body. Thus, when one stays within this relation the soul comes under the consideration of the natural philosopher. However, when considered apart from that accident it comes under the consideration of the metaphysician. Thus this very same soul according to its concrete being partially is the concern of natural philosophy and partially of metaphysics. Just as a triangle in a certain respect is a quantity, and for this reason comes under the study of geometry; however, with regard to the form from which the term is applied, the triangle is a quality and is not the concern of the discipline of geometry.

22. Perhaps someone might suggest that it is the task of the theologian to deal with the soul.

Against this: It is the task of the theologian to investigate the way in which the soul obtains or loses merit, and what leads to salvation and what to punishment. It is not his task to enquire into what the soul is, and in which category it lies, nor how it is infused into the body. From which it follows that to know these things is rather the concern of another expertise. Therefore, it follows that the task of the theologian is only to teach how one gains or loses merit; it is not his proper task to teach what the soul is nor what its essence is.

III. *On the opinion of various authors concerning the soul*

23. The next thing which should be known is that various authors had diverse opinions concerning the soul. For some said that the soul was blood, some that it was nature, some indeed that it was air.

24. They held that blood was the soul, through this observation: As a result of the extraction of blood the soul is necessarily drawn out of the body. Moreover, all things are procreated out of heat and hu-

et humore procreantur omnia. Sed sanguis est calidus et humidus; propter hoc dicebant quod ex sanguine provenit vite productio; et ita sanguis est anima.

25. Contra. Quanto aliquid magis participat de sua causa, tanto est magis in effectu sue cause; ergo si sanguis sit anima, quanto quis magis 5 participat sanguinem, tanto est magis in effectu ipsius anime; ergo tanto magis vivificatur; quod falsum est: quoniam tanta potest esse sanguinis habundantia quod tendit ad corruptionem et destructionem animalis.

26. Qui autem dicebant animam esse naturam, hoc iudicio illud 10 habebant: sicut ignis de sui natura movetur sursum, et ponderosum de sui natura tendit deorsum, ita ponebant animalia habere suam propriam naturam que fuit causa motus ipsorum.

27. Sed obicitur. Omne illud quod movetur a motore naturali in unicum certum terminum et definitum de sui natura habet moveri, ut 15 patet in igne et aliis elementis. Sed motus animalis non est in unicum terminum tantum; ergo motus animalis non est a motore naturali, et ita cum animal moveatur tum sursum tum deorsum tum in latus, et ille motus sit ab anima motore, anima non est natura.

28. Qui autem ponebant animam esse aera, hac ratione inducti erant 20 ad illud probandum. Videbant enim quod quamdiu animal inspirat et exspirat aera, vivit; quamcito deficit inspiratio et exspiratio aeris, moritur. Et ideo ponebant aera esse animam, quia illud quod corpus vivificat; et ita dicebant animam esse aera.

29. Sed obicitur. Aer in inspiratione ab aliquo attrahitur et in 25 exspiratione ab aliquo emittitur. Et licet virtus horum actuum sit ipsius pulmonis, tamen ipsa virtus ab aliquo alio est quod totum corpus vegetat. Et illud non est aer, cum nihil seipsum attrahat; et quicquid illud sit, illud est anima. Ergo anima non est aer.

30. Item. Non est virtutis corporee apprehendere res incorporeas, 30

---

5 cause: anime V   6 ipsius: sue V   7 tanta: tantum P   materia *add.* C
esse *om.* V   8 ad: in P   11 de *om.* P   12 de *om.* CP   ita: et *add.* P
13 ipsorum: illorum P   14 in: et V   18 animal: animalis V   20 animam
om. P   23 esse: animam, quia *om.* CV   illud: id CV   24 ita: ideo V
esse aera *om.* P   25 attrahitur: trahitur V   26 ipsius *om.* V   28 illud: ita
CV   28–29 quicquid illud sit *om.* C; et est anima P   30 Item: amplius V

---

10 Cf. Alfredus Anglicus *De Motu Cordis* vii (28–29).
14 Cf. Alfredus Anglicus ibid. xi (53).
17–19 Cf. supra, n. 9.
18 Cf. Arist. *De An.* I. ii (405[a]21); Avic. *De An.* I. ii (2[va]A).
21–24 Cf. Zeno Stoicus et Chrysippus apud Calcidium *In Platonis Timaeum* ccxx (232).
26–28 Cf. Nemesius ibid. (544–5).
30–p. 9, l. 4 Cf. Avic. *De An.* v. i (22[va]B).

mour. But blood is warm and moist; for this reason they said that the production of life arose out of blood; and thus blood is the soul.

25. Against this: The more something participates in its cause, the more there is of the cause in the effect; therefore if blood were the soul, the more someone had of blood, the more there would be of the soul itself in the effect; therefore it would be more vivified; which is false, because an abundance of blood can be so great that it leads to the destruction and corruption of the animal.

26. Those, however, who said that the soul is nature held it to be so for the following reason: just as fire moves upwards by its own nature, and something heavy moves downwards by its own nature, in the same way they posited that animals have their own nature which would be the cause of their movements.

27. But it is objected: Everything which is moved by a natural mover is moved by its own nature towards one precise and definite end, as is clear in the case of fire and the other elements. However, the movement of an animal is not towards one end only, therefore, the movement of an animal does not come from a natural mover. And so, since an animal is moved sometimes upwards, then downwards, then sideways, and that movement is from the soul as a mover, the soul is not a nature.

28. Those, however, who posited that the soul is air tried to prove this by means of the following argument. They saw, indeed, that for as long as an animal breathes in and breathes out, it lives; as soon as it no longer breathes in and out, it dies. And so they posited that air is the soul, since it is that which vivifies the body; and thus they said that the soul is air.

29. But it is objected: when someone breathes in they draw air in, and when they breathe out they emit air. And even though the power for these acts belongs to the lungs, yet this power is from something else which enlivens the whole body. And that is not air since it draws in nothing to itself; and whatever that power is, it is the soul. Therefore, the soul is not air.

30. Again, a corporeal power is unable to grasp incorporeal things

quia res incorporee neque subiacent sensui neque ymaginationi. Sed
in anime comprehensionem cadunt universalia et res incorporee. Ergo
anima non est corporea. Et est corporea vel incorporea; ergo anima est
incorporea.

31. Preterea. In hoc est similitudo inter animam et intelligentiam, 5
quod intelligentia intelligit ea que simplicia sunt, et extra vim sensi-
tivam et ymaginativam cadunt. Similiter et anima. Sed intelligentia
secundum suam propriam essentiam intelligit ea que intelligit, et est
substantia res separata, per se existens. Ergo et anima, cum ipsa simili
modo apprehendat, erit res per se existens; et ita anima erit sub- 10
stantia.

32. Preterea ad idem. Ab Avicenna habemus in commento *Metaphysice*, quod omnis substantia aut est corpus, aut est non-corpus. Si
est non-corpus, aut est pars corporis, aut non est pars corporis. Si est
pars corporis, vel est materia vel forma. Si non est pars corporis, vel 15
est tale quid quod habet ligationem cum corporibus, ut illa moveat, et
dicitur esse anima; vel est tale quid quod non habet ligationem cum
corporibus ut illa moveat, et dicitur esse intelligentia; et ita habemus
a precedentibus quod anima est substantia et incorporea.

33. Preterea. Omnis corruptio est secundum contrarium. Sed cum 20
anima sit res incorporea, nullam habet in se causam contrarietatis.
Ergo anima non potest corrumpi; ergo est incorruptibilis. Quod bene
concedimus.

34. Multas alias opiniones de anima ponit tam Aristoteles quam
Augustinus, et alii auctores; et ostendunt quod anima non est ex tra- 25
duce; quas opiniones non oportet hic tangere, quoniam ipsi sufficienter
eas ponunt et improbant.

---

1 sensui: visui CV   3 est¹: res *add.* P; est *om.* V   et est . . . vel
incorporea *om.* P   5 in hoc: hec C   intelligentiam: intellectum P
8 suam propriam essentiam: sui naturam P   9 ipsa *om.* CP   10 apprehendat: apprehendit V   12 Ab *om.* P   habemus: dicit P   13–14 si est
non-corpus *om.* P   14 corporis: corporalis V   15 vel–vel: aut V   16 ligationem: longitudinem P *etiam* 17   19 substantia et *om.* P   22 non *om.* V
24–27 Multas . . . improbant *om.* P   25 non est: non *om.* V   27 eas: et
eas V

12–18 Avic. *Metaphys.* II. i (75ʳᵃʙ).
24 Arist. *De An.* I. i–v.
25 Aug. *De Natura et Origine Animae* III–IV (512–14). alii auctores: cf. Macrobius *In Somnium Scipionis* I. xiv. 19–20; Nemesius ibid. ii (536–89); Avic. *De An.* I. ii
(2ʳᵇ–3ᵛᵃ); Gundiss. *De An.* ii (37).
25–26 non ex traduce: cf. Guill. de Conchis *Dragm.* VI (306): 'Cum Augustino igitur
credo et sentio quotidie novas animas creari non ex traduce, non ex aliqua materia, sed
ex nihilo, solo iussu eas a creatore creari'; Hugo de Ribodimonte *Epistola de Anima*
(833–6); Gundiss. *De An.* vi (52); Alex. Nequam *De Nat. Rer.* II. clii (233).

since incorporeal things are not grasped by either the senses or the imagination. But universals and incorporeal things do come under the grasp of the soul. Therefore the soul is not corporeal. And it is either corporeal or incorporeal; therefore the soul is incorporeal.

31. Moreover, the similarity between the soul and an intelligence consists in the following: an intelligence understands those things which are simple, and these fall outside of the sensitive and imaginative powers. The same is true of the soul. But an intelligence according to its own proper essence understands those things which it understands, and it is a substance, a separate thing existing in itself. Therefore so also the soul, since it apprehends in a similar way, will be a thing existing in itself; and thus the soul will be a substance.

32. Again, regarding the same: from Avicenna in his commentary on the *Metaphysics* we have it that every substance is either a body or not a body. If it is not a body, it is either part of a body or is not part of a body. If it is part of a body it is either matter or form. If it is not part of a body, either it is such that it has a link with bodies, so that it moves them, and is said to be a soul; or it is such that it does not have a link with bodies in order to move them, and is said to be an intelligence; and thus we know from what has been discussed that the soul is a substance and is incorporeal.

33. Moreover: all corruption is because of a contrary. But since the soul is an incorporeal thing, it does not have any cause of contrariety in itself. Therefore the soul cannot be corrupted; therefore it is incorruptible. Which we readily concede.

34. Both Aristotle and Augustine as well as other authors put forward many other views concerning the soul; and they show that the soul does not derive its existence from propagation; but it is not the place to touch on them here, since these authors have adequately put forward these views and refuted them.

IV. *Utrum anima vegetabilis, sensibilis et rationalis sint eadem*
anima

Sequitur videre utrum anima vegetabilis, et sensibilis, et rationalis
sint in homine eadem anima an diverse.

35. Quod sint eadem anima sic ostenditur.   5

Hoc ipsum 'animatum' predicatur univoce de corpore animato, et
de animali, et de homine. Sed dicitur esse animatum ab anima. Ergo
sicut animatum secundum eandem intentionem nominis de unoquoque
eorum predicatur, ita hoc ipsum 'anima' secundum unam et eandem
intentionem unicuique convenit, et ita eadem est anima corporis ani- 10
mati, et animalis, et hominis.

36. Item. Hoc ipsum 'substantia' univoce predicatur de corpore, et
corpore animato, et de unoquoque suorum inferiorum; et hoc ipsum
'substantia' per adictionem differentie in descendendo gradatim plus
et plus specificatur, ut hoc ipsum corporeum additur substantie, et 15
postea animatum, et tertio superadditur sensibile, et sic deinceps.
Simili modo hoc ipsum 'anima' per adictionem quod est vegetabile
trahitur ad speciem, et postea superadditur hoc ipsum sensibile, et
tertio rationale. Qua ratione ergo substantia est genus ad ea que specificantur per differentias in descendendo appositas, et anima genus erit 20
ad ea que specificantur per differentias in descendendo gradatim appositas. Sed non dicetur in aliqua specie substantie quod in ea sint
plures substantie, immo una sola substantia specificata per differentias.
Pari ratione dicendum est quod in homine non sunt tres anime, immo
una sola anima specificata per differentias tres, ut per vegetabile, sensi- 25
bile, rationale; et ita anima vegetabilis, sensibilis, rationalis non sunt
tres anime in homine, immo una sola anima.

37. Preterea. Si sint diverse anime, contingit hominem habere tres

---

1–2 Utrum . . . anima *om.* PV   3 Sequitur videre: restat ut videamus V   et
rationalis *om.* P   4 in homine *om.* CP   an: aut CV   5 eadem: idem V
anima *om.* P   10–11 unicuique . . . hominis *om.* V, *sed add.*: anime vegetabili,
et sensibili (sensibilis *cod.*) anime, et anime rationali convenit. Ergo cum anima
rationalis sit substantia incorporea, sed omnis substantia incorporea est incorruptibilis,
quod constat esse falsum si non sit in homine eadem anima et vegetabilis et sensibilis
et rationalis. Relinquitur ergo quod eadem est anima in homine et rationalis et vegetabilis et sensibilis.   12 Item: amplius V   13 de *om.* V   13 suorum
inferiorum: de suo inferiori C   15 additur: adicitur V   est *om.* P   21 in
descendendo *om.* P   22 appositas: oppositas P   aliqua: alia P   26 sensibilis: et *add.* V   27 in homine *om.* CP

1–2 Cf. Avic. *De An.* v. vii (26^vbA–27^vbE); Gundiss. *De An.* iv (44–47).
6 Cf. Arist. *Cat.* i (1^a6–12) (47).
12 Cf. Arist. *Cat.* v (3^a33–^b9) (51–52); Boeth. in loc. et *In Isagogen* II ed. III. 4
(208–9).

IV. *Whether the vegetative soul, the sensitive soul, and the rational soul are the same soul*

Next we must see whether the vegetative soul, the sensitive soul, and the rational soul are the same soul in human beings or are many.

35. That they are the same soul is shown in the following way. This same word 'animated' is predicated univocally of an animated body, both that of an animal and that of a human being. But something is said to be animated by a soul. Thus just as 'animated' is predicated according to the same meaning of the word in each one of them, so this same word 'soul' applies to each with one and the same meaning,[4] and so the soul of an animated body, both of an animal and of a human being, is the same.

36. Moreover, this same word 'substance' is predicated univocally of a body and of an animated body as well as of each one of its inferiors; and this same word 'substance' by the addition of differences in descending order specifies more and more, as when the same word 'corporeal' is added to 'substance', and followed by 'animated', and thirdly 'sensitive' is then added, and so on. In the same way, this word 'soul' by means of the addition of what is vegetable is applied to the species and then the word 'sensitive' is added, and thirdly 'rational'. Therefore, for this reason, substance is a genus in respect of those things which are specified by means of differences applied in descending order, and the soul will be a genus with respect to those which are specified by means of differences applied in a descending order. But it will not be said of any substance that there are many substances in it, rather one sole substance specified by means of differences. For the same reason, it should be said that there are not three souls in a human being, rather only one soul specified by three differences, namely by 'vegetative', 'sensitive', 'rational'; and thus the vegetative, sensitive, and rational soul are not three souls in a human being but rather one soul only.

37. Moreover, if there were various souls, it would mean that in effect a human being would have three souls, which is against what

---

[4] V adds: to the vegetative soul, and the sensitive soul and the rational soul. Therefore, since the rational soul is an incorporeal substance, and every incorporeal substance is incorruptible, which must be false if the vegetative, sensitive, and rational soul is not the same in man, it thus follows that the rational, vegetative, and sensitive is the same soul in man.

animas in effectu, quod est contra Avicennam, qui dicit quod ab anima rationali est in homine vegetatio, sensibilitas, rationalitas.

38. Quod autem sint diverse anime sic ostenditur. Si anima vegetabilis et anima rationalis sint eadem anima, ergo cum anima rationalis sit incorruptibilis, et anima vegetabilis erit incorruptibilis; quia eadem 5 sunt per se accidentia rerum eiusdem speciei. Ergo destructo corpore animato, ut arbore aliqua vel animali aliquo, ut asino, remanet eius anima perpetua habens esse necessario. Et ita secundum hoc, sicut anima rationalis potest separari a corpore habens perpetuitatem essendi, et ita anima vegetabilis, et sensibilis similiter.   10

39. Item. In genere non est perfectio, immo in specie; unde non est invenire aliquod animal quin ipsum sit sub aliqua specie animalis, et nullum genus parificatur sue speciei: quia omne genus in plus est quam aliqua suarum specierum. Ergo cum omnis differentia habeat specificare suum genus, differentia specificans primo suum genus habet differentiam 15 sibi oppositam, que differentie primo loco dividunt ipsum genus. Ergo cum hec differentia 'vegetabilis' primo loco specificet hoc genus 'anima' habet aliquam differentiam coequevam et sibi oppositam, illa differentia nec est hec differentia 'sensibilis', nec hec differentia 'rationalis', quoniam utraque illarum differentiarum posterior est vegetabili; et si 20 hoc, tunc est aliqua species anime que non est anima vegetabilis, quod non videtur posse esse. Sed vegetabile est proprium anime conveniens omni anime; unde non est differentia divisiva anime, nec est eius perfectiva, immo est concomitans differentiam perfectivam anime. Quoniam si vegetabile est differentia perfectiva essentie anime, tunc est hic 25 nugatio: anima vegetabilis, sicut hic: homo rationalis. Si autem vegetabile esset differentia divisiva anime, tunc esset aliqua anima in corpore que non vegetaret corpus, et aliquod corpus vegetatum quod non

---

1 Avicennam: Abicennam P   2 sensibilitas: sensualitas P   5 quia: et V
5–6 quia . . . speciei om. P   6 speciei: sunt anima vegetabilis et anima sensibilis in bruto animali et in homine, quia sensibile et vegetabile in eadem significatione hinc et inde sumuntur add. V   7 eius om. P   9 corpore: consequenter permanere add. P   10 et¹ om. C   11 Item: amplius V   immo: scriptum, sed deletum et corr. in margine: nisi C   13 in om. P   14 habeat: habet V   15 differentia . . . habet om. propter homoeoteleuton sed add. post genus: item V   15–16 differentia . . . genus om. propter hom. C   17 specificet: specificat CV   19 nec hec differentia om. P   22–p. 12, l. 2 Sed . . . ab anima om. PV (sed vid. infra p. 12, l. 4)   28–p. 12, l. 1 vegetatum . . . animatum: esset animatum quod non esset vegetatum V

---

1 Avic. De An. I. iii (3ᵛᵃ–4ʳᵃ); v. vii (27ᵛᵇE).
11 Cf. Porphyr. Isagoge De diff. (16–17); De comm. gen. et diff. (21–22), et Boeth. In Isagogen II ed. IV. 5–6 (253–4); v. 2 seq. (289 seq.); Abaelard. Glossae super Porphyrium in loc. (72); Avic. De An. I. i (1ʳᵇE).
24 Cf. Abaelard. Glossae super Praedicamenta (134–5).

Avicenna says, that vegetation, sensitivity, and rationality are in the human being from the soul.

38. However, that there are various souls can be shown in this way. If the vegetative soul and the rational soul are the same soul, since therefore the rational soul is incorruptible, so also the vegetative soul will be incorruptible, since in themselves the accidents of things of the same species are the same. Therefore once the animated body is destroyed, such as a certain tree or a certain animal such as an ass, its perpetual soul remains because it has a necessary existence. And thus in this way, just as the rational soul can be separated from the soul while having a perpetual existence, so also can the vegetative soul and similarly the sensitive soul.

39. Again, perfection is not to be found in the genus but rather in the species; so it is that no animal is to be found such that it does not come under some species of animal, and no genus can be made the equal of its species, because every genus is something more than any of its species. Therefore, since each difference has the effect of specifying its genus, a difference which first specifies its genus has a difference which is opposite to it, differences which in the first place divide the genus itself. Therefore, since this difference 'vegetative' specifies in the first place this genus 'soul', it has a certain coeval difference, one opposite to itself. That difference is neither this difference 'sensitive' nor this difference 'rational', since both of these differences come after the vegetative. And if this is the case, then there is another species of soul which is not the vegetative soul, which, it seems, cannot exist. But the vegetative is proper to the soul, applying to each soul; whence it is not a difference divisive of the soul, nor is it perfective of it; rather it accompanies the perfective difference of the soul. For if the vegetative is the perfective difference of the essence of the soul, then this statement 'vegetative soul' is a tautology, like this 'rational man'. For if the vegetative were a divisive difference of the soul, then there would be a certain soul in the body which does not enliven the body, and a certain enlivened body which would not be animated. And this

esset animatum. Quod etiam videtur esse consonum ei quod dicitur a quibusdam, scilicet stellas et planetas moveri ab anima.

40. Solutio. Dicimus quod hoc nomen 'anima' significat genus anime vegetabilis et anime sensibilis et rationalis; et in homine est una sola anima a qua est vegetatio, sensus et ratio. Et anima sensibilis est 5 genus subalternum, quia anima sensibilis est genus anime rationalis et species anime vegetabilis.

41. Ad hoc quod queritur ab aliquo, utrum anima secundum sui generalem intentionem sit substantia corporea vel incorporea; Dicendum est quod est substantia incorporea et incorruptibilis, quia omne 10 universale est incorruptibile. Nec tamen quelibet anima est incorruptibilis, ut est anima vegetabilis que est in arboribus, et anima sensibilis que est in brutis: hee enim secundum suum esse speciale corrumpi possunt, sed non secundum suum generale esse secundum quod sunt substantie incorporee. Et hoc contingit quia suum esse 15 speciale salvari non potest nisi in corpore organico; unde destructo organico corpore destruitur et esse et esse ipsius anime, que simplex est, vegetabilis vel sensibilis.

42. Item queritur. Cum intellectus et ymaginatio sint virtutes anime sicut ratio, quare potius rationalitas sit differentia perfectiva speciei 20 anime quam intelligibilitas vel ymaginabilitas.

43. Ad hoc dicendum est quod ymaginatio et intellectus a rationalitate procedunt, et ymaginatio nihil apprehendit nisi quod prius fuit in sensu; et ideo contingit quod neque potentia ymaginandi neque potentia intelligendi differentie perfective sunt specierum, quia sunt concomi- 25 tantes ipsas differentias.

44. Item. In genere non est perfectio, ut dictum est. Ergo cum

---

3 anima: anime CV   4 vegetabilis et anime om. CV   rationalis: Sed vegetabile . . . ab anima add. V (vid. supra p. 11, l. 22)   4–5 et in homine . . . ratio om. P   5 est² om. C   7 vegetabilis om. CV   8 ab aliquo om. P   10 est² om. C   10–11 et incorruptibilis . . . incorruptibile om. CP   11–12 incorruptibilis: quia omne universale est incorruptibile add. C (vid. supra ll. 10–11)   13 enim om. P   speciale: singulare V   16 speciale: singulare V   17 organico corpore: organo corporis P   et esse¹ om. CV   simplex: simpliciter CV   19 Item: Rursus V   anime: corporis CP   22 Ad hoc: et ad hoc V   ymaginatio: ratio CV   23 procedunt: procedere P   25 differentie om. CV   perfective sunt: perfectiva est CV   specierum: speciei C   27 Item: amplius V   ut dictum est om. P

1–2 Vid. supra, n. 9.
4–5 Cf. supra, n. 37.
6–7 Cf. Avic. De An. I. v (4^{vb}A): 'Cum autem volueris altius scrutari, utique melius est ut vegetabilis accipiatur genus sensibilis et sensibilis genus humane'.
9 Cf. Abaelard. Glossae super Porphyrium (48–49); Glossulae super Porphyrium (526–7).
27 ut dictum est: vid. supra, n. 39.

seems to agree with what some people say, namely that the stars and the planets are moved by the soul.

40. Solution: We state that this word 'soul' signifies the genus of the vegetative soul as well as the sensitive soul and the rational; and in a human being there is one soul only, from which comes vegetation, sensation, and reason. And the sensitive soul is a subalternate genus, because the sensitive soul is a genus of the rational soul and a species of the vegetative soul.

41. Regarding that which is asked by someone, whether the soul according to its general meaning is a corporeal or incorporeal substance: it should be said that it is an incorporeal and incorruptible substance, because every universal is incorruptible. Yet neither is just any soul incorruptible, such as the vegetative soul which is in trees and the sensitive soul which is in brute animals: for these souls according to their specific being can be corrupted, but not according to their general being, by virtue of which they are incorporeal substances. And this happens because its specific being cannot be preserved except in an organized body; whence if the organized body is destroyed both its being and that of its soul are destroyed, which is simple whether vegetative or sensitive.

42. Again it is asked: Since the intellect and the imagination are powers of the soul just like reason, why should rationality be more of a perfective difference of the species of soul than the ability to understand or imagine.

43. In this regard, it should be said that the imagination and intellect proceed from rationality, and the imagination apprehends nothing that was not first in the senses. Thus, it is necessary that neither the power of imagining nor the power of understanding is a perfective difference of species, since they are found together with these differences.

44. Again, as has been said, perfection is not to be found in the

anima vegetabilis sit genus, non erit aliqua anima quin ipsa sit sub aliqua specie specialissima ipsius generis. Ergo cum in arboribus et plantis sit anima vegetabilis, est ibi aliqua species anime vegetabilis habundans ab anima vegetabili per aliquam differentiam substantialem. Sed que sit illa differentia non videtur posse ostendi, cum illa neque sit hec diffe- 5 rentia rationalis neque sensibilis. Et ita videtur quod anima vegetabilis non sit genus.

45. Ad hoc dicendum quod anima vegetabilis genus est et dividitur in duas differentias, quarum altera est sensibilis, altera insensibilis, sicut anima sensibilis dividitur in rationale et irrationale.   10

v. *De viribus anime vegetabilis*

46. Consequenter sciendum est quod anime vegetabilis tres sunt vires: vis nutritiva, vis augmentativa, vis generativa. Vis nutritiva est vis convertens nutrimentum a corporalitate in qua erat in similitudinem corporis in quo est, et unit illud ei propter restaurationem eius quod 15 resolvitur ex illa; et est illa vis ad conservationem individui per restaurationem deperditorum. Vis augmentativa est vis augens corpus in quo est ex corpore quod ipsa assimilat illi augmento proportionali in omnibus dimensionibus, que sunt longitudo, latitudo, spissitudo; et est hec vis ad conservandum perfectionem quantitatis determinate 20 ipsi individuo a natura. Vis generativa est vis accipiens a corpore in quo est partem similem illi in potentia, et operatur in ea per attractionem omnium membrorum, et assimilat illi generationem et complexionem.

47. Sed videtur posse ostendi quod vis augmentativa et vis generativa 25 et vis nutritiva sunt una vis et non diverse vires hoc modo. Generatio est a vi generativa. Sed vis nutritiva spoliat nutrimentum a sua propria forma substantiali conferens materie formam rei nutriende; et ita est

---

2 in *om.* P   3 vegetabilis *om.* P   5 ostendi: ostendere V   5–6 hec differentia *om.* P   6 vegetabilis *om.* CV   8 vegetabilis *om.* CV.   11 De . . . vegetabilis *om.* PV   12 est *om.* P   15 unit: vivit V   propter: per P 15–17 eius . . . restaurationem *om. per hom.* V   16 illa[2]: ita V   17 augens: agens CP   18 assimilat: assimulat C   21 a: in P   22 partem *om.* P   23 illi *om.* P   25 Sed: Item P   26 sunt: sint V   hoc modo *om.* CP 26–27 Generatio . . . Sed *om.* CV   28 nutriende: nutriendam C

3–4 Cf. Porphyr. *Isagoge* De diff. (17, 16): 'Differentia est qua abundat species a genere'. Et Boeth. in loc. II ed. (262).
8–10 Vid. supra, n. 40, ll. 6–7.
11 Cap. V: Vid. Arist. *De An.* II. iv; III. xii (434[a]20–26).
12–24 Cf. Avic. *De An.* I. v (4[vb]B); Algazel *Metaphys.* II. iv. 1 (162–3); Gundiss. *De An.* ix (66).
16–17 Vid. p. xviii.
26 una vis: cf. Arist. *De An.* II. iv (416[a]19).
27–28 Cf. Avic. *De An.* II. i (6[ra]A).

genus. Therefore, since the vegetative soul is a genus, there will not be any soul such that it will be under a certain, most specific species of this genus. Therefore, since the vegetative soul is in trees and plants, there is a certain species of the vegetative soul there, flowing from the vegetative soul by means of a certain substantial difference. But it does not seem possible to show what that difference is, since it is neither this rational difference nor this sensitive difference. Thus, it seems that the vegetative soul is not a genus.

45. Regarding this, it should be said that the vegetative soul is a genus and it is divided into two differences, of which one is the sensitive and the other is insensate, just as the sensitive soul is divided into the rational and irrational.

v. *On the powers of the vegetative soul*

46. The next thing which should be known is that there are three powers of the vegetative soul: the nutritive power, the augmentative power, and the generative power. The nutritive power is a power which converts the food from the material condition in which the food was into a likeness of the body in which the nutritive power is. It unites the food to itself in order to restore that which has been lost from its body, and it is a power which preserves the individual by restoring what was lost. The augmentative power increases the body in which it is by means of another body which it assimilates to proportional growth in every dimension, namely, length, breadth, and depth. This is a power which preserves the perfection of quantity determined by nature to this individual. The generative power is a power which takes a part from the body in which it is, which is potentially similar to the body, and it works on it by means of the attraction of all of the bodily organs, and produces something similar to the body in generation and bodily constitution.

47. However, it seems that it can be shown as follows that the augmentative power, the generative power, and the nutritive power are one power and not various powers. Generation is from the generative power, yet the nutritive power strips the nourishment of its own substantial form, conferring the form of matter of the thing which is to be nourished. Thus there is the corruption of one substantial form

ibi corruptio unius forme substantialis et generatio alterius forme substantialis. Sed illud est a vi nutritiva, et omnis generatio est a vi generativa; ergo secundum hoc vis nutritiva est vis generativa.

48. Eodem modo potest ostendi quod ipsa est vis augmentativa;
quoniam vis nutritiva in spoliatione forme nutrimenti dat eius materie 5
formam novam similem forme rei nutriende et unit illam rei nutriende,
et ita adicit quantum quanto. Ergo inducit augmentum; quoniam, ut
dicit Aristoteles in libro *de Generatione et Corruptione*, augmentum est
magnitudinis additamentum. Sed omne augmentum est a vi augmentativa, et istud augmentum est a vi nutritiva. Ergo vis nutritiva est vis 10
augmentativa, et ita tres vires predicte sunt una et eadem vis.

49. Item. Augmentum et diminutio sunt motus oppositi, et ab
oppositis viribus sunt oppositi motus, et diminutio et augmentum sunt
motus oppositi; ergo fluunt ab oppositis viribus. Ergo vis augmentativa
et vis illa a qua fluit diminutio, scilicet vis diminutiva, sunt diverse 15
vires. Ergo cum habeamus vim diminutivam sicut augmentativam,
plures sunt vires anime vegetabilis quam tres preassignate.

50. Eodem modo potest ostendi quod habemus vim corruptivam
oppositam in generatione.

51. Solutio. Distinguendum est quod hoc nomen 'vis' quandoque 20
sumitur abstractive quandoque concretive: abstractive, secundum quod
supponit tantum pro potentia ipsius anime relata ad aliquem eius
effectum; concretive, secundum quod supponit simul animam et
potentiam ipsius anime in universali. Et propter hoc dicunt plures
auctores quod omnes vires anime sunt una vis, quia ipse omnes sunt 25
eedem in subiecto. Cum igitur vis generativa secundum quod ipsa est
vis anime vegetabilis sit abscindens a corpore in quo est partem similem
illi in potentia, et virtutis nutritive non sit ita abscindere, alia est vis
nutritiva et alia est vis generativa. Quod autem alia sit vis augmentativa
et alia vis nutritiva patebit inferius, ubi ostenditur quod nutrimentum 30
potest fieri sine augmento. Ultimo autem obiectum solvendum est per

2 illud: illa C   4 vis *om.* P   6 et . . . nutriende *om. per hom.* V
7 adicit: additum P   9 omne *om.* V   10 et: sed V   istud: illud PV
12 Item: amplius V   15 illa *om.* V   vis²: illa *add.* C   17 vegetabilis: vegetabiles P   quam: quoniam P   24 ipsius *om.* V   Et *om.* C   26 eedem:
eadem C   in *om.* V   27 sit: sic CP   28 ita *om.* P   29 et *om.*
P   vis² *om.* P   31 Ultimo: Ultimum V

5–7 Avic. *De An.* II. i (6^{rb}A).
8 Arist. *De Gen. et Corr.* I. v (320^{b}30–31).
24–26 Cf. Arist. *De An.* II. iv (416^{a}19); Avic. *De An.* I. iv. (4^{rb}B).
26–29 Cf. supra, n. 47.
29–31 Cf. supra, n. 48.
35 inferius: cf. infra, nn. 52–54.
31–p. 15, l. 2 Ultimo autem obiectum: vid. supra, n. 49. Cf. Avic. *De An.* I. iv (4^{ra}A).

and the generation of another substantial form. However, that is from the nutritive power, and all generation is from the generative power; therefore it follows that the nutritive power is the generative power.

48. In the same way it can be shown that the nutritive is the augmentative power, since the nutritive power in stripping away the form of the nourishment gives it a new form of matter similar to the form of the thing to be nourished and unites that to the thing to be nourished and thus adds a quantity to a quantity. Therefore, it introduces growth because, as Aristotle says in the book *On Generation and Corruption*, growth is the increase of size. However, all growth is from the augmentative power, and this growth is from the nutritive power. Therefore, the nutritive power is the augmentative power, and so the aforementioned three powers are one and the same power.

49. Moreover, growth and decrease are opposite movements, and from opposite powers come opposite movements. Now decrease and growth are opposite movements, therefore they come from opposite powers. Therefore, the augmentative power and the power from which decrease comes, namely, the diminutive power, are diverse powers. Therefore, since we have a diminutive power just like an augmentative, there are more powers of the vegetable soul than the three already assigned.

50. In the same way it could be shown that we have a corruptive power opposite to generation.

51. Solution: A distinction should be made. This word 'power' is sometimes taken abstractly and sometimes concretely: abstractly, in so far as it stands for the power of the soul itself in relation to some or other of its effects only; concretely, according as it stands for the soul and the power of the soul itself in general at the same time. For this reason, many authors say that all of the powers of the soul are one power, because all of these are the same in a subject. Therefore, since the generative power in so far as it is a power of the vegetative soul is something which separates a part which is potentially similar to it from the body in which it is, and the nutritive power does not separate, in this way the nutritive power is one thing and the generative power is another. However, that the augmentative power is one thing and the nutritive power is another will be clear below, where it will be shown that nourishment can take place without growth. However, the final objection is solved as follows: decrease does not flow from any power,

hoc quod diminutio non fluit ab aliqua vi, immo exit in effectum per
inpotentiam et virtutis nutritive debilitatem. Similiter dicendum est de
corruptione.

52. Item. Videtur posse ostendi quod nutrimentum non possit fieri
sine augmento hac ratione. Vis nutritiva spoliat nutrimentum a sua 5
propria forma et confert ei formam similem forme rei nutriende, et unit
illud nutriendo: ut si caro sit nutrienda, spoliatur nutrimentum a sua
propria forma et confertur ei carnalitas, et unitur caro nova cum carne
antiqua nutrienda. Sed in illa unitione est aditio quanti cum quanto.
Ergo ibi est augmentum, et non contingit aliter fieri nutrimentum. 10
Ergo nutrimentum non potest fieri sine augmento. Ab Aristotele etiam
habetur in libro *de Generatione et Corruptione* quod ad esse augmenti
tria exiguntur: quoniam aliquo adveniente fit augmentum; et re salvata
que augmentatur; et unamquamque partem magnitudinis augmentate
necesse est maiorem esse. Sed hec tria inveniuntur in carne nutrita, 15
cum unaqueque pars nutriatur proportionaliter, et ita nutrimentum, ut
prius dictum est, non fit sine augmento.

53. Preterea. Motus ipsius nutrimenti est motus augmenti, quia in
unitione nutrimenti cum nutrito est motus secundum omnem dimensionem, scilicet in longum, in latum, in spissum. Sed ille motus 20
est motus augmenti; ergo idem est motus nutrimenti et augmenti.
Quod si concedatur; habetur instantia in senibus et decrepitis qui
nutriuntur non tamen augmentantur.

54. Solutio. Dicimus quod aliud est nutrimentum et aliud augmentum, et unum fit sine reliquo. Quoniam ubi caro vel huiusmodi 25
nutritur non est ibi substractio materie in materiam, nec continuatio
nutrientis cum nutrito, sed adiccio quedam, ut cum nutrimentum sit
complexionatum per proprietates ipsius augentur proprietates nutriende, scilicet frigiditas, caliditas, humiditas et siccitas: simile enim

---

4 Item: Rursus V   possit: posset V   5 nutrimentum: augmentum C
7 ut *om.* C   8 carnalitas: carnelitas P; carneitas V   9 aditio: adiectio C
11–12 etiam habetur: autem habemus P   12 quod *om.*P   13 re salvata:
resolvata P   15 in: cum C   18 est motus augmenti *om.* P   quia: cum P
in *om.* P   19–20 secundum omnem dimensionem: in omni dimensione P
20 longum . . . spissum: longitudine, latitudine, spissitudine P   22 habetur:
habet P   23 non: nec V   25 vel huiusmodi *om.* P   26 substractio:
subintratio V   28 ipsius: rei P   augentur proprietates *om.* P   29 et *om.* P

2–3 De corruptione: vid. supra, n. 50.
9–10 Cf. Arist. *De An.* II. iv (416ᵇ11–12).
11–15 Arist. *De Gen. et Corr.* I. v (321ᵃ19–22).
18–20 Cf. Avic. *De An.* II. i (6ʳᵇA). Cf. supra, n. 46.
22–23 Cf. Arist. *Topic.* II. iv (111ᵇ25–26): 'Nutriuntur enim semper animalia,
augentur autem non semper' (cod. Coll. Trin. Oxon. 47, fol. 23ᵛ).
27–p. 16, l. 3 Cf. Avic. *De An.* II. i (6ʳᵃ–ᵇA).

rather it goes into effect by means of the feebleness and weakness of the nutritive power. The same should be said of corruption.

52. Again, it seems that it can be shown that nourishment cannot take place without growth for the following reason. The nutritive power strips the nourishment of its own form and confers a form on it similar to the form of the thing to be nourished, and it adds to it by nourishing. So if there is flesh to be nourished, the nourishment is stripped of its own form and the form of flesh is conferred on it, and the new flesh is added to the original flesh which was to be nourished. However, in this adding there is the addition of one quantity to another. Therefore, there is growth there and nourishment does not happen in any other way. Therefore nourishment cannot take place without growth. Indeed, we know from Aristotle in the book *On Generation and Corruption* that in order for growth to be present three things are required: because if something is added growth happens; the thing which grows is preserved; any part of the increased-size magnitude must be bigger. However, these three are to be found in the flesh which has been nourished, since each part is nourished proportionately, and thus nourishment, as has already been said, does not take place without growth.

53. Moreover, the movement of the nourishment itself is a movement of growth, because in the uniting of the nourishment with the one who is nourished there is a movement in every direction, namely, in length, width, and depth. But that movement is a movement of growth, therefore the movement of nutrition and growth is the same. However, if this is conceded, we would have an exception in the old and decrepit, who are nourished and yet do not grow.

54. Solution: We state that nourishment is one thing and growth another, and one happens without the other. For where flesh or something similar is nourished there is no entering in[5] there of matter into matter, nor a continuity between the nourisher and the nourished, but rather a certain addition, so that when the nourishment is constituted by its own properties it increases the properties which nourish, namely the cold, hot, moist, and dry: for what is like increases like, so

---

[5] Reading *subintratio* with V.

suum simile augmentat, ut si cibus sit calide complexionis vel frigide semper nutrit suum simile. Et sicut nutrimentum recipitur in corpore nutriendo, ita eius tota materia facta digestione eicitur. Illi autem qui dicunt quod cum fit digestio separatur purum ab impuro, et retinetur purum et eicitur impurum, et quod puro retento et spoliato a sua 5 propria forma datur forma rei nutriende, dicunt quod contingit quandoque nutrimentum fieri sine augmento, eo quod maior est deperditio per resolutionem quam sit per nutrimentum restauratio.

vi. *De anima sensibili*

Dictum est de anima vegetabili et eius viribus. Sequitur ut agamus de 10 anima sensibili et de eius viribus.

55. Anima sensibilis est anima movens corpus voluntarie. Huius autem anime due sunt vires generales, quarum una dicitur esse vis motiva, alia vis apprehensiva. Vis motiva alia est vis movens et imperans, alia est movens et efficiens motum. Vis movens et imperans motui que- 15 dam est vis concupiscibilis, quedam est irascibilis. Vis concupiscibilis est vis appetitiva boni et eorum que appropinquant ad summam necessitatem, et est hec vis imperans motui cum appetitu delectamenti. Vis irascibilis est vis respuitiva eius quod putatur esse nocivum, vel quod putatur esse corrumpens spem vincendi. Et ab Aristotele in *Topicis* 20 quandoque hec vis appellatur animositas. Animositas enim est cum spe habendi victoriam. Vis autem motiva et efficiens motum est illa vis que operatur in nervis et in musculis, et contrahit cordas et ligamenta membrorum ad invicem. Vis autem apprehensiva est vis anime qua ipsa potest percipere ea que putantur esse nociva vel delectabi- 25 lia vel medio modo se habentia. Huiusmodi autem quedam est vis

---

1 simile *om.* P   2 recipitur: recipit P   2–3 in corpore nutriendo *om.* P
3–5 Illi . . . eicitur *om. propter hom.* P   4 impuro: puro V   5 et² *om.* P
6 rei *om.* V   10 Sequitur ut agamus: Dicendum est V; dicamus P   11 viribus: Animam autem sensibilem dicunt multi esse substantiam, que est usiosia, et corruptibilis, sicut igneitas est corruptibilis. Illud vero potest inprobari per hoc: anima
rationalis est subusia (*sic*). Si ergo anima sensibilis sit usyosis, anima simpliciter sumpta
in universali nec est usya nec usyosis, immo partim sub hoc genere partim sub illo
add. V   13 una: vis *add.* V   14 apprehensiva: apprehensitiva, *et sic deinceps* C
15 est: vis *add.* P   movens¹: manens P   17 appropinquant: appropinquantur V
18 est *om.* V   25 potest percipere: primo percipit V   26 vel medio: et P

---

1 Cf. Arist. *De An.* ii. iv (416ᵃ29–31).
12–p. 17, l. 3 Cf. Avic. *De An.* i. v (4ᵛᵇB); Gundiss. *De An.* ix (67–68).
20–21 Arist. *Topic.* iv. v (126ᵃ8) (iuxta versionem Boethii): 'timor in animositate',
et (126ᵃ10): 'ira in animositate'. Ita legitur in cod. Coll. Trin. Oxon. 47, fol. 30ᵛ,
ut cl. Dr. L. Minio-Paluello mihi humanissime indicavit. Cod. 253 Coll. Balliol. Oxon.
habet: 'timor in furoris specie . . . ira autem in furiosi animositate' (fol. 121). Occurrit
etiam apud Macrobium, *In Somnium Scipionis* i. xii. 14: 'in Martis animositatis
ardorem, quod θυμικόν nuncupatur'.

that if food is of a hot or cold constitution it will always nourish what is like itself. And just as the food is received in the body to be nourished, so all of its matter is ejected after digestion has taken place. Indeed, those who say that when digestion happens the pure is separated from the impure, and the pure is retained and the impure is ejected, and that when the pure is retained and stripped of its own form and given the form of the thing to be nourished, also say that it happens sometimes that nourishment takes place without growth, because the loss is greater through dispersal than it is through the replenishment of nourishment.

VI. *On the sensitive soul*

Having spoken about the vegetative soul and its powers, it follows that we should next deal with the sensitive soul and its powers.

55. The sensitive soul is the soul which moves the body voluntarily. There are, however, two general powers of this soul, one of which is said to be the motive power and the other the perceptive power. One motive power is a motive and directing power and another is a motive motion-making power. A certain motive and directing power is the concupiscible power and another is the irascible power. The concupiscible power is a power which desires the good and those things which come near to the Highest Necessity, and this is a power directing motion together with the desire of pleasure. The irascible power is one which rejects what it regards as harmful, or that which it regards as destructive of its hope of winning. Sometimes this power is called 'spiritedness' by Aristotle in the *Topics*. Indeed, spiritedness is found together with the hope of obtaining victory. The motive and motion-making power is that power which operates in the nerves and in the muscles, and it pulls the tendons and the ligaments together. Now the perceptive power is a power of the soul by means of which it can perceive those things which are regarded as being harmful or pleasurable or those which lie in between. However, a certain part of this power is

apprehensiva deforis, quedam est vis apprehensiva deintus. Vis autem apprehensiva deforis est sensus, qui dividitur in quinque species, scilicet, visum, gustum, tactum, olfactum, auditum.

56. Potest autem queri utrum visus, gustus, tactus, olfactus, auditus sint eiusdem speciei vel diversarum specierum. Quod sint diverse 5 videtur per hoc quod gustus et tactus et visus et olfactus et auditus sunt diverse vires et diversa habent propria obiecta.

57. Quod sint eiusdem speciei sic ostenditur. Instrumentum operis nec est de essentia operis nec est de essentia opificis. Ergo non facit diversitatem in specie quantum est ad motorem, quia propter diversitatem 10 instrumentorum non sunt motores diversi in specie. Ergo licet oculus et auris et alia instrumenta sensuum sint diversarum specierum non erunt virtutes ipsius anime apprehensive per illa instrumenta diversarum specierum.

58. Preterea. Si fiat percussio per lapidem vel per baculum vel per 15 alia, propter hoc non erit alterius speciei percussio una quam alia. Pari ratione si fiat sensus anime vel per tactum, vel per auditum, propter hoc non erit sensus hinc et inde alterius speciei, et ita de consimilibus.

59. Solutio. Distinguendum est sicut Aristoteles distinguit in libro de Intellectu et Intellecto, quod est intellectus agens et intellectus 20 materialis, et intellectus formalis sive adeptus, et intellectus exiens in effectum, ut patebit inferius. Similiter est visus agens, scilicet, vis ipsius anime secundum quam apprehendit res visas secundum quod ipse afficiuntur coloribus; et est visus materialis, scilicet oculus, qui est corpus politum et planum recipiens inmutationes a corpore lucido 25 agente in ipsum; et est visus formalis sive adeptus, scilicet inmutatio recepta in oculo, que est ymago et similitudo rei extra, que apprehenditur. Visus exiens in effectum est visus formalis in visu materiali exiens secundum quod visus agens apprehendit in effectu rem illam

---

1 autem *om.* V   4 Questio *add. in marg.* V   5 specierum *om.* P   7 obiecta: sensata PV   8–9 nec est de essentia operis *om.* P   16 alia: aliud P; aliquod aliud V   17 tactum: oculum P   18 hinc: hic C; huic V   ita: iterum V 22 effectum: effectu V   24 ipse *om.* P   28 exiens: existens PV   effectum: effectu V   in visu: in *om.* P   29 exiens: existens PV

4–5 Avic. *De An.* I. iv (4ʳᵇA): 'An exteriores sint ex una vi que facit diversis instrumentis actiones diversas'.
5–7 Cf. Avic. *De An.* v. vii (27ʳᵃB): 'Dicemus igitur ex premissis manifestum esse actiones diversas ex diversis virtutibus esse; et quod omnis virtus, ex eo quod est virtus, non est sic nisi ob hoc ad quod ex ea provenit actio quam principalem habet. ... Impossibile est autem duas virtutes esse unam'. Gundiss. *De An.* ix: 'Diversitas enim actionum provenit ex diversitate virium' (64; cf. 65). Cf. Avic. ibid. I. iv (4ʳᵃ).
19–20 Aristoteles: potius Alex. Aphrod. *De Intell.* (74 seq.).
22 inferius: cf. n. 337, ubi haec opinio adscribitur 'Avicenne et multis aliis auctoribus'.

the external perceptive power and another is the internal perceptive power. The external perceptive power is sensation, which is divided into five kinds, namely, sight, taste, touch, smell, and hearing.

56. It can, however, be asked whether sight, taste, touch, smell, and hearing are of the same kind or of differing kinds. It seems that they are of differing kinds by the fact that taste, touch, sight, smell, and hearing are different powers and have different proper objects.

57. That they are of the same kind can be shown in the following way. The tool for a job is neither of the essence of the job nor of the essence of the workman. Therefore, a difference in kind does not come about due to a mover since different movers in kind do not come about because of different tools. Therefore, even if the eye and the ear and the other organs of sense are of different kinds, the powers of the same perceptive soul will not be of differing kinds because of those organs.

58. Moreover, if a blow is made with a stone or a stick or something else, it is not because of this that one blow is of a different kind from another. For the same reason, if a sensation of the soul occurs, be this through touch or hearing, it is not because of this that this sensation and that are of differing kinds, and so also with similar cases.

59. Solution: One should distinguish, as Aristotle does in the book *On the Understanding and the Understood*, between the active understanding and the material understanding and between the formal or obtaining understanding and the understanding which goes out into an effect, as will be seen below. Similarly, there is active sight, namely a power of the soul according to which it perceives the things which are seen according as these are endowed with colours. Again, there is material sight, namely the eye, which is a smooth and even body that receives modifications from a bright body acting upon it. Again, there is formal or obtaining sight, namely the modification received in the eye, which is an image and a likeness of the external thing which is perceived. The sight which goes into an effect is the formal sight in the material sight 'going out' according as the active sight perceives in the effect that thing of which the formal sight is the likeness. A similar

cuius visus formalis est simulacrum. Similiter distinguendum est de gustu et olfactu et de aliis. Et propter sensum existentem in effectu dicit Aristoteles in libro *de Anima* quod sentire perficitur a duobus, scilicet a pati et a iudicare: pati vero in corpore est, iudicare vero in anima. 5

60. Si ergo queratur utrum omnes sensus sint eiusdem speciei, secundum quod hoc nomen 'sensus' sumitur concretive, pro iudicio anime cum inmutatione recepta in instrumento corporeo, dicendum est quod omnes sensus sunt eiusdem speciei quoad apprehensionem anime; quoniam omnes sensus, secundum quod agentes sunt, eiusdem speciei 10 specialissime sunt. Quo autem ad diversa instrumenta corporea, ut ad oculum et auditum, et quoad inmutationes in ipsis receptas, diversarum sunt specierum. Unde secundum hoc posset dici quod diverse sunt species tactus propter diversa instrumenta tactuum et propter diversas impressiones in ipsis receptas, ut est caliditas, frigiditas, asperitas et 15 lenitas, et gravitas et levitas, et consimilia. Sed quoad tactum agentem, tactus est species specialissima.

61. Vis autem apprehensiva deintus est sensus communis, et vis ymaginativa, et vis estimativa, et vis memorialis et reminiscibilis. De istis dicetur inferius secundum ordinem. 20

VII. *De viribus anime sensibilis secundum ordinem*

62. Sequitur ut agamus de viribus anime sensibilis secundum ordinem. Primo igitur queritur, utrum vis concupiscibilis et vis irascibilis sint una et eadem vis in essentia, vel diverse.

63. Quod sint diverse vires sic ostenditur. Ab Aristotele habemus in 25 *Topicis*, quod duo contraria sunt in eadem vi, ut cum amor sit in vi concupiscibili, odium erit in eadem vi; et per hoc ostendit quod odium

---

2 et olfactu *om.* P   4 iudicare: videre P   in: a P   vero² *om.* P   in²: ab P   7 concretive: et *add.* C   8 anime: vel pro vi anime a qua est iudicium *add.* V   10 quod *om.* V   11 ut *om.* P   12 et auditum: vel aures V   inmutationes: diminutiones V   14 propter *add. eadem manu in marg.* C; *om.* P   15 et *om.* P   16 et levitas *om.* P   consimilia: similia P   18 apprehensiva: apprehensitiva C   19 et vis estimativa *om.* P   21 VII . . . ordinem *om.* PV   24 una *om.* V   25 vires *om.* P   27 ostendit: ostendendum est P

---

4 pati: Arist. *De An.* II. xii (424ᵃ18); cf. II. vii (418ᵇ26); etc. iudicare: II. vi (418ᵃ7-25); III. ii (426ᵇ8-427ᵃ16).
13-17 Cf. Arist. *De An.* II. xi (422ᵇ17 seq.); Avic. *De An.* I. v (5ʳᵃC); Gundiss. *De An.* ix (68-69).
18-20 Cf. Avic. *De An.* I. v (5ʳᵇDE); Gundiss. *De An.* ix (71). Inferius: cap. XVII-XVIII.
23-24 Cf. Avic. *De An.* I. iv (4ʳᵇB); Gundiss. *De An.* ix (65).
25-26 Arist. *Topic.* II. vii (113ᵃ34-ᵇ3): 'Nam idem contrariorum susceptibile' (cod. Coll. Trin. Oxon. 47, fol. 24ᵛ).

distinction is to be made concerning taste, smell, and the others. And because of the sensation existing in the effect, Aristotle says in the book *On the Soul* that sensing is made up out of two things, namely, undergoing and judging: undergoing is in the body, judging, however, is in the soul.

60. Therefore, if it is asked whether all of the senses are of the same kind, according as this word 'sense' is taken concretely as regards the judgement of the soul together with the change received in the bodily organ, it should be said that all the senses are of the same kind as regards the perception of the soul; for all senses inasmuch as they are active are most specifically of the same species. However, with regard to the various bodily organs such as the eye and the ear, and regarding the modifications received in them, they are of differing kinds. Thus, for this reason it can be said that there are different kinds of touch because of the various organs of touching and because of the various impressions received in them such as heat, coldness, sharpness and smoothness, heaviness and lightness, and others like these. However, with regard to active touch, this touch is the most specific kind.

61. However, the internal perceptive power is the central sense, the imaging power, the estimative power, the memorizing and recollective power. These will be discussed below in turn.

VII. *On the powers of the sensitive soul taken in order*

62. Next we shall deal with the powers of the sensitive soul in order. Thus, the first question which arises is whether the concupiscible power and the irascible power are one and the same power in essence or are different powers.

63. That they are different powers can be shown in the following way. From Aristotle in the *Topics* we know that there are two contraries in the same power, such that when love is in the concupiscible power, hate will be in the same power; and in this way he shows that

non est in vi irascibili: et ita secundum eum diverse sunt vires, scilicet
vis concupiscibilis et vis irascibilis.

64. Preterea. Dicit Aristoteles in libro *de Anima* quod diverse sunt
vires vis concupiscibilis et vis irascibilis. Et Avicenna illud idem dicit
in commento super librum *de Anima*.   5

65. Item. Ab alia vi est quod aliquid ascendit naturaliter et ab alia
vi quod descendit. Sed sicut ascendere et descendere sunt oppositi
actus, ita respuere et desiderare sunt actus oppositi, et respuere est a vi
irascibili et desiderare est a vi concupiscibili. Ergo vis concupiscibilis
et vis irascibilis sunt diverse vires.   10

66. Preterea. Contrarii actus a contrariis potentiis sunt. Sed respuere
et desiderare sunt actus oppositi; ergo sunt a contrariis potentiis; et
respuere est a vi irascibili et appetere a vi concupiscibili. Ergo vis
irascibilis et concupiscibilis sunt diverse vires.

67. Quod sint eadem vis sic ostenditur. Medicina una et eadem est 15
disciplina sani et egri, nec est alia disciplina ad inducendum sanitatem
et alia ad expellendum egritudinem; pari ratione non erit alia vis qua
appetitur delectabile et alia vis qua respuitur eius oppositum. Sed vis
appetitiva boni sive delectabilis est vis concupiscibilis et vis respuitiva
eius oppositi est vis irascibilis; ergo eadem est vis irascibilis et vis 20
concupiscibilis.

68. Item. Contraria sunt in eadem vi. Sed respuere et appetere sunt
contraria; ergo sunt in eadem vi. Sed unum illorum est in vi concupiscibili, reliquum in vi irascibili. Ergo vis concupiscibilis et vis irascibilis
sunt una et eadem vis.   25

69. Item queritur, Quid sit dictu, quod aliquid sit in vi concupiscibili
vel in vi irascibili quia non est sicut accidens in subiecto.

70. Preterea. Odium est id quod putatur esse nocivum; ergo est in
vi irascibili: et est contrarium amori. Ergo, cum amor sit in vi concupiscibili, odium est in eadem vi. Sed nihil unum et idem est in 30
diversis viribus. Ergo vel est una et eadem vis vis irascibilis et vis
concupiscibilis, vel odium non est in vi irascibili.

---

4 Et *om.* CV   Avicenna: Abicenna *fere semper habet* P   dicit *om.* V   5 super
librum *om.* C   6 Item: Amplius V   ascendit: descendit V   7 descendit:
ascendit V   8 et desiderare . . . et respuere *om. propter hom.* V   est *om.* C
10 vis *om.* C   11 Preterea . . . sunt: Ea sunt contraria opposita que sunt
sub contrariis oppositis potentiis C   12 oppositi: contrarii V   a *om.* V
15 eadem: una P   ostenditur: potest ostendi P   19–21 et vis . . . concupiscibilis
om. V   20 eius: et eius P   vis³ *om.* C   22 Item: Amplius V   26 Item
queritur: Preterea potest queri V   sit in: est in V   27 vel: et P   est *om.* C
29 cum *om.* P   30 est: erit V   nihil . . . est: nihil manens unum et idem V
31–32 vis concupiscibilis: vis *om.* P   32 vel *om.* P

3 Arist. *De An.* III. ix (432ᵃ22–ᵇ8); cf. x (433ᵇ1–5).
4–5 Avic. *De An.* I. v (4ᵛᵇB).

hate is not in the irascible power; and so, according to him they are different powers, namely the concupiscible power and the irascible power.

64. Moreover, Aristotle says in the book *On the Soul* that the concupiscible power and the irascible power are different powers. Again, Avicenna says the same thing in the commentary on the book *On the Soul*.

65. Again, that something naturally rises is from one power and that it falls is from another power. However, just as rising and falling are opposite acts, so rejecting and desiring are opposite acts, and to reject is from the irascible power and to desire is from the concupiscible power. Therefore, the concupiscible power and the irascible power are different powers.

66. Moreover, contrary acts are from contrary potencies. However, rejecting and desiring are opposite acts, therefore they are from contrary potencies; and to reject is from the irascible power and to desire is from the concupiscible power. Therefore, the irascible power and the concupiscible power are different powers.

67. That they are the same power can be shown in the following way. Medicine is one and the same science of health and sickness, nor is there one science to induce health and another to expel sickness. For the same reason there will not be one power by which something pleasing is desired and another through which its opposite is rejected. However, the power of desiring the good or the pleasing is the concupiscible power and the power of rejecting its opposite is the irascible power; therefore the irascible power and the concupiscible are the same.

68. Again, contraries are in the same power; but to reject and to desire are contraries, therefore they are in the same power. However, one of them is in the concupiscible power and the other is in the irascible power. Therefore, the concupiscible power and the irascible power are one and the same power.

69. Again, it is asked what it means when it is said that something is in the concupiscible power or in the irascible power, since it is not like an accident in a subject.

70. Moreover, hate relates to that which is regarded as being harmful, therefore it is in the irascible power and is contrary to love. Therefore, since love is in the concupiscible power, hatred is in the same power. However, nothing is one and the same in different powers. Therefore, either the irascible power and the concupiscible power are one and the same power, or else hatred is not in the irascible power.

71. Solutio. Ad primum. Dicimus quod hoc nomen 'vis' quandoque
sumitur concretive quandoque abstractive. Concretive, quando dat
intelligere in concretione quadam simul animam et illud accidens quod
copulatur per hoc verbum *vireo, vires*, et secundum hoc vis concupiscibilis et vis irascibilis sunt unum et idem subiecto. Secundum autem quod 5
hoc nomen 'vis' sumitur abstractive, eadem sunt, et diverse tamen sunt
vires vis concupiscibilis et vis irascibilis prout ipse ad diversos actus
referuntur, quoniam dicitur vis esse concupiscibilis secundum quod
anima dicitur esse appetitiva rei delectabilis; dicitur autem esse vis
irascibilis ab eo quod anima est respuitiva nocivi et habet spem vin- 10
cendi.

72. Ad aliud. Dicendum est quod aliquid dicitur esse in aliqua vi
vel ut consequens operationem illius virtutis, vel ut antecedens operationem illius virtutis; ut consequens, secundum quod amor est in anima
secundum vim concupiscibilem; ut antecedens, in quantum temperantia 15
est in anima regens operationes virtutis concupiscibilis, et dicitur esse
in vi concupiscibili; et fortitudo est in anima regens vim irascibilem, et
dicitur esse in vi irascibili; prudentia autem est in anima secundum vim
rationabilem. Et distinguimus preterea quod dupliciter dicitur aliquid
esse in eadem vi, vel per se vel per accidens: ut medicina est disci- 20
plina sani et egri, sani per se, egri per accidens, quoniam per se est ad
inducendum sanitatem et conservandum eandem; per accidens autem
expellit egritudinem. Similiter amor est per se in vi concupiscibili et
odium in eadem vi per accidens.

73. Item. Videtur posse ostendi quod vis concupiscibilis et vis irascibi- 25
lis sunt vires apprehensive, cum tamen dicat Avicenna eas esse motivas
et non apprehensivas, hoc modo. Vis appetitiva secundum quod appetit
id quod putatur esse delectabile apprehendit illud, quoniam non est
appetere sine apprehendere. Ergo vis concupiscibilis, ex quo ipsa est

1 dicimus: dicendum P   2 quando: secundum quod PV   dat: sumit V
4 vires: es CV   5 idem: in *add*. P   autem *om*. P   6 eadem: vis in
essentia *add*. V   eadem sunt et *om*. P   tamen sunt *om*. P   7 vires *om*. V
7–8 prout . . . referuntur *om*. P   8 referuntur: refertur C   9 anima *om*. V
12 est *om*. P   15 ut: secundum P   in quantum: quantum V   16–17 et
dicitur . . . concupiscibili *om*. P   17 est *om*. P   17–18 et dicitur . . . irascibili
om. P   19 Et *om*. P   20 vel² *om*. V   21 se² *om*. V   22 eandem:
eam P   23 expellit egritudinem: ad expellendum V   per se *add*. *in marg*. C;
om. V   25 Item: Rursus V   26 apprehensive: apprehense C   Avicenna:
Aristoteles P   27 quod *om*. V   29 sine: eo quod est *add*. V   apprehendere: apprehensione P   29–p. 21, l. 2 ex quo . . . concupiscibilis *om. per hom*. V

1 Ad primum: cf. supra, nn. 63–68.
2 concretive, abstractive: cf. supra, n. 51.
12 Ad aliud: cf. supra, nn. 69–70.
26 Avic. *De An*. I. v (4vb B).

71. Solution: Regarding the first objection, we state that this word 'power' is sometimes taken concretely and sometimes abstractly. Concretely, when it conveys a concrete understanding simultaneously of the soul and of that accident which is linked to it through the verb 'to be vigorous', and in this way the concupiscible power and the irascible power are one and the same in the subject. However, inasmuch as this word 'power' is taken abstractly, they are the same, and yet the concupiscible power and the irascible power are different powers inasmuch as these refer to different acts, since the concupiscible power is said to exist inasmuch as the soul is said to desire something which is pleasant; the irascible power, however, is said to exist from the fact that the soul rejects something which is harmful and which it hopes to overcome.

72. With regard to the next objection, it should be stated that something is said to be in the same power either as something following on from the operation of that power, or as something coming before the operation of that power: as following from it as when love is in the soul according to the concupiscible power; or as coming before it inasmuch as temperance is in the soul, directing the operations of the concupiscible power, and is said to be in the concupiscible power. Again, fortitude is in the soul directing the irascible power, and is said to be in the irascible power; prudence, however, is in the soul according to the rational power. Moreover, in addition, we make the distinction that something can be in the same power, either in itself or in an accidental manner: For example, medicine is the study of health and sickness; of health in itself, and of sickness accidentally, since medicine in itself is concerned with inducing health and preserving the same, but accidentally it drives out sickness. Similarly, love in itself is in the concupiscible power and hate is in the same power in an accidental way.

73. Again, it seems that it can be shown as follows that the concupiscible power and the irascible power are perceptive powers, even if Avicenna says that they are motive and not perceptive powers. The appetitive power, in so far as it desires what it regards as pleasing, perceives it, since there cannot be any desiring without perceiving. Therefore, the concupiscible power, inasmuch as it is appetitive of a

appetitiva rei delectabilis, ipsa est apprehensiva eiusdem rei. Ergo vis
concupiscibilis est vis apprehensiva, et est vis motiva, ut dictum est
prius. Non ergo hec divisio est per opposita: virium anime sensibilis
alia est vis motiva alia vis apprehensiva.

74. Solutio. Dicendum est quod vis appetitiva non illud apprehendit 5
quod anima secundum eam appetit, immo per aliam vim illud anima
apprehendit, ut per sensum vel per ymaginationem.

75. Item. Videtur posse ostendi quod vis concupiscibilis et vis irascibilis sunt partes prudentie hac ratione. Prudentia est vis discretiva boni
et mali cum electione boni et detestatione mali, et ita a prudentia est 10
electio boni et detestatio mali. Sed electio boni est a vi concupiscibili,
eo quod ipsa est appetitiva boni, et detestatio mali est a vi irascibili, eo
quod ipsa est respuitiva mali. Ergo tam vis concupiscibilis quam vis
irascibilis est pars prudentie; quoniam unus et idem effectus non est
a diversis ita quod ab utroque per se. Sed prudentia est in vi rationabili; 15
ergo vis concupiscibilis et vis irascibilis sunt in vi rationabili ex quo
sunt partes prudentie.

76. Solutio. Dicendum est quod electio boni que est pars prudentie
et a ratione procedit et a collatione boni cum malo; unde ipsa non est
a vi concupiscibili: vis enim concupiscibilis est circa sensibilia, et non 20
cum collatione et discretione.

77. Item. Ab eodem est appetere et velle. Sed appetere est a vi
concupiscibili; ergo velle est a vi concupiscibili. Sed illud quod est
a vi concupiscibili est in vi concupiscibili; ergo voluntas est in vi
concupiscibili; quod est contra Aristotelem dicentem in *Topicis* quod 25
omnis voluntas est in vi rationabili.

78. Solutio. Distinguendum est quod duplex est voluntas, voluntas
sensualis et voluntas rationalis. Voluntas a sensualitate procedens est
in vi concupiscibili, voluntas a ratione procedens est in vi rationali,
et de hac dicitur quod omnis voluntas est in vi rationali; et de hac 30
etiam dicitur: 'habe caritatem et fac quicquid vis', scilicet voluntate
rationali et non voluntate sensuali.

4 vis² *om.* P   5 est *om.* P   non illud: illud non P   6–7 quod . . . apprehendit *om. propter hom.* V   6 illud: illum P   anima *om.* P   7 per² *om.* C
8 Item: Rursus V   9 sunt: sint V   10 ita *om.* C   11–12 est a vi . . . boni
om. V   15 rationabili: irascibili P   13–16 quam . . . ergo *om.* V   16 rationabili: irascibili P   quo: ipse *add.* V   18 est¹ *om.* P   19, 21 collatione: collecti-
one P   20 est: non est nisi P   22 Item: Amplius V   est *om.* P   24–25 ergo
. . . concupiscibili *om.* V   26 rationabili: rationali PV   27 est² *om.* P   volun-
tas¹ *om.* C   28 est *om.* C   29 in: a P   30 vi *om.* P   31 habe: habeas V

2–3 ut dictum est prius: cf. supra, n. 55.
25 Arist. *Topic.* IV. v (126ª13).
31 habe caritatem et fac quicquid vis: attribuitur Augustino. Cf. *In Ep. Iohan. ad
Parthos* VII. 8 (2033): 'Dilige et quod vis fac'.

pleasant thing, is itself perceptive of the same thing. Thus, the concupiscible power is a perceptive power and is a motive power, as has
been said above. This division is not, therefore, due to opposites: of
the powers of the sensitive soul, one is the motive power and another
the perceptive power.

74. Solution: It should be stated that the appetitive power does not
perceive that which the soul desires by means of it; rather it is through
another power that the soul perceives, either through a sense or by
means of the imagination.

75. Again, it seems that it can be shown for the following reason that the concupiscible power and the irascible power are parts
of prudence. Prudence is a power which distinguishes between good
and evil by choosing good and hating evil, so the choosing of good
and the hating of evil are from prudence. However, the choice of the
good is from the concupiscible power because it is the desire of the
good, and the hatred of evil is from the irascible power because it is
that which rejects evil. Therefore, both the concupiscible power and
the irascible power are parts of prudence, since one and the same effect does not come from different powers such that the effect comes
from both. However, prudence is in the rational power, therefore the
concupiscible power and the irascible power are in the rational power
since they are parts of prudence.

76. Solution: It should be stated that the choice of the good which
is part of prudence comes both from reason and from the comparison of good with evil. Thus, this is not from the concupiscible power,
for the concupiscible power is concerned with sensible things and not
with comparing and distinguishing.

77. Again, to desire and to will have the same source. However, to
desire is from the concupiscible power, so to will is from the concupiscible power. Yet, that which is from the concupiscible power is in the
concupiscible power, therefore the will is in the concupiscible power,
which is against what Aristotle says in the *Topics*, that all of the will
is in the rational power.

78. Solution: A distinction should be made: the will is twofold, a
sensual will and a rational will. The will which comes from sensuality
is in the concupiscible power, and the will which proceeds from reason
is in the rational power. Moreover, concerning the latter (the rational
will and not the sensual will), it is said that all of the will is in the rational power. It is also said of it (namely with regard to the rational
will and not the sensual will), 'have love and do what you wish'.

79. Item. Queritur in qua virium anime sit iustitia. Si in vi rationali, potest ostendi quod est in vi irascibili, quia non est consequi iustitiam sine fortitudine, et quicquid consequitur anima per fortitudinem consequitur per vim irascibilem. Pari ratione erit in vi concupiscibili, quia non est consequi iustitiam sine temperantia, et temperantia est in vi 5 concupiscibili.

80. Solutio. Dicendum est quod iustitia est in anima non secundum unam solam vim, sed secundum omnes vires anime. Iustitia enim in effectum fluit post temperantiam, fortitudinem, prudentiam; qui enim iustus est oportet ipsum esse temperatum, prudentem et fortem.   10

81. Item. Dolor et gaudium sunt opposita. Sed gaudium est in vi concupiscibili; ergo dolor est in eadem vi. Contra. Dolor est id quod putatur esse nocivum. Ergo dolor est in vi irascibili. Ad hoc dicendum quod dolor est in vi irascibili per se, et est in vi concupiscibili per accidens; quoniam duorum contrariorum que sunt in eadem vi, unum est 15 in aliqua vi per se, reliquum in eadem vi per accidens.

82. Item. Sicut est ratiocinari sic est intelligere: unde habemus rationem et intellectum. Sed ratio est vis apprehensiva, vis rationabilis est vis motiva et imperans; ergo pari ratione cum intellectus sit vis apprehensiva, erit vis intelligibilis vis motiva et imperans: et ita sicut 20 anime sensibilis sunt due vires motive et imperantes, scilicet vis concupiscibilis et vis irascibilis, pari ratione anime rationalis sunt due vires motive et imperantes, scilicet vis rationalis et vis intelligibilis, sed diverse in comparatione. Si autem dicatur quod una et eadem vis est vis rationalis et vis intelligibilis, sed diverse comparationes; pari ratione 25 una et eadem vis sunt in essentia ratio et intellectus, sed diverse in accidente. Quod bene concedimus.

VIII. *De viribus anime sensibilis apprehensivis*

83. Sequitur de viribus anime sensibilis apprehensivis. Primo dicendum est de vi apprehensiva deforis, ut de sensu. Primo igitur inquiren- 30 dum est quid sit sensus et que sunt species sensus.

1 Item: preterea *om.* V   in qua *om.* V   rationali: rationabili V   3–4 consequitur: assequitur P   7 Dicendum: Distinguendum P   11 Item: Preterea V 12 id: illud PV   13–14 Ad . . . quod *om.* P   14 dolor *om.* V   per se *om.* C est *om.* P   15 sunt *om.* V   est *om.* C   16 vi² *om.* P   17 Item: Rursum V 17–18 unde . . . intellectum *om.* V   18 rationabilis: rationalis P   22 anime: anima V   vires: virtutes P   23 scilicet *om.* CV   vis² *om.* P   23–24 sed diverse in comparatione *om.* P   24–25 si . . . comparationes *om.* CV   26–27 accidente: actione P   28 *Tit. om.* PV   29 Sequitur: Consequenter agendum est V Primo: Et primo V   vi apprehensiva: viribus apprehensivis P   31 et . . . sensus *om.* P   sunt: sint V

8–9 Gregorius M. *Moral.* XXII. 2 (212D): 'Nec vera iustitia est quae prudens, fortis et temperans non est'.

79. Again, it is asked in which of the powers of the soul is justice. If it is in the rational power it can also be shown that it is in the irascible power, since justice cannot be attained without fortitude, and whatever the soul attains through fortitude it attains through the irascible power. For the same reason it will also be in the concupiscible power, since justice cannot be attained without temperance, and temperance is in the concupiscible power.

80. Solution: It should be stated that justice is in the soul not according to one power alone but according to all of the powers of the soul. For justice comes into effect, following temperance, fortitude, and prudence; for he who is just must be temperate, prudent, and strong.

81. Again, sorrow and joy are opposites. However, joy is in the concupiscible power, therefore sorrow is in the same power. On the contrary, sorrow is that which is regarded as being harmful, therefore sorrow is in the irascible power. To this it should be stated that sorrow is in the irascible power in itself and is in the concupiscible power in an accidental manner, since when two contraries are in the same power, one is in a certain power as such, and the other is in the same power in an accidental way.

82. Again, it is the same to reason as it is to understand—and so from this we have reason and understanding. However, reason is a perceptive power and the rational power is a motive and directing power. Therefore following the same argument, since the understanding is a perceptive power, the intellective power will be a motive and directing power. And thus, just as there are two motive and directing powers of the sensitive soul, namely the concupiscible and the irascible power, for the same reason there are two motive and directing powers of the rational soul, namely the rational power and the intellective power, but they are different when compared. If, however, it were to be said that the rational power and the intellective power are one and the same power, but different when compared, by the same token reason and intellect are one and the same in essence but differ in an accidental way.—Which is something we willingly concede.

VIII. *On the perceptive powers of the sensitive soul*

83. What comes next concerns the perceptive powers of the sensitive soul. First of all, we shall discuss the external perceptive power, namely sensation. Therefore, we should investigate first of all what sensation is and what types of senses there are.

84. Sensus est vis apprehensiva rei presentis in quantum ipsa est presens per inmutationes receptas in instrumento corporeo ab extrinseco advenientes. Potest autem in primis queri quorum sit sensus, utrum scilicet solum singulare habeat sensum apprehendendi vel et singulare et universale. Quod solum singulare sentiatur testantur omnes auctores de sensu 5 loquentes. Sed quod sensus apprehendat universalia videtur posse ostendi.

85. Ens est prima anime impressio, ut dicit Avicenna, unde per se cadit in intellectum. Quicquid autem in sensum cadit est ens. Quod autem in intellectum cadit, tum est ens tum est non-ens. Ergo potius dicetur quod ens, secundum quod ens cadit in sensum quam in intel- 10 lectum et ens dicetur prima esse anime impressio secundum sensum et non secundum intellectum. Sed ens est primum universale aggregans omnia in sua intentione universali; ergo universale cadit in sensum.

86. Preterea. Lucidum per se et proprie videtur. Sed lucidum, secundum quod lucidum, est universale. Ergo universale per se et 15 proprie videtur.

87. Solutio. Dicendum quod solum singulare videtur. Et dicitur ens esse prima impressio anime quoad intellectum, quia intellectus in componendo inchoat ab ente et in resolvendo stat in ente universali, nec ulterius in resolvendo procedit; cum ens sit intentio impartibilis, et 20 nunquam cadit ens, secundum quod ens, in sensum, sed in quantum est sub aliquo accidente sensibili, ut calore et albedine, et consimilibus. Nec valet hec argumentatio: lucidum proprie, secundum quod lucidum, per se et proprie in visum cadit, et lucidum est universale; ergo universale in sensum cadit. Est enim ibi fallacia accidentis; quoniam in 25 sensum cadit singulare quod est lucidum secundum quod tale; postea predictus redditur universale gratia universitatis. Sed quia dicit Boethius in secunda editione super Porphirium, quod idem est singulare quod universale, singulare cum sentitur, universale cum cogitatur, potest concedi quod id quod est universale sentitur, non tamen universale.   30

1 rei presentis *om.* V   4 solum *om.* C   et[1] *om.* C   7 prima *om.* P   ut: unde P   9 est[2] *om.* C   10 quod[1] *om.* CP   cadit: cadere C; habet cadere V in *om.* P   11 ens *om.* P   esse *om.* C   12 est *om.* P   13 in[1]: de P 15 Ergo universale *om.* P   17 Dicendum: est *add.* V   19–20 nec . . . procedit *om.* P   20 in resolvendo *om.* V   22 est *om.* CP   sub: informatur P 23 proprie *om.* P   25–26 quoniam . . . singulare: quod illud P; quando in lucidum cadit quod est singulare V   27 predictus *om.* P; predicatus V   universitatis: universalitatis V   29 cogitatur: cognoscitur P

1–3 Cf. Avic. *De An.* I. i (2^{ra}T). Vid. p. xix.
5–6 Cf. Arist. *De An.* II. v (417^{b}20–28); *Anal. Post.* I. xviii (81^{b}6); I. xxxi (87^{b}28–30); Avic. *De An.* II. ii (6vb–7^{ra}A).
7 Avic. *Metaphys.* I. vi (72^{rb}A).
17–22 Cf. Arist. *De An.* II. v (417^{b}20–28).
27–29 Boeth. *In Isagogen* II ed. I. 11 (167).

84. Sensation is a power which perceives a thing which is present inasmuch as it is present by means of some modifications received in a bodily organ which come to it from outside. One of the first things we can ask is what sensation is of—in other words, whether a sense perceives only the particular or both the particular and the universal. All of the authors who talk about sensation hold that only the particular is sensed. However, it seems that it can be shown that sensation does perceive universals.

85. Being is the first impression on the soul, as Avicenna states. Thus in itself it comes under the understanding. However, what comes under the sense-power is being. What comes under the understanding, however, is both being and non-being. Therefore, it would be preferable to say that being as such comes under sensation rather than the understanding, and being should be said to be the first impression on the soul as regards sensation and not from understanding. Yet being is the first universal which gathers everything together in a universal intention; therefore the universal comes under sensation.

86. Moreover, a bright object is seen properly and in itself. However, to be 'bright' as such is a universal. Therefore a universal is seen properly and in itself.

87. Solution: It should be stated that only something which is particular is seen. Again, being is said to be the first impression on the soul as regards the understanding, because when the understanding combines it begins from being, and when it divides it remains in universal being, nor does it go further in dividing since being is an indivisible intention. Moreover, being as such never comes under sensation except inasmuch as it does under some sensible accident such as heat or whiteness and others of the same kind. Nor is this reasoning valid: '"Bright thing" inasmuch as it is bright properly and in itself comes under sight and "bright thing" is a universal, therefore a universal comes under sensation.' For here there is a fallacy of the accident because the particular which is bright as such comes under sensation, and afterwards the aforementioned becomes universal thanks to universality. However, because Boethius says in his second commentary on Porphyry that the particular is the same as the universal, particular when sensed, universal when thought, it can be conceded that that which is universal is sensed but not the universal as such.

88. Sensus autem dividitur in quinque, scilicet in visum, in tactum, gustum, olfactum, auditum. Utrum autem sint species an non, prius dictum est.

IX. De visu

89. Primo autem dicendum est de visu. Visus est vis ordinata in nervo 5 concavo ad apprehendendam formam eius quod formatur in humore cristallino ex similitudine corporum coloratorum per radios venientes in effectum ad superficies corporum tersorum. Hanc descriptionem ponit Avicenna in commento *de Anima*. 'In nervo concavo', dicit ad differentiam aliorum nervorum qui non sunt concavi quibus com- 10 parantur alii sensus; 'in humore cristallino', dicit, quia illud quod inmutationes recipit in oculo que sunt similitudines rerum extra est humor cristallinus.

90. Sed videtur posse ostendi secundum hoc quod quodlibet visum debeat apparere rotundum; quoniam signum rotundum indicat causam 15 rotundam, et ideo per rotunditatem umbre probatur rotunditas terre. Inmutatio autem existens in oculo representans rem extra est rotunda, quia suum subiectum in quo est [est] rotundum, scilicet oculus, et illa inmutatio representat rem extra. Ergo cum anima iudicet secundum representationem illius inmutationis, quodlibet videtur apparet ro- 20 tundum.

91. Item. Videtur posse ostendi quod nulla res posset videri minor quam sit diameter secans oculum et transiens per eius centrum. Quoniam quicquid videtur, videtur tanquam sub duabus lineis exeuntibus a duobus terminis diametri ipsius oculi. Sed cum ipse linee sint equedistantes, 25 et quelibet linea secans paralellogrammum ad pares angulos sit equalis linee sibi opposite, res visa secundum equedistantiam a termino diametri non videbitur maior quam sit ipse diameter; et ita nihil quod sit

2 gustum, olfactum, auditum: etc. CV   species *om.* V   an: vel PV   prius: superius
4 *Tit. om.* PV   6 humore: humido C; humano P; (cf. 11, 13)   7 ex: et PV
corporum coloratorum: corporararum coloratarum P   8 corporum tersorum: *lacuna*
V   10 qui... concavi *om.* P   10–11 comparantur: operantur V   11 quia:
quod V   13 cristallinus: cristallini P   15 signum: lignum P   indicat: inducit P
19 representat: presentat P   20 apparet: apparebit P   20–21 rotundum: est
ipsum iudicare habet quodlibet esse rotunde forme *add.* V   22 Item: Amplius V
posset: possit C   23 diameter: diametrum V   24 exeuntibus: extendentibus
P   25 diametri: diametur V   27–28 diametri: diameter V   28 ipse: ipsa C

1–3 Cf. supra, n. 55.
2–3 prius dictum est: cf. supra, nn. 56–60.
9 Avic. De An. I. v (4^{vb}B).
18 oculus rotundus: Guill. de Conchis *Dragm.* VI (280): 'Oculus est quaedam orbiculata substantia'.
22–p. 25, l. 2 Cf. Alex. Nequam *De Nat. Rer.* II. cliii (235).

88. Sensation, however, is divided into five, namely, into sight, touch, taste, smell, hearing. However, whether each is a type of sensation or not should be discussed first.

ix. *Concerning sight*

89. To begin with, we should talk about sight. Sight is a power located in the concave nerve in order to grasp the form of what is shaped in the vitreous humour from a likeness of coloured bodies. This happens by means of rays coming from the surfaces of polished bodies. Avicenna proposes this description in his commentary *On the Soul*. He says 'on the concave nerve' as distinct from other nerves which are not concave and which are related to other senses. He says 'in the vitreous humour' because that which receives the impressions in the eye which are likenesses of external things is the vitreous humour.

90. However, it seems that following from this it can be shown that whatever is seen should appear as round, since a round impression indicates a round source, for the curvature of the earth is proved by means of the roundness of shadows. Indeed, an impression which is in the eye and which represents something external is round since what it is in is round, namely the eye, and that impression represents an external thing. Therefore, since the soul judges according to a representation of that impression, whatever is seen appears as round.

91. Moreover, it seems that it can be shown that no thing can be seen which is smaller than a diameter dividing the eye and passing through its centre. For whatever is seen is seen as it were within two lines coming from the two ends of the diameter of the eye itself. However, since these lines are equidistant, and any line dividing an equilateral parallelogram at right angles is equal to a line opposite to it, a thing which is seen as being equidistant from the end of a diameter will not be seen as greater than the diameter itself; and thus nothing

equedistans ab oculo potest videri maius quam sit ipse oculus, quod
patet esse falsum.

92. Solutio. Ad primum. Dicendum est quod humor cristallinus in
oculo est aqueus, unde de facili recipit quamcumque inpressionem; sed
eam non retinet, sicut nec aqua suam retinet inpressionem. Unde 5
humor cristallinus in oculo de facili recipit inpressionem ad similitudinem forme agentis in ipsum. Unde non omnis inmutatio recepta in
oculo est rotunda, quia non est iudicium secundum inmutationem
existentem in superficie oculi.

93. Ad aliud. Dicendum est quod quicquid videtur, videtur sub 10
angulo quasi inter duas lineas concurrentes in oculo facientes ibi
angulum. Res autem visa est basis et opponitur illi angulo; unde est
ibi triangulus: et quanto linee sic intellecte extense ab oculo magis
distant ab oculo tanto maior est distantia ipsarum ad invicem; quanto
autem magis accedunt ad oculum magis ad concursum accedunt. Cum 15
ergo non protendantur linee secundum equam distantiam in quantum
intelliguntur extendi ab oculo, manifestum est quod res maior oculo
potest videri.

94. Item. Hec est propositio per se nota in principio intellectus:
unum et idem secundum subiectum in numero ex eadem parte sui et in 20
eadem parte sui simul et semel non habet diversas impressiones diversarum specierum, ut ymaginem hominis et ymaginem leonis, et capre
et consimilium, ut patet in cera molli. Ipsa enim simul et semel non
recipit quoad eandem partem in se impressionem representantem
hominem et asinum, et sic de consimilibus. Sed si aer inmutatur 25
mediante quo videtur res, eo quod visus habet fieri per medium, contingit unum et eundem aera inmutari a forma asini et a forma hominis
et, capre, ut si homo et asinus et capra videantur a diversis, et unus et
idem aer intercipiatur inter res visas et oculos videntes. Sed unum et
eundem aera in numero a formis diversarum specierum simul inmutari 30
est impossibile per premissam propositionem; ergo relinquitur quod
aer non inmutatur.

1 maius: maior V   ipse *om.* CV   3 est *om.* P   4–6 sed eam . . . inpres-
rius sionem *om. propter hom.* C   7 inmutatio *om.* P   7–8 recepta . . . rotunda: in
oculo est recepta rotunda P   8 quia: et V   iudicium: inmutatio *sed deletum* V
10 est *om.* P   videtur² *om.* P   13 sic *om.* V   14 ab oculo: et P   maior:
magis V   15 magis²: ma . . . *lacuna* V   16 equam: eque V   20 secundum
om. PV   21 sui *om.* P   24 partem: artem V   26–27 contingit: continget
P; contingat V   27 inmutari: haberi in se impressionem a forma hominis et a
forma asini V   29 videntes: videntur V   32 aer: anima V   inmutatur:
mutatur V

3 Ad primum: cf. supra, n. 90.
5 nec aqua suam retinet inpressionem: cf. Avic. *De An.* I. v (5rbD).
10 Ad aliud: cf. supra, n. 91.

which is equidistant from the eye can be seen as greater than the eye itself, which is clearly false.

92. Solution: In reply to the first objection, it should be said that the vitreous humour in the eye is aqueous and so can easily receive any impression; but it does not retain it, just as water does not retain an impression made on it. Thus the vitreous humour in the eye easily receives the impression which is a likeness of the form acting upon it. Thus, not every impression received in the eye is round, since no judgement is made according to the impression which is on the surface of the eye.

93. With regard to the other objection, it should be stated that whatever is seen, is seen as it were within an angle between two lines converging in the eye which form an angle there. However, the thing which is seen is at the base ⟨of the triangle⟩ and is opposite to that angle; thus there is a triangle there: and the more the lines understood to be extended thus from the eye become distant from the eye, so much greater is the distance from each one to the other; however, the more they come closer to the eye, the more they come closer to meeting. Since, therefore, the lines are not extended according to an equal distance inasmuch as they are understood to be extended from the eye, it is clear that a thing which is bigger than the eye can be seen.

94. Again, here is a proposition which is known in itself at the beginning of understanding: one and the same thing in number according to subject, from the same part of itself and in the same part of itself at one and the same time, does not have different impressions of different kinds, such as the image of a man and the image of a lion, and of a goat, and others of a similar kind, as is plain in soft wax. For, with regard to the same part, the wax does not receive an impression in itself at one and the same time representing a man and a donkey, and the same is true in similar cases. However, because sight occurs through a medium, if the air by means of which the thing is seen is changed, it is necessary that one and the same air is changed by the form of a donkey and by the form of a man and of that of a goat, so that if a man and a donkey and a goat are seen by different people, then one and the same air between the things seen and the eyes seeing is grasped. However, it is not possible that one and the same air in number can be changed at the same time by the forms of the various species, in accordance with the proposition mentioned above; therefore it follows that the air is not changed.

95. Preterea. A causis contrariis fluunt inmutationes contrarie. Ergo cum album et nigrum sint contraria ab eis fluunt inmutationes contrarie in effectum. Ergo, si aer inmutetur ab albo et nigro, in aere sunt simul duo contraria. Sed hoc est impossibile; ergo aer non inmutatur ab albo et nigro: pari ratione nec ab alio colore. Ergo aer non inmutatur ab 5 aliquo colore. Queritur ergo qualiter ab albo cum ipsum videtur hoc generetur inmutatio in oculo nisi aer intermedius inmutetur. Forte dicet aliquis: quod si album et nigrum videantur, aer intermedius inmutatur ab albo et nigro, et generatur in oculo color medius vel mixtus.

96. Contra. Aer inmutatur et eius color est medius color; ergo 10 generat sibi consimilem colorem in oculo; ergo inmutatio recepta in oculo representabit medium colorem. Ergo secundum hoc non videtur nisi medius color, et apparebit res visa medio colore colorata; quod est falsum. Relinquitur ergo quod aer non inmutatur a colore rei vise nec ab eius figura, quod prius ostensum est. Quod bene concedimus, 15 dicentes quod sicut contingit quod artifex aliquis ad similitudinem forme precogitate facit formam in materia sue actioni subiecta, nec tamen est instrumentum quo ipse operatur sub illa forma quam ipse imprimit materie sibi subiecte, ut si ipse faciat ymaginem draconis propter hoc non est suum instrumentum quo ipse agit sub forma 20 draconis; ita contingit quod lucidum agens facit inmutationem in oculo ad similitudinem rei vise mediante ipso aere, nec oportet quod ipse aer sit sub eadem inmutatione. Est autem res visa agens, et lux est dans formam oculo patienti. Hoc habetur ab Alpharabio. Item ab Aristotele in libro de Anima habetur quod magnitudo, numerus, motus, sunt 25 sensata sensus communis.

97. Sed potest ostendi quod magnitudo est sensatum visus proprie; quia quicquid videtur, videtur sub angulo et res visa est basis opposita illi angulo. Sed si caret magnitudine non ei opponetur; unde nec videtur. Sed videtur; ergo magnitudo est adminiculum rei quare ipsa videatur 30 sicut color. Qua ratione ergo color est proprium sensatum visus, magnitudo est proprium sensatum visus.

3–5 in aere . . . et nigro *om.* PV   5 ab[1] *om.* CP   colore: corpore P   6 colore *om.* V   5–6 Ergo . . . colore *om.* P   6 ab *om.* P   hoc *om.* P   8 aliquis *om.* CV   inmutatur: inmutetur V   9 oculo: eo C   9–10 vel mixtus . . . color *om.* V   11 sibi *om.* P   12 videtur: immutabitur V   17 forme: rei P   precogitate: prerogate P   actioni: actionem P   subiecta: subiectam P   18 quam: quoniam P   21 lucidum *om.* P   24 Hoc . . . Alpharabio *om.* P   Item: Amplius V   28–29 et res visa . . . angulo *om.* V   30 ergo: quod *add.* P

1–9 Cf. Guill. de Conchis *Dragm.* VI (285).
15 prius ostensum est: cf. supra, nn. 94–95.
24 ab Alpharabio: locum non inveni.
24–26 Arist. *De An.* II. vi (418[a]17–19); III. i (425[a]14–20).
30–31 Cf. Arist. *De An.* III. i (425[b]8–9).

95. Moreover, from contrary causes contrary changes flow. There-fore, since white and black are contraries, from them contrary changes flow into an effect. Therefore, if the air is changed by white and black, then there are two contraries at the same time in the air. But this is impossible; therefore the air is not changed by the white and the black, and for the same reason neither by any other colour. Therefore, air is not changed by any colour. Thus, the question arises: how is the air changed by white since when this is seen a change is produced in the eye, unless the air in between is changed? Perhaps someone might say that if white and black are seen, the air in between is changed by the white and black, and an intermediate or mixed colour is produced in the eye.

96. Against: The air is changed and its colour is an intermediate co-lour, therefore it produces a colour similar to itself in the eye. There-fore the impression received in the eye will represent the intermediate colour, and so according to this only the intermediate colour is seen, and the thing seen will appear as coloured by the intermediate co-lour, which is false. Therefore, it follows that the air is not changed by the colour of the thing seen nor by its shape (which has already been shown). This is something which we readily concede, stating that in the same way as it happens that any workman makes a shape in the matter which he is working upon according to the likeness of a form which he had already thought of, nevertheless, the instrument with which he works is not part of that form which he impresses upon the matter which he is working on, so that if he were to make a picture of a dragon it does not follow that the tool which he uses is part of the form of a dragon. Thus it follows that a bright object which acts makes an impression upon the eye similar to the thing seen by means of the air itself, nor is it necessary that the air itself is subject to that same impression. For the thing seen is active, and the light gives the form to the eye of the receiver. This is what we know from al-Fārābī. Again, we know from Aristotle in *On the Soul* that size, number, and movement are sensed by the central sense.

97. Yet it can be shown that size is a proper sensible of sight, since whatever is seen is seen within an angle and the thing seen is the base opposite to that angle. However, if it is lacking in size it cannot be opposite to it; and so neither would it be seen. But it is seen; therefore size is a support to the thing whereby it is seen just like colour. There-fore by the same reason that colour is a proper object of the sensation of sight, size is a proper object of what is sensed by sight.

98. Ad hoc dicendum est quod sicut res non videtur nisi mediante suo colore, similiter non videtur nisi mediante magnitudine; unde dicit [Aristoteles] quod magnitudo est sensatum visus sicut color. Dicitur tamen ab Aristotele quod magnitudo est sensatum sensus communis, quia magnitudo apprehenditur per plures sensus particulares, ut per 5 visum, gustum et tactum, et in sensu communi percipitur iudicium rei sensate.

99. Item. Queritur quare ita sit quod tactus apprehendat inmutationem aliquam causa illius omnino destructa, et visus non; ut si calor generetur in tactu ab igne, destructo igne totaliter, adhuc sentitur 10 calor, sed si ab albo generetur inmutatio in visu, albo penitus destructo non videbitur album.

100. Solutio. Dicendum est quod hoc contingit propter ineptitudinem materie que non potest retinere inmutationem generatam a forma rei vise. Generatur enim in humore cristallino qui propter liquiditatem sui 15 non potest retinere formam, cum defecerit causa imprimens.

101. Item. Omnis motus localis fit in tempore. Ergo si videamus extra mittentes radios in emissione radii non adhuc est terminus radii in termino rei vise. Ergo cum magna sit distantia inter rem visam et rem videntem erit tempus commensurans motum radii venientis ab 20 oculo ad rem visam, eo quod ipsum radium necesse est pertransire omnes partes intermedias, et prius pertransit propinquiorem quam remotiorem partem. Non ergo res visa videtur quamcito ei opponitur oculus; quod quilibet experimento percipere potest esse falsum. Ergo illud est falsum ex quo istud sequitur. Sed hoc sequitur dato quod 25 videmus extra mittentes; ergo non videmus extra mittentes. Quod bene concedimus, dicentes quod inmutatio generatur a lucido subito ex quo non invenit aliquod resistens, sicut testatur tam Avicenna quam Algazel in commento prime philosophie, quod illuminatio subito et non

2 suo . . . mediante *om.* PV   4 tamen: autem C   6 percipitur: participatur P; perficitur V   8 Item: Rursus V   9 aliquam: a materia C
10 sentitur: sentietur C   11 ab *om.* C   albo: illo V   17 Item: Amplius V
18 extra mittentes: extramitates P   20 tempus: aliquod *add.* V   22 pertransit: transit V   24 esse *om.* V   25 illud: id C   sequitur: queritur
V   26 videmus: vidiatur P; videamus V   extra mittentes[1]: extramitates P   extra
mittentes[2]: extremitates P   27 subito: subiecto V   28 aliquod: aliquid C
tam . . . quam: ab Avicenna et ab P   29 philosophie: scilicet *add.* V

1–3 Cf. Arist. *De An.* loc. cit.
3 Aristoteles: cf. supra, p. 26, ll. 24–25 et infra, l. 4.
4–7 Arist. *De An.* II. vi (418ª17–20); III. i (425ª14–20).
17 Cf. Arist. *Physic.* IV. xiv (223ª30–ᵇ22).
17–28 Cf. Arist. *De An.* II. vii (418ᵇ20–26); *De Sensu* vi (446ª21–447ª11); Aug. *De Genesi ad Litteram* IV. xxxiv. 54.
28 Avic. *De An.* III. vii (15ʳᵇA); Algazel *Metaphys.* II. iv. 3 (168).

98. Regarding this it should be stated that just as a thing is not seen except by means of its colour, similarly it is not seen except by means of size. For this reason, Aristotle says that size is the proper object of sight just like colour. For it is stated by Aristotle that size is a proper object of the sensation of the central sense, because size is perceived by means of many individual senses such as sight, taste, and touch, and in the central sense the judgement of the thing sensed is perceived.

99. Again, it is asked why it is that touch perceives some impression the cause of which has been completely destroyed, whereas sight does not. For example, if heat is produced in touch by fire, if the fire has been completely destroyed, the heat is still felt. However, if an impression is produced in sight by a white object, as soon as the white object is destroyed the white object will not be seen.

100. Solution: It should be stated that this happens because of the unsuitability of the matter which cannot retain the impression produced by the form of the thing seen. For the impression is generated in the vitreous humour, which because of its liquidity cannot retain the form when the cause of the impression is lacking.

101. Again, all local motion takes place in time. Therefore, if we were to see by sending out rays, when a ray is emitted the position of the ray is not yet at its end point in the thing seen. Therefore when the distance is great between the thing seen and the one seeing, there will be a time corresponding to the movement of the ray coming from the eye towards the thing seen, because it is necessary for this ray to pass through all of the intermediate parts, and firstly to pass through the nearer rather than the more remote part. The thing seen is not, therefore, seen as soon as an eye is located opposite it, which anyone can see is false through experience. Therefore, the latter is false inasmuch as the former ensues. However, this follows if it is the case that we see by sending out rays, therefore we do not see by sending out rays. This is something which we readily concede, stating that an impression is produced by a bright body immediately in so far as it does not encounter something resisting it, as both Avicenna and al-Ghazālī hold in their commentaries on the *Metaphysics*: illumination is pro-

gradatim generatur et pereunt tenebre: tenebre enim nihil aliud sunt quam privatio lucis. Nec illuminatur citius pars propinquior quam remotior, ex quo illuminatio subito est facta. Et potest hoc probari simili modo quo probat Aristoteles in *Physicis*, quod vacuum non est, quia si vacuum esset, motus localis fieret in instanti, quia res mota non inveniret sibi 5 resistens. Et ponit hoc principium ibi: que est proportio resistentie ad resistentiam, eadem est proportio motus ad motum.

102. Item. Quod sub maiori angulo videtur, maius videtur, et quod sub minori minus, et quanto res visa magis elongatur ab oculo, tanto minor videtur. Unde quanto aliquid a remotiori videtur tanto minus 10 videtur, quanto a propinquiori tanto maius. Ergo quanto periferia a remotiori videtur, tanto minor videtur. Sed quanto periferia est minor, tanto angulus contingentie est maior. Ergo quanto periferia a remotiori videtur, tanto angulus contingentie maior videtur: non ergo quanto aliquid a remotiori videtur tanto minus videtur. 15

103. Solutio. Hec argumentatio non valet: quanto periferia est minor tanto angulus contingentie est maior; ergo quanto periferia a remotiori videtur tanto angulus contingentie minor videtur. Quia sicut decrescit angulus sub quo videtur circulus per elongationem ipsius ab oculo, ita decrescit angulus sub quo videtur angulus contingentie per elongati- 20 onem anguli contingentie ab oculo, quia quanto plus et plus recedit angulus contingentie ab oculo tanto magis minoratur angulus linearum concurrentium in oculo, ut ipse recessus sit proportionalis diminutioni ipsius anguli.

104. Dictum est superius quod quanto aliquid a propinquiori videtur, 25 tanto maius videtur, quanto a remotiori tanto minus. Sed obicitur.

105. Describatur aliquis semicirculus et ducatur diameter ab uno terminali puncto secundum circumferentiam ad aliud punctum termi- nalem, et postea in medio puncto semicircumferentie ponatur oculus

2 pars *om.* P   propinquior: rei lucide *add.* PV   6 que: quod P   est *om.* V resistentie: resistentis C   7 eadem: vel *add.* P   est *om.* C   8 Item: Amplius V   maius: magis C   10 minor: minus V   11 periferia: per inferia P   12 tanto *corr. in marg.* C   quanto: quantum C   13 contingentie: contingentis P; contingenti V   13–14 tanto . . . videtur¹ *om.* V   14 maior: minor P   15 aliquid *om.* P   16 est *om.* CP   18 minor: maior V 19 elongationem: elongitudinem P   21 quia: et C   25 aliquid *om.* P 26 maius: maior P   a remotiori: videtur *add.* P   minus: minor CP   28 secun- dum circumferentiam: semicircumferentie PV   29 semicircumferentie: circum- ferentie P

1–2 Cf. Arist. *De An.* II. vii (418ᵇ18–19).
4–7 Arist. *Physic.* IV. viii (215ᵃ20–22; 216ᵃ8–12); ix (216ᵇ24–26).
8–15 Fere ad litt. apud Alex. Nequam *De Nat. Rer.* II. cliii (234). Cf. Algazel *Metaphys.* II. iv. 3 (167).
16–24 Alex. Nequam ibid. (235).
25 superius: cf. n. 102.

duced immediately and not gradually and the darkness ends: darkness is nothing other than the absence of light. Nor is the nearer part illuminated more quickly than the remoter part inasmuch as illumination is produced immediately. And this can be proved in a way similar to that by which Aristotle proves in the *Physics* that a vacuum does not exist. For if a vacuum existed local motion would take place in an instant, since a moving thing would not encounter something resisting itself. And he posits this principle there: that which is the proportion of a resistance to a resistant is the same proportion of a movement to a moved.

102. Again, that which is seen within a greater angle is seen more, and that which is seen within a lesser angle is seen less, and the more something which is seen moves away from the eye, the less it is seen. Thus the more something is seen at a distance so much less is it seen, and the more from close up the more it is seen. Therefore, the more the circumference is seen from a distance the less it is seen. But the smaller the circumference the greater is the angle of incidence. Therefore, the further away the circumference is seen, the greater is the angle of incidence seen; and it is not therefore the case that the further away something is seen the less it is seen.

103. Solution: This argumentation is not valid: the less the circumference the greater the angle of incidence; therefore the more a circumference is seen from afar, the angle of incidence is seen less. Because just as the angle decreases within which a circle is seen because of its elongation from the eye, in the same way the angle decreases in which the angle of incidence is seen because of the elongation of the angle of incidence from the eye, since the more the angle of incidence recedes from the eye the more the angle of the lines meeting in the eye become less, so that this receding is proportional to the diminution of this angle.

104. It has been stated above that the more something is seen from close up the bigger it is seen, and the further away the smaller. This, however, is disputed.

105. Let a semicircle be described and let a diameter be drawn from one end point on the circumference to another end point, and afterwards let an eye be located at the midpoint on the circumference which sees all of the diameter, and let another eye be located on the circum-

videns totum diametrum, et ponatur alius oculus in circumferentia in alio
puncto propinquiori termino circumferentie quam sit prior oculus, et
ille oculus videat similiter totum diametrum. Inde ille oculus est propinquior diametro quam sit alius oculus. Ergo per premissam proportionem
diameter videtur maior hoc oculo propinquiori quam oculo remotiori. 5

106. Contra. Probatum est in Geometria in tertio Euclidis quod
omnes anguli existentes in semicircumferentia habentes diametrum pro
basi eis opposita sunt equales. Sed anguli sub quibus duo predicti
oculi vident diametrum sunt consistentes in semicircumferentia. Ergo
ipsi sunt equales, et quecumque videntur sub equalibus angulis videntur 10
equalia. Ergo cum diameter videatur sub equalibus angulis non videtur
maior uni oculo quam alii, et tamen propinquior est uni oculo quam alii.

107. Solutio. Dicendum est quod diameter non videtur maior uni
oculo quam alii, et non est unus oculus propinquior totali diametro
quam alius; quia licet unus oculus propinquior sit termino diametri 15
quam alius, tamen non est propinquior totali diametro, quia quanto
maior est appropinquatio ad unum terminum tanto est maior elongatio
ab alio termino.

108. Item. Piscis qui videtur in aqua maior videtur in fundo aque
quam in superficie aque vel in superficie aeris supra. Sed propinquior 20
est oculo quando est in superficie aque quam quando est in fundo aque.
Non ergo quanto aliquid a propinquiori videtur, videtur maius.

109. Solutio. Dicendum est quod hoc contingit propter inmutationem
receptam in aqua fluxili. Fluit enim inmutatio a pisce in aqua quasi sub
duabus lineis non equedistantibus quando piscis ab oculo videtur in 25
aqua, sed in quantum accedunt ad superficiem aque plus et plus magis
distant, quia intelliguntur ibi quasi due linee protracte a duobus terminis ipsius piscis usque ad superficiem aque.

110. Item. Queritur quare recta virga appareat curva si eius medietas
sit in aqua. Ad hoc dicendum est quod illa pars virge que est in aqua 30

---

1 diametrum: diametrem C   3 ille: iste V   similiter: sic P   ille: iste V
4 diametro . . . oculus *om.* V   Ergo *om.* P   4 proportionem: propositionem
V   5 hoc *om.* V   quam: in *add.* V   7 pro: et P   8 eis: ei P; sibi V   duo *om.* V
10–11 videntur . . . angulis *om.* V   12 alii: alio P   13–14 Solutio . . . quam
alii *om.* V   13 uni: uno P   14 totali diametro: cuidam termino diameter V
15–16 quam alius . . . alius *om. propter hom.* V   18 ab *om.* V   19 Item:
Amplius V   maior: magis P   20 aeris: aere P   21 quam . . . aque *om.* V
22 Non *om.* P   24 a pisce: inquantum piscis ab oculo videtur P   25 duabus:
rectis *add.* V   25–26 quando . . . plus et plus: plus et plus distantibus, sed inquantum accedunt ad superficiem aque P   in aqua sed *om.* V   26–27 magis
distant *om.* C   27 ibi *om.* CV   29 virga: linea V   30 est *om.* P

---

6 Euclides *Geometria* III: locum non inveni.
19–22 Cf. Alex. Nequam ibid. (234).
29–30 Cf. Guill. de Conchis *Dragm.* III (69); Alex. Nequam ibid. (235).

ference at another point nearer to the end of the circumference than
the first eye, and the second eye similarly will see all of the diameter.
Then that eye is nearer to the diameter than the other eye is. Therefore, by means of the aforementioned relation, the diameter appears
greater to the nearer eye than to the eye which is further away.

106. Against: It is proved in Geometry in the third book of Euclid
that all angles existing in a circumference and having a diameter as a
base opposite to them are equal. But the angles within which the two
aforementioned eyes see the diameter are present together on the circumference. Therefore these are equal, and whatever is seen within
equal angles is seen as equal. Therefore since the diameter is seen
within equal angles it is not seen as bigger to one eye than to the other,
and yet it is nearer to one eye than to the other.

107. Solution: It should be stated that a diameter does not appear
greater to one eye than to another and one eye is not nearer to all of
the diameter than another eye; because even if one eye is nearer to the
end of the diameter than the other, yet it is not nearer to all of the diameter because the nearer it is to one end, the further is the distance
from the other end.

108. Moreover, a fish which is seen in the water seems bigger in
the depths of the water than at the surface of the water or in the air
above the surface of the water. Yet the fish is nearer the eye when it is
at the surface of the water than when it is in the depths of the water.
Therefore it is not the case that the more something is seen from close
up the bigger it appears.

109. Solution: It should be stated that this happens because the impression which is received in water is liable to change. For the impression flows out from the fish in the water as between two lines which
are not equidistant when the fish is seen by the eye in the water, yet
inasmuch as the lines come closer to the surface of the water they become more and more distant, because they are grasped there like two
lines drawn out from two ends of the fish itself up to the surface of
the water.

110. Moreover, the question arises as to why a straight stick appears curved if half of it is in the water. In answer to this it should
be stated that the part of the stick which is in the water is not really

revera non videtur, sed eius inmutatio que generatur in aqua, et oculus
videns effectum credit se videre causam illius effectus; et ideo cum illa
inmutatio sit propinquior oculo quam illa pars virge que est in aqua,
ideo illa virga apparet curva. Si autem queritur quare potius fiat inmutatio figure in aqua quam in aere, dicendum est quod hoc contingit 5
propter habilitatem materie, eo quod aqua non est rara sicut aer nec
adeo fluxilis.

111. Item. Queritur quare quoddam animal perspicacius videat de
nocte quam de die cum lux exigatur ad visum.

Ad hoc dicendum est quod sicut potest esse diminutio lucis ad visum 10
perficiendum, ita et superhabundantia lucis; et sicut diminutio lucis
impedit visum, ita lucis superfluitas impedit visum. Et ideo quia illud
animal quod melius videt de nocte quam de die illuminat rem visam
suo proprio instrumento visus, et etiam illuminatur a sole, eo quod dies
est sol lucens super terram, est lucis habundantia impediens visum 15
animalium talium. De nocte autem illuminant huiusmodi animalia per
claritates oculorum suorum rem visam.

112. Item. Queritur quare contingat audire sonum venientem ab
anteriori parte et a posteriori, et a dextra et a sinistra, et similiter non
contingat videre ante et retro. Similiter dextrorsum et sinistrorsum.   20

113. Similiter. Queritur quare contingat auditu percipere sonum
pariete existente medio, et non contingat similiter videre per medium
parietem.

114. Item. Queritur quare oculi inspicientes oculos lippos incurrant
labem consimilem potius quam oculi egri aspicientes oculos sanos 25
incurrant sanitatem.

115. Solutio. Ad primum. Dicimus quod aer inmutatur a sono, et
inmutatus pertransit corpus porosum; unde, cum auris et caput sint
corpora porosa, pertransit aer inmutatus ad nervum audibilem, et sic
fieri auditum contingit; et etiam aer inmutatus inmutat sibi aerem 30

---

1 in aqua: videtur *add.* V   4 quare *om.* P   6 rara: tam rara V; rarum C
8 Item queritur: Similiter potest queri V   10 sicut *om.* V   11 ita et: ita quod
P   sicut: sic P   diminutio lucis: lucis *om.* C   16 autem: et V   17 claritates:
claritatem V   visam: ab eis *add.* V   18 Item: Rursus V   quare *om.* P   audire
lacuna V   19 posteriori: parte *add.* P; in posteriora V   similiter *om.* V
20 dextrorsum: deorsum V   21 auditu percipere: audire V   22 medio:
et sono audito *add.* P   24 Item: Similiter V   oculos *om.* V   24, 26 incurrant:
incurrunt P

8–17 Cf. Avic. *De An.* III. vii (15^{va}B); Adelard. Bath. *Quaest. Nat.* xii (15); Guill.
de Conchis *Dragm.* VI (290); Alex. Nequam *De Nat. Rer.* II. cliii (236); *De Laudibus
Divinae Sapientiae* D. ix (490); Anon. *Quaestiones Phisicales* 37 (172): 'Cur nocte
magis discernere visu possit murilegus' (Ex Urso *Aphorism.*). Vid. p. xix.
24–26 Cf. Guill. de Conchis ibid. (287–8).
27 Ad primum: cf. supra, n. 112.

seen, but rather its impression which is produced in the water, and when the eye sees the effect it believes that it is seeing the cause of the effect; and so since that impression is nearer to the eye than that part of the stick which is in the water, therefore that stick appears to be curved. If, however, it is asked why the impression of a shape occurs more readily in water rather than in air, it should be stated that this happens because of the capacity of the matter, since water is not as thin as air nor consequently is it as changeable.

111. Again, it is asked why some animals see more clearly at night rather than during the day since light is required for sight.

In reply to this it should be stated that just as a lessening in light can improve sight, the same is true of a lot of light, and just as a lessening of light can hinder sight, so too much light can impede sight. And so because that animal which sees better at night than during the day illuminates the thing seen by means of its own organ of sight, and if it is also illuminated by the sun, since daytime is when the sun shines on the earth, the excess of light impedes the sight of such animals. At night-time, however, such animals illuminate the thing seen by means of the rays from their eyes.

112. Again it is asked, why does it happen that we hear sound coming from the place in front and from behind, from the right and from the left, but it does not similarly happen that we see in front and behind and likewise things to the right and to the left?

113. Similarly, it is asked why does it happen that one perceives sound by hearing when there is a wall in between and it does not similarly happen that one can see through such a wall?

114. Again it is asked, why is it that eyes looking into inflamed eyes incur an identical blemish more so than the eyes of the sick looking into the eyes of the healthy incur health?

115. Solution: In answer to the first question, we state that air is affected by sound, and having been affected it passes through a porous body. Thus since the ear and the head are porous bodies, the affected air penetrates to the auditory nerve and in this way it happens that hearing takes place. Moreover, when the air is affected it affects the

proximum usque perveniat ad concavitatem auris, et in illa concavitate fit reflexio inmutationis ad nervum audibilem, et fit auditus. Aer autem, ut supra ostensum est, non inmutatur a forma visibili, immo oculus tantum secundum quod ipse opponitur recte rei vise.

116. Per hoc autem solvitur secunda questio. Quia aer inmutatus 5 a sono pertransit poros parietis. Nisi enim esset corpus porosum non fieret auditus per medium. Si enim aliquid includeretur in corpore spherico non habente poros, illa res inclusa non audiret aliquem sonum extra illud corpus. Illud autem quo existente medio fit visus est pervium et translucens; et istud est primum. Nec habet queri propter quid 10 hoc sit, scilicet quare visus habeat fieri per translucens; paries autem non est pervius nec translucens, et ideo per medium parietem non potest fieri visus.

117. Ad ultimum. Dicendum est quod cum sanus oculus inspicit egrum oculum inmutatur a forma oculi egri, et inficitur ab illa inmu- 15 tatione. Res enim visa agit in oculum videntem, et oculus est patiens recipiendo inmutationem a re visa et ita incurrit labem.

x. *De luce*

118. Quoniam dictum est superius quod visus non potest stabiliri in effectu nisi mediante luce, ideo videndum est quid sit lux, cum ad esse 20 visus exigatur esse lucis. Queritur ergo quid sit lux.

119. Si color; Contra. Aliquis color est alicui colori contrarius. Ergo si idem sit color quod lux, aliqua lux alicui luci erit contraria, quod non videtur esse verum.

120. Preterea. Lux et tenebra privative opponuntur, et lux est color, 25 ergo color et tenebra privative opponuntur. Ergo nihil quod est tenebrosum est coloratum; et preterea secundum hoc colores non sunt de nocte.

121. Item. Si lux est color, aut lux est eiusdem intentionis in universali cuius est color, aut est intentio particularis contenta sub colore. 30

1 auris . . . concavitate *om.* P   2 auditus: auditum P   6 Nisi: Ubi V
10 istud: illud V   11 quare: quod P   habeat: habet V   11–12 paries
. . . translucens *om.* PV   15 inficitur: invenenatur P   16 enim: autem V
18 *Titulum om.* PV   22 contrarius: contrarium V   23 lux² *om.* P   25 tenebra:
tenebre *bis* PV   27 sunt: erunt V   29 Item: Amplius V

3 supra: cf. n. 96.
5 secunda questio: cf. supra, n. 113.
14 Ad ultimum: cf. supra, n. 114.
19 superius: cf. nn. 96, 111. Vid. Arist. *De An.* II. vii; Avic. *De An.* III. i (10ra−b^AB).
22–24 Cf. Avic. ibid. iii (11^ra^A).
25 Cf. Arist. ibid. (418^b^18); Avic. loc. cit.
27–28 Cf. Avic. loc. cit.

air next to it until it arrives at the cavity of the ear, and in that cavity a reflection of this impression occurs and hearing happens. For air, as has been shown above, is not affected by a visible form; rather only the eye is, inasmuch as it is placed directly opposite the thing which is seen.

116. In this way the second problem is solved, since air which is affected by sound passes through the pores of a wall. For unless there is a porous body hearing will not occur through the medium. For if something is enclosed within a spherical body which does not have pores, the thing enclosed will not hear any sound outside of that body. However, what exists as a medium when sight happens is pervious and translucent, and this is a requirement. Nor should it be asked why this is the case, namely why does sight take place by means of something translucent, for a wall is neither pervious nor translucent and so sight cannot happen through a wall.

117. With regard to the last question, it should be stated that when a healthy eye looks into a sick eye it is affected by the form of the sick eye, and it is harmed by that affect. For the thing which is seen acts upon the eye which is seeing and the eye undergoes the reception of an impression from the thing seen and thus incurs a blemish.

x. *Concerning light*

118. Since it was said above that sight cannot actually take place except by means of light, it follows that we should examine what light is, since in order for sight to be, there must be light.

119. Is light colour? Against: any colour is contrary to another colour. Therefore if colour were the same as light, any light would be contrary to another light, which does not appear to be true.

120. Moreover, light and darkness are privative opposites. Therefore, nothing which is dark is coloured; and again because of this there are no colours at night.

121. Again, if light is a colour, either light is of the same universal understanding that light is, or it is a particular concept contained within the term 'colour'. If it is of the same universal understanding,

Si est eiusdem intentionis in universali, ergo cum illa intentio que est
color sit genus albedinis et nigredinis, lux erit similiter genus albedinis
et nigredinis. Sed omne genus predicatur in quid de sua specie; ergo
hec est vera: nigredo est lux. Si autem lux sit species coloris, ergo vel
est unum extremorum vel aliquod mediorum. Si unum extremorum, 5
ergo lux vel est albedo vel nigredo. Non nigredo, quia nigredo impedit
potius lucem quam sit inducens eandem. Similiter non est albedo, quia
aliquis lapis est totaliter albus, et ita intrinsecus est albus, nec tamen
lux est intra ipsum, quoniam lux solummodo est in pervio vel in
superficie. 10

122. Si lux sit medius color;

Contra. Omne medium conficitur ex extremis. Ergo si lux sit color
medius ipsa conficitur ex albedine et nigredine, cum albedo et nigredo
sint extrema mediorum. Sed hoc esse non potest, quoniam nigredo
potius destruit lucem quam eam procreet. Forte dicet aliquis quod lux 15
nihil aliud est quam detectio coloris; sed detectio coloris et occultatio
coloris sunt opposita, et detectio coloris est lux; ergo occultatio coloris
est tenebra. Sed affectio generatur ex rubeo in re sibi opposita, que
affectio occultat colorem illius rei. Ergo illa affectio est tenebra. Sed illa
affectio est lucida. Ergo tenebra est lucida, quod esse non potest. 20

123. Solutio. Dicimus quod idem subiecto est lux et color; sed
diversitas est inter ea in accidente. Dicitur enim lux in comparatione
ad rem translucentem, et color in comparatione ad rem coloratam. Lux
enim, secundum quod lux, est qualitas que ex essentia sua est perfectio
translucentis. Sed illa qualitas que est color solum dicitur esse lux in 25
quantum ipsa ex se generat lumen, et est perfectio translucentis; et
dicitur illa qualitas esse color in quantum ipsa colorat suum subiectum.
Distinguitur autem in commento inter lucem et lumen et splendorem.
Lucem appellat Commentator perfectionem translucentis; lumen vero

2–3 lux erit similiter genus albedinis et nigredinis *om. propter hom.* C   4 Si . . .
coloris: Si species sit species coloris V   5 aliquod: aliquid V   7 inducens
om. P   eandem: sibi *add.* P; eam V   Similiter *om.* V   8 albus: albedo P
9 est intra ipsum *om.* P   10 superficie: solidi *add.* P   12–13 Contra . . .
medius *om.* V   15 procreet: provocet C   lux *om.* V   18 ex: in V
19 illa[1]: illius P   21 subiecto: in subiecto P   23 translucentem . . . rem *om.* V
25 Sed: Unde P   solum: solis V   26–27 et dicitur: sed dicitur P   28 in
commento *om.* P   29 vero *om.* V

15–16 Cf. Avic. ibid. (11[ra]A).
21 Solutio: cf. Avic. loc. cit.
23–25 Cf. Avic. *De An.* III. iii (11[rb]C).
25–27 Avic. loc. cit.
28 in commento: i.e. in libro *De Anima* Avicennae, loc. cit., et c. i (10[ra]A).
29 Commentator: i.e. Avicenna, loc. cit.

then when the concept which is 'colour' is of the genus of white and black, similarly light will be of the genus of white and black. But every genus is predicated of all as regards its species, therefore this is true: 'blackness is light'. If light then is a species of colour, therefore it is either one of the opposites or something of those in the middle. If it is one of the opposites, then light is either whiteness or blackness. Not blackness since blackness impedes light more than it produces it. Similarly it is not whiteness since any stone can be totally white, and so internally is white, and yet there is no light within it, since light is only where it can pass through or is at the surface.

122. If we take it that light is a middle, then it can be argued against this that everything in the middle is made up from the opposite extremes. Therefore, if light were a middle colour, it would be made up of whiteness and darkness since whiteness and darkness are the extremes of those in the middle. But this cannot be since blackness more destroys light than produces it. Perhaps someone might say that light is nothing other than the revealing of colour; yet the revealing and the covering up of colours are opposites, and the revealing of colour is light; therefore the covering up of light is darkness. However, an impression is produced by a red thing on something which is opposite to it, and this red impression obscures the colour of the thing. Therefore, that red impression is dark. However, that impression is bright; therefore darkness is light—which is something that cannot be.

123. Solution: We state that both light and colour fall under the same subject; but there is a difference in accident between them. For light is said with reference to something which is translucent, and colour with reference to something which is a coloured thing. For light, inasmuch as it is light, is a quality which by its very essence is the perfection of a translucent thing. However, that quality which is colour is only said to be light inasmuch as it generates brightness from itself and is the perfection of a translucent thing; and that quality is said to be colour inasmuch as it colours its subject. In the commentary ⟨of Avicenna⟩ a distinction is made between light, brightness, and brilliance. The Commentator (Avicenna) calls light the perfection of the translucent; brightness, however, he calls the modification produced

appellat passionem generatam in translucente, ut in aere; splendorem autem dicit esse passionem generatam ex colore aliquo in re translucente, ut ex rubore vel aliquo consimili.

124. Ad predictas obiectiones dicendum est quod in ipsis est fallacia accidentis; ut hic: idem est color quod lux; sed color est genus albedinis 5 et nigredinis; ergo est lux: ab alio enim dicitur lux et ab alio color. Similiter dicendum est ad alias obiectiones.

125. Item. Queritur utrum nomen lucis dicatur univoce de luce supercelestium, ut solis et stellarum, et de luce rerum naturalium, ut ignis et flamme et consimilium. Quod eiusdem speciei sit hinc et inde 10 videtur per effectum, quia luce solis generatur lumen in aere, a luce ignis generatur idem effectus in aere, et idemptitas effectus ab idemptitate cause est; ergo lux ignis et lux solis sunt eiusdem speciei, et nomen lucis dicitur de ipsis secundum eandem intentionem.

126. Contra. Lux ignis res nature est, lux solis est res supra naturam; 15 sed a diversis in specie exeunt diversa in specie; ergo, cum natura inferior sit alterius speciei quam res procreans passiones superiorum, alterius speciei est lux in inferioribus et alterius speciei est lux in superioribus. Ergo nomen lucis non est eiusdem intentionis secundum quod dicitur de luce solis et secundum quod dicitur de luce ignis; ergo 20 nomen lucis non predicatur univoce de ipsis.

127. Solutio. Bene concedimus quod lux solis et lux ignis non sunt eiusdem speciei specialissime, et tamen hoc nomen 'lux' predicatur de illis secundum unam et eandem intentionem universalem, quoniam lux in abstractione universali est genus ad illas duas species, quarum una 25 est ab operatione nature, ut in rebus inferioribus, altera autem a re non naturali, immo a re supra naturam, ut supercelestium lux; et licet nomina illarum specierum non sint inventa, non ideo minus sunt ille species.

128. Item. Queritur quomodo opponantur lux et tenebra. Non secundum affirmationem et negationem, ut lux et non-lux; non secun- 30 dum viam contrariorum, quia tenebre non proveniunt nisi ex privatione lucis: ex privatione autem nihil relinquitur in re; unde nomen tenebre non significat accidens contrarium luci. Non opponuntur secundum viam relative oppositorum; hoc manifestum est.

1 in translucente: translucentem C   ut in aere *om.* P   2 translucente: translucentem C   5 hic: hoc V   6 lux² *om.* V   8 Item: Amplius V   10 sit: sint V   11 luce: a luce V   17 sit . . . superiorum *om.* V   18–19 et alterius . . . in superioribus: quam lux in inferioribus V   20 et secundum . . . ignis *om.* V   23–24 de illis *om.* C   24 unam et *om.* C   25 abstractione *scriptum sed deletum* V   29 Item: Amplius V   tenebra : tenebre P *passim*   31 proveniunt: veniunt P   nisi *om.* V   33 opponuntur: opponitur P   34 hoc *om.* P

4 Ad predictas obiectiones: cf. supra, nn. 119–22.

in the translucent, such as in air; brilliance, however, he states to be an effect produced by any colour in a translucent thing, such as by redness or something similar.

124. It should be replied to the aforementioned objections that in them there is a fallacy of the accident, such as this: colour is the same as light; but colour is of the genus of whiteness and blackness; therefore it is light: but light is said from one thing and colour from another. The other objections are replied to in the same way.

125. Moreover, it is asked whether the word 'light' is said univocally of the light of the heavenly bodies such as the light of the sun and the stars, and of the light of natural things such as the light of fire and of flames and other similar things. That light is of the same kind both there and here can be seen through the effect, since by the light of the sun brightness is produced in the air, from the light of fire the same effect is produced in the air, and an identical effect is produced by an identical cause. Therefore the light of fire and the light of the sun are of the same kind, and the word 'light' is said of them according to the same meaning.

126. Against: The light of fire is a natural thing, the light of the sun is a thing which is above nature; but from things which are also diverse in kind come forth things which are diverse in kind. Therefore since the lower nature is of a different kind from the thing producing the actions undergone by the superior bodies, light in inferior bodies is of one kind and light in superior bodies is of a different kind. Therefore, the word 'light' does not have the same meaning inasmuch as it is said of the light of the sun and inasmuch as it is said of the light of fire; therefore the word light is not predicated univocally of these.

127. Solution: We readily concede that the light of the sun and the light of fire are not of the same most specific kind, and yet this word 'light' is predicated of them according to one and the same universal understanding, since light in a universal abstraction is a genus in respect of those two species, one of which is from the workings of nature such as in inferior bodies; the other, however, is not from a natural activity—rather it is from something which is above nature, such as heavenly light, and whereas the names of these species are not to be found, none the less they are of those species.

128. Again, it is asked in what way light and darkness are opposed. Not according to affirmation and negation, such as light and non-light; not according to way of contraries, since shadows do not occur except by the privation of light; by privation, however, nothing is left in reality. Thus the word 'darkness' does not signify an accident contrary to light. Nor are they opposed according to the relative way of opposite, and this is clear.

129. Si opponantur secundum privationem et habitum; Contra. Ut dictum est superius, aliud est lumen et aliud est lux. Lumen enim est passio generata ex luce, lux vero est qualitas innata, nec potens est relinquere suum subiectum, ut ignem non potest relinquere lux; unde subiectum lucis non potest privari luce; ergo non potest esse tene- 5 brosum. Sed privatio et habitus habent esse circa idem subiectum, sed tenebre et lux non habent esse circa idem subiectum; ergo non opponuntur privative.

130. Preterea. Si est ibi oppositio privationis et habitus inter tenebras et lucem, ratio tenebrosi est aptum natum esse lucidum non ens luci- 10 dum. Sed nihil potest esse tale in quo est lux; ergo inter lucem et tenebras non est privativa oppositio.

131. Item. Inter privationem et habitum est ordo irregressibilis, ut habetur ab Aristotele in *Predicamentis*. Ergo si lux et tenebre privative opponuntur, a tenebris in lucem non potest fieri regressus; ergo si ali- 15 quod corpus sit tenebrosum non potest fieri lucidum quamvis lucidum prius fuerit.

132. Solutio. Dicendum est quod nomen tenebrarum duplicis est intentionis, in una enim intentione opponitur lumini, et secundum hoc luminosum et tenebrosum privative sunt opposita; unde aer tum est 20 tenebrosum tum luminosum, et habent fieri lumen et tenebre circa idem subiectum, et est hic ordo regressibilis. Et quod dicit Aristoteles quod inter privationem et habitum non est ordo regressibilis, intelligit hoc auctor tantum de illo habitu qui est ab aptitudine naturali intrinseca; lumen autem non est qualitas innata, immo est ab extrinseco. In alia 25 significatione nomen tenebrarum significat privationem lucis, et secundum hoc si ipsum non ponat aptitudinem habendi lucem posset concedi, quod quodlibet corpus in quo non est lux est tenebrosum; et ita aer, cum in eo non sit lux, immo lumen, quando est illuminatus posset dici tenebrosus a privatione lucis et non a privatione luminis; et 30 secundum hanc intentionem nominis non tenet hec argumentatio: est

1 opponantur: opponatur P   2 dictum: habitum P   3 vero: autem P   potens: potest V   9 ibi *om.* C   et habitus *om.* C   tenebras: privationem V   10-11 non ens lucidum *om.* P   10 ens: et eius V   13 Item: Preterea V   15 si *om.* C   18 tenebrarum: tenebrosum V   19 intentione: intentione P   significatione P   lumini: lumine P   20 aer: cum *add.* V   22 hic: ibi V   23 hoc *om.* P   27 ponat *om.* C   posset: potest CV   si . . . lucem: hoc nomen tenebrosum non compleat aptitudinem habendi lucem V   30 lucis . . . a privatione *om.* CV

1 Cf. Arist. *De An.* II. vii (418b18-20).
2 superius: cf. n. 123.
13 ordo irregressibilis: cf. Abaelard. *Glossae super Praedicamenta* (272-4).
14 Arist. *Cat.* x (13a31-34) (73).
22 dicit Aristoteles: cf. supra, n. 131.

129. What if they are opposed in the manner of privation and possession? Against: As has been said above, brightness is one thing and light is another. For brightness is a modification produced from light; light, however, is an innate quality, nor is it able to leave its subject—for example, light cannot leave fire. Thus, the subject of light cannot be deprived of light, therefore it cannot be dark. However, a privation and a possession have their existence with regard to the same subject, but shadows and light do not have their existence regarding the same subject, therefore they are not opposed in a privative manner.

130. Moreover, if there is an opposition of privation and possession there between shadows and light, the understanding of a dark thing is that it is capable of being bright and is not a bright thing. But nothing in which there is light can be such, therefore between light and shadows there is not a privative opposition.

131. Moreover, between privation and possession there is an irreversible order, as we know from Aristotle in the *Categories*. Therefore, if light and darkness are opposed privatively, no return could happen from darkness to light; therefore if any body was dark it could not become bright even if it had been bright before.

132. Solution: It should be stated that there are two ways of understanding the word 'darkness'. In one meaning it is opposed to light, and in this way brightness and darkness are opposed privatively; and so air is now dark, now bright, and light and darkness occur with regard to the same subject, and here there is a reversible order. And when Aristotle says that between privation and possession there is not a reversible order, he means only that possession which arises from an intrinsic natural aptitude. Brightness, however, is not an innate quality, rather it is something from without. In another meaning the word 'darkness' means the privation of light. Thus, if this does not imply an aptitude to have light, it can be conceded because any body in which there is no light is dark. Therefore, since there is no light in air but rather brightness, when air is illuminated it can be said to be dark from the privation of light and not from the privation of brightness. Thus, according to this understanding of the word, this argument does not hold: 'it is dark, therefore it is not bright'; rather

tenebrosus, non ergo est luminosus; immo sic deberet intelligi: ergo non est lucidus. Aer enim secundum intentionem lucidi non est lucidus, a luce enim est aliquid lucidum et a lumine luminosum. Secundum hoc etiam bene potest concedi quod tenebre et lumen simul sunt in eodem, quia lux et lumen simul in eodem subiecto vel aere non sunt. Dicendum 5 est etiam quod lux et tenebre secundum hoc non possunt esse circa idem subiectum; et hoc est quia lux est accidens inseparabile a suo subiecto sicut ab igne et a sole, et in quocumque subiecto est lux et lumen, et non convertitur.

133. Sequitur videre utrum tenebre possint videri. Quod possint, sic 10 videtur posse ostendi. Omne quod visu percipitur videtur; sed tenebra visu percipitur; ergo tenebra videtur. Ergo non solum lucidum videtur.

134. Item. Facilius est unumquodque agere cum ipsum sit liberum ab omni passione quam quando passione opprimitur. Sed oculus in tenebris liber est ab omni passione; in luce autem non est liber a pas- 15 sione, quoniam patitur inmutationes receptive. Ergo potius est animam iudicare penes visum in tenebris quam in luce.

135. Item. Plato dicit tenebras videndo nihil videmus; sed per hoc quod dicit 'tenebras videndo' ponit tenebras videri. Et ita secundum eum tenebre videntur. 20

136. Solutio. Hoc ipsum videre secundum quod perficitur a pati et a iudicare dicendum est quod tenebre non videntur; quia a tenebris non fit aliqua inmutatio in oculo; et ideo non patitur oculus cum ipse tenebre non generent aliquam inmutationem in oculo: manet enim oculus vacuus ab inmutatione in tenebris. Nam habetur ab Aristotele in primo 25 *Topicorum*, quod sentire dicitur multipliciter, secundum corpus et secundum animam. Similiter dicendum est de videre. Et de illo, scilicet videre quod est secundum animam, dicit Plato: tenebras videndo nihil videmus. Et secundum hoc bene potest concedi quod tenebre videantur.

---

1 ergo *om.* V   est *om.* P   deberet intelligi: debet inferri CV   2 Aer . . . lucidus *om.* V   enim: non C   4 simul: solis P   5 quia: quando CV   simul *om.* C   vel aere *om.* P; non sunt ut in aere V   9 et non convertitur *om.* P   10 Sequitur videre: Restat ut dicamus V   11–12 tenebra . . . percipitur: tenebre percipiuntur P   12 tenebra videtur: tenebre videntur P   13 Item: Preterea V   ipsum *om.* V   sit: est CV   14–16 quando . . . quoniam *om. per hom.* C   14 quando: ipsum *add.* V   16 patitur *om.* P   inmutationes: inmutationem C   17 iudicare: videre P   18 Item: Preterea V   20 eum: hoc C   22–23 non fit . . . ideo *om.* P   23 oculus *om.* V   tenebre *om.* P   24 generent: generet P   inmutationem: similitudinem P   manet: movet C   25 Nam: secundum quod C   27 Similiter: Et similiter V   dicendum *om.* CV   29 videmus: Et dicit Plato melius videmus tenebras videndo, quia tenebra non est aliquid, sed privatio et absentia lucis *add.* V   videantur: videntur P

---

18 Cf. Calcidius *In Platonis Timaeum* cccxlv (337–8).
21–22 a pati et a iudicare: cf. supra, n. 59.
25–26 Arist. *Topic.* I. xv (106^b23).   28 dicit Plato: cf. supra, n. 135.

this is how it should be understood: 'therefore it is not shining'. Air according to the meaning of the word 'shining' is not shining, for it is from light that something becomes shining and it becomes bright from brightness. Therefore, it can even be readily conceded that darkness and brightness are in the same thing at the same time, since light and brightness are not in the same subject or air at the same time. It should also be stated that light and darkness taken in this way cannot exist with respect to the same subject; and this is because light is an accident which is inseparable from its subject such as from fire or the sun. Again, in any subject whatever there is light and brightness and they are not interchangeable.

133. We should next examine whether darkness can be seen. That it can seems capable of being shown as follows: Everything which is perceived through sight is seen; but darkness is perceived by sight; therefore darkness is seen. Therefore, it is not only brightness which is seen.

134. Moreover, it is easier for anything to act when it is completely free from being acted upon than when it is burdened by being acted upon. Yet an eye in darkness is completely free from being acted upon; in light, however, it is not free from being acted upon since it undergoes modifications receptively. Therefore the soul is better able to judge by the power of sight in darkness than in light.

135. Again, Plato says that in seeing darkness we see nothing; yet by that which he says, 'in seeing darkness', he maintains that darkness can be seen. And thus according to him darkness is seen.

136. Solution: In respect of seeing itself, inasmuch as it is perfected by being acted upon and by judging, it should be stated that darkness is not seen, because from darkness no impression happens in the eye. Thus, the eye is not acted upon since darkness itself does not produce any modification in the eye: for the eye remains empty of impressions in darkness. Indeed, we know from Aristotle in the first book of the *Topics* that sensing is said in many ways, according to the body and according to the soul. The same is to be said about seeing. And it is concerning that, namely seeing which is according to the soul, that Plato says 'in seeing darkness we see nothing'. Again, in this regard, it can be readily conceded that darkness is seen. However, inasmuch

Secundum autem hoc quod dicitur lucidum videri, vel luminosum, hoc dicit solum de illo videre quod in pati consistit quoad instrumentum videndi.

XI. *De umbra*

137. Consequenter sciendum est quod quedam sunt propositiones de 5 umbris satis note apud intellectum per sensus experimentum. Possunt tamen huiusmodi propositiones demonstrari geometrice. Propositiones autem hee sunt.

Si corpus luminosum et corpus tenebrosum sunt equalia, corpus luminosum proicit umbram equedistantium laterum, et dicitur illa 10 umbra chilindreidos. Si corpus tenebrosum fuerit minus corpore luminoso proicit umbram conoydos. Si corpus tenebrosum sit maius corpore luminoso proicit umbram calatoydes. Item. Si corpus tenebrosum fuerit minus corpore luminoso, quanto corpus tenebrosum magis accedit ad corpus luminosum tanto minorem proicit umbram, 15 quanto magis recedit tanto maiorem proicit umbram. Si corpus tenebrosum sit maioris quantitatis quam corpus luminosum, quanto corpus tenebrosum magis accedit ad corpus luminosum tanto maiorem proicit umbram, quanto magis recedit tanto minorem proicit umbram. Si fuerit corpus tenebrosum equale corpori luminoso non dicitur umbra maiorari 20 vel minorari propter accessum vel recessum corporis tenebrosi a corpore luminoso, quia cum umbra sit equedistantium laterum ipsa procedit in infinitum.

138. Hiis visis fiant hee tres argumentationes.

Prima est hec: A et B sunt duo quanta eiusdem forme et eiusdem 25 quantitatis, et in principio diei sunt equalia, et tam A quam B continue movebitur per hunc totum diem a maiori quantitate in minorem et uniformiter, et A per hunc totum diem ita diminuitur quod in quolibet

---

1 Secundum: *om.* C   videri: videtur P   hoc *om.* C   2 dicit solum: dicitur C
5 est *om.* P   sunt *om.* P   9 Si . . . equalia: Si corpus tenebrosum fuerit equale corpori luminoso CV   9–10 corpus luminosum *om.* CV   10 umbram: umbra C   et *om.* CV
11–12 Si . . . conoydos *om.* P   11 fuerit: sit V   12 proicit *om.* C   umbram: umbra C   sit: fuerit P   maius: minus P   13 calatoydes: colatoydes C; calatioydos V   Item *om.* V   13–14 tenebrosum: tenebrarum C   16 quanto . . . umbram *om. per hom.* V   17 sit . . . luminosum: fuerit maius corpore luminoso P
17–18 corpus tenebrosum *om.* P   19 quanto . . . umbram *om.* V   22 laterum: latus C   24 visis: multis P   hee *om.* P   28 et A *om.* P   diminuitur *om.* P

9–23 Cf. Calcidius *In Platonis Timaeum* lxxxix-xc (141–3); Guill. de Conchis *Dragm.* IV (145); *Phil. Mundi* II. xxxii (73C–74D).
11 chilindreidos: κυλινδροειδής.
12 conoydos: κωνοειδής.
13 calatoydes: καλαθοειδής.

as it is said that a shining object is seen or a bright one, this refers only to that seeing which consists in being acted upon as regards the organ of sight.

XI. *Concerning shadows*

137. Next it should be known that there are some propositions regarding shadows which are already sufficiently known to the intellect through the experience of sensation. None the less, propositions of this kind can be demonstrated geometrically. Now these propositions are the following.

If a bright body and a dark body are the same size, the bright body projects a shadow whose sides are equidistant, and that shadow is called 'cylindrical'. If the dark body is smaller than the bright body it will cast a conical shadow. If the dark body is greater than the brighter body it casts a basket-shaped shadow. Again, if a dark body is smaller than a bright body, the more the dark body comes closer to the bright body the smaller the shadow it casts, and the more it recedes the greater the shadow it casts. If the dark body is of a greater quantity than the bright body, the more the dark body comes closer to the bright body the greater the shadow it casts, and the more it recedes the smaller the shadow it casts. If the dark body is equal to the bright body one would not say that the shadow would become bigger or smaller because of the dark body's coming closer or receding from the bright body, because since the shadow would have sides which are equidistant, it would go on to infinity.

138. Having seen the above, these three arguments follow.

The first is this: A and B are two quantities of the same form and the same size, and at the beginning of the day they are equal, and both A and B are moved continually throughout all of the day uniformly from a greater quantity to a lesser, and A throughout the whole day is

tempore huius diei erit minus quam B. Ergo in fine diei A erit minus
quam B.

139. Secunda argumentatio est hec: A et B sunt duo corpora equalia
et eiusdem forme, et movebuntur in quolibet tempore huius diei continue et uniformiter a minori quantitate in maiorem, et A in quolibet 5
tempore huius diei magis augmentabitur quam B. Ergo in fine diei erit
maius quam B.

140. Tertia argumentatio est hec: A B C sunt tria quanta, et A et B
pariter accepta sunt minora quam C, et per hunc totum diem continue
et uniformiter tam A quam B minorabitur, et C erit immobile per hunc 10
totum diem ita quod neque crescet neque decrescet. Ergo A et B pariter
accepta in fine diei erunt minora quam C.

141. Instantia prime argumentationis. Sint A et B duo corpora
sperica eiusdem quantitatis et tenebrosa, et opponantur corpori luminoso
et sperico minoris quantitatis quam sit utrumque illorum, et decrescet 15
tam A quam B continue et uniformiter per hunc totum diem ita quod
utrumque illorum in fine diei sit equale corpori luminoso. B maneat
immobile quoad locum, et corpus luminosum similiter. Sed A per
hunc totum diem continue et uniformiter recedat a corpore luminoso.
Umbra A corporis dicatur A, et umbra B corporis dicatur B. Hic ergo 20
contingit quod A umbra et B umbra sunt duo quanta et equalia in principio huius diei; et A in quolibet tempore huius diei magis decrescit
quam B; tamen in fine diei A erit equale B, A non magis decrescit, quia
quantum est ad decrementum A corporis A umbra decrescit equaliter
B umbre, sed per recessum A corporis a corpore luminoso magis de- 25
crescit A umbra quam B umbra, quia quanto magis recedit A corpus
a corpore luminoso, cum sit maius eo, tanto minorem proicit umbram,
et tamen A umbra in fine diei erit equale B umbre, quia utraque umbra
est umbra chilindreidos, et ab equalibus corporibus procedunt ille
umbre.   30

142. Instantia secunde argumentationis. Sit opposito modo ut sint
A et B duo corpora tenebrosa equalia sperica minora corpore luminoso
sperico, et crescant per hunc totum diem continue et uniformiter ita
quod in fine diei sint equalia corpori luminoso, et B maneat immobile

---

1 huius *om.* C   erit minus: magis diminuetur P   diei: die P   3 est *om.* C
sunt: quanta *add.* V   5 in² *om.* P   6 erit: A *add.* V   8 tria: duo P   et²
om. C   9 sunt: sint P   11 crescet: crescit C   decrescet: decrescit C
13 Sint: Sit V   15 et¹ *om.* CV   decrescet: decrescit C; decrescat P   17 diei:
die P   B: et C per hunc totum diem *add.* P   22 quolibet: quocumque
V   23 A¹ *om.* CP   erit . . . magis: erit equale B magis AP   24 A
corporis: quod A corpus P   decrescit: decrescet P   26–27 quia . . . umbram
om. C   27 proicit: proicet P   28 quia: que P   umbra² *om.* C   31 opposito: oppositi P   33 crescant: crescunt P   et *om.* P   34 sint: sunt CV

lessened in such a way that at any time of this day it is lesser than B. Therefore, at the end of the day A will be less than B.

139. The second argument is this: A and B are two equal bodies and of the same form and they are moved continually and uniformly at any time of this day from a lesser quantity to a greater quantity, and A at any time of this day will increase more than B. Therefore, at the end of the day A will be greater than B.

140. The third argument is this: A, B, and C are three quantities, and A and B taken together are less than C and throughout the whole day both A and B are continually and uniformly lessened, and C will remain unchanged throughout all of this day such that it neither increases nor decreases. Therefore, A and B taken together at the end of the day will be less than C.

141. An objection to the first argument. Let A and B be two dark spherical bodies of the same quantity, placed opposite a bright and spherical body which is of a lesser quantity than both of them, and both A and B decrease continually and uniformly throughout all of this day so that both of them at the end of the day are equal to the bright body. B remains unchanged as regards its position and the bright body similarly. A, however, throughout all of this day, continually and uniformly moves away from the bright body. Let the shadow of A be called A and the shadow of B be called B. Here, therefore, it happens that the shadow A and the shadow B are two quantities and are equal at the beginning of this day, and A at any time of this day is decreasing more than B. However, at the end of the day A will be equal to B. A does not decrease more, since inasmuch as there is a decrease in the body A, the shadow A decreases to the same extent as the shadow B. However, through the moving away of body A from the bright body, shadow A will decrease more than shadow B, since the more A moves away from the bright body, since it is bigger than it, the smaller the shadow it casts. Yet at the end of the day shadow A will be equal to shadow B, since both shadows are cylindrical shadows, and those shadows come from equal bodies.

142. An objection to the second argument. Take the opposite situation such that A and B are two equal dark spherical bodies which are smaller than the bright spherical body and they increase continually and uniformly throughout all of the day such that at the end of the day they are equal to the bright body. And let B remain in the same place

localiter, et A recedat a corpore luminoso continue et uniformiter per
hunc totum diem. Umbra A corporis dicatur A umbra, umbra B corporis dicatur B. Hic ergo contingit quod A et B in principio huius diei
sunt equalia, et A per hunc totum diem magis crescit quam B, et tamen
A in fine diei erit equale B; quia utraque umbra est equedistantium 5
laterum in fine diei, et a corporibus equalibus exeunt.

143. Instantia tertie argumentationis. Sint A et B due semispere
equales et tenebrose, et utraque sit minor corpore luminoso, et C sit
una spera tenebrosa equalis corpori luminoso sperico; et tam A quam B
sit corpus tenebrosum maius medietate corporis luminosi, et decrescat 10
tam A quam B continue et uniformiter per hunc totum diem, ita tamen
quod A et B pariter accepta sint in fine diei maiora corpore luminoso,
et opponantur A et B corpori luminoso ita quod diametri A et B eque
distent, et si protendantur diametri continue in directum procedant ille
due linee a diametris protense usque ad duos terminos extremos cor- 15
poris luminosi, et moveatur A continue et uniformiter accedendo ad B,
et B continue et uniformiter moveatur accedendo ad A, et recedant in
illo motu a corpore luminoso ita quod in fine diei coniungantur in unam
speram.

144. Hic contingit instantia ad predictam argumentationem. Ut si 20
dicatur A umbra A corporis et B umbra B corporis, quoniam tam A
quam B est umbra conoides, et pariter accepte sunt minores quam
umbra C corporis, quoniam illa est umbra chilindreidos, et ideo infinita;
et tamen A et B per hunc totum diem decrescunt, et C umbra erit
immobilis, et A et B pariter accepte in fine diei maiores erunt quam C 25
umbra: quia A et B in fine diei erunt una umbra calatoydes et illa umbra
proicietur a corpore facto ex duabus semisperis quod est maius C corpore, et ideo eius umbra est maior; et tamen in quolibet instanti ante
finem diei, quando illa umbra fuit divisa in duas partes, ille due umbre
partiales fuerunt umbre conoides et minores quam umbra C corporis. 30
Sed continuatio A corporis cum B corpore fuit facta subito non gradatim
et sine successione; et ita privatio luminis ex altera parte corporis in
instanti facta fuit, et ita augmentum umbre factum fuit in instanti: et

4 B: et C add. P   6 et om. P   7 Sint: Sunt C   semispere: spere P
8 et tenebrose . . . luminoso om. P   9 equalis: equaliter C   10 maius
medietate: magis magis medietate medietatis P   11 continue et uniformiter om. P   12 in fine diei om. P   13 opponantur: opponitur C   14 distent: distunt C   diametri continue om. P   15 a om. V   a diametris
protense om. P   extremos om. P   16–17 accedendo . . . uniformiter om. per
hom. C   18 motu: loco C   coniungantur in: coniungant P   21 A² om.
P   et B . . . corporis om. P   tam A om. V   24 et tamen A et B: tam A
quam B P   decrescunt: decrescent V   26 calatoydes: calatoides C; caladides P
27 ex: a C   duabus: coniunctis add. V   30 fuerunt: fuerint V   31 Sed: Et C
32 in om. CV

and let A move away continually and uniformly from the luminous body throughout the whole of this day. Let the shadow of body A be called shadow A, the shadow of body B be called shadow B. Here, therefore, it happens that A and B at the beginning of this day are equal, and A throughout the whole of this day increases more than B, and yet A at the end of the day will be equal to B, since each shadow has equidistant sides, at the end of the day, and they came from equal bodies.

143. An objection to the third argument. Let A and B be two equal and dark hemispheres, and let both of them be smaller than the bright body, and let C be a dark sphere equal to the bright spherical body. And let both A and B be dark bodies greater than a half of the bright body, and let both A and B decrease continually and uniformly throughout all of this so that A and B taken together are greater at the end of the day than the bright body. And let A and B be placed opposite the bright body such that the diameters of A and B are equally distant, and if the diameters are continuously extended these two lines will be extended in a straight line, extended from the diameters as far as the two furthest ends of the bright body. And let A be moved continually and uniformly in coming closer to B, and let B be moved continually and uniformly in coming closer to A, and let them move away from the bright body in that movement such that at the end of the day they will be joined into one sphere.

144. Here an objection arises regarding the above argument. Suppose that one calls A the shadow of body A and B the shadow of body B, since both A and B are basket-shaped shadows and when taken together are smaller than the shadow of body C, since the latter has a cylindrical shadow, and therefore it is infinite, yet, when A and B decrease throughout all of this day, the shadow of C will be unchanging, and A and B taken together at the end of the day will be greater than the shadow C, because A and B at the end of the day will be one basket-shaped shadow and that shadow is cast from a body made up from two hemispheres which is greater than the body C, and therefore its shadow is greater. However, at any given instant before the end of the day, when that shadow was divided into two parts, those two partial shadows were basket-shaped shadows and smaller than the shadows of body C. However, the joining together of body A with body B happened immediately and not in stages and without one following upon the other. Therefore, the privation of light from the other part of the body happened in an instant, and thus the increase of the shadow occurred in an instant. Thus, because of this sudden change

propter huius subitam mutationem sine successione factam quoad
crementum et decrementum in umbris non tenent argumentationes
superius assignate.

XII. *De auditu et sono et eorum dispositionibus*

Sequitur ut dicamus de auditu. Unde primo loco videndum est et 5
quid sit auditus et quo mediante habeat fieri auditus, et cum illud sit
sonus, dicendum est de sono.

145. Ab Avicenna in commento *de Anima* habetur quod auditus est
vis ordinata in nervo expanso ad apprehendendum formam eius quod
sibi advenit ex commotione aeris, qui constringitur constrictione vio- 10
lenta inter percutiens et percussum. Nisi enim sit constrictio violenta,
non erit ibi sonus qui est proprium sensatum auditus; et cum non sit
resistentia non erit violenta constrictio, unde si percutiatur lana non est
ibi sonus.

146. Videtur autem inprimis posse ostendi quod auditus sit species 15
tactus, et hac ratione. Ex constrictione aeris inter percutiens et percussum fit sonus in aere, et ille aer inpulsus et informatus impellit aera
sibi proximum et eum inmutat consimili inmutatione, et sic procedit
continua inmutatio donec aer inmutatus recipiatur in concavitatibus
aurium, et contingat instrumentum auditus et ipso contactu fit auditus. 20
Qua ratione ergo in contactu calidi si sentiatur calidum dicitur esse
tactus; pari ratione, cum in contactu ipsius aeris inmutati sentiatur ipse
sonus, dicendum est quod huiusmodi sensus est tactus, et huiusmodi
sensus est auditus; et ita auditus est tactus. Sed tactus universalior est
quam auditus et predicatur in quid. Ergo tactus est genus auditus, et ita 25
auditus est species tactus.

---

2 in umbris *lacuna* P   tenent: tenet P   4 *Titulum om.* PV   5 loco *om.* C
6 quid: quod V   9 formam: formas P   11 Nisi . . . violenta: nisi enim
ibi sit violentia P   12 ibi *om.* C   qui: quod C   qui . . . auditus *om.* V
et: unde P   13 resistentia: residentia V   lana: simplex *add.* P   17 ille:
iste V   19 concavitatibus: concavibus C   20 et . . . auditus *om.* C
21 dicitur esse: est V   23-24 tactus . . . est *om.* P   24 est tactus: et
tactus V

---

4 Cf. Arist. *De An.* II. viii (419^{b}4–421^{a}6).
6–7 Cf. Avic. *De An.* II. vi (9^{ra}A).
8 Avic. *De An.* I. v (5^{ra}B).
12 proprium sensatum: cf. Arist. *De An.* II. vi (418^{a}11–13).
12–14 Cf. Arist. *De An.* II. viii (419^{b}6, 15); Avic. *De An.* II. vi (9^{rb}A).
15–16 Cf. Arist. *De An.* II. viii (419^{b}19 seq.).
17–20 Cf. Boeth. *De Instit. Musica* I. xiv (200); Abaelard. *Glossae super Praedicamenta* (176); *Dialectica* (67); Adelard. Bath. *Quaest. Nat.* xxi (25); Guill. de Conchis
Dragm. VI (292).
24 Cf. Arist. *De An.* II. ii (413^{b}5); iii (414^{b}3); III. xii (434^{b}9–24); etc.
25 predicatur in quid: cf. Arist. *Topic.* I. v (102^{a}31); Porphyr. *Isagoge* De gen. (7).

with respect to the increase and decrease in the shadows without one thing following upon another, the arguments put forward above do not hold.

XII. *On hearing and sound and their organization*

It follows that we should now say something concerning hearing. Thus, first of all we should look at both what hearing is and by what means it takes place, and since the latter is through sound, we should talk about sound.

145. We know from Avicenna in the commentary *On the Soul* that hearing is a power located in an extended nerve so as to grasp the form of what comes to it from a disturbance of air which is compressed in a violent compression between what strikes and what is struck. For unless there is a violent compression, no sound which is proper to the sense of hearing will arise there. Again, when there is no resistance there will not be a violent compression, so if wool is struck there is no sound there.

146. It seems, however, first of all that it can be shown that hearing is a kind of touch, and for the following reason. Sound is produced in air out of the compression of the air between something striking and something struck, and when that air has been stimulated and informed it strikes the air near to it and changes it with a similar modification. Thus a continual modification goes forward so that the affected air is received into the cavities of the ears, and it touches the organ of hearing and in this contact hearing occurs. Therefore, for the same reason that if heat is felt through contact with a hot thing touch is said to exist, so also because sound itself is sensed when in contact with the air which has been affected, it should be said that this kind of sensation is touch and this kind of sensation is hearing, and thus hearing is touch. However, touch is more general than hearing and is predicated of other things, therefore touch is the genus of hearing, and hearing is a species of touch.

147. Preterea. Secundum aliquod membrum facilius animal apprehendit tactu quam auditu, et citius vel tardius apprehendere non impedit
quin ibi sit tactus; ergo licet secundum instrumentum auditus cum
facilitate in contactu aeris apprehendat sonum, et per aliud membrum
non possit sentiri sonus, propter hoc non impedietur quin ille sensus sit 5
tactus; et ille sensus est auditus; ergo auditus est tactus.

148. Item. Proprium sensatum visus est color, proprium sensatum
auditus est sonus. In re autem visa dicitur esse color; similiter in re
gustata dicitur esse sapor, et sic de aliis sensatis propriis sensuum. Ergo
simili modo in re audita erit sonus; ergo cum es sonat, in ere est sonus, 10
et sic de aliis ex percussione quorum fit semper sonus.

149. Item. In qua proportione se habet lux ad colorem in eadem
proportione se habet collisio sive percussio ad sonum; quoniam lux est
manifestatio et detectio coloris, percussio autem est manifestatio soni:
cuiusmodi enim sonus sit ab hoc corpore vel ab illo non potest perpendi 15
nisi per collisionem, sicut cuiusmodi color sit in hoc corpore vel in illo
non potest perpendi nisi per detectionem lucis. Quo ergo modo dicitur
quod color est in corpore luce non existente, debet dici quod sonus
est in aere nulla collisione existente; et ita sicut aer non est proprium
subiectum corporis, ita nec aer est proprium subiectum soni.   20

150. Item. Si aer sit proprium subiectum soni, sicut hec conceditur:
paries videtur, quia proprium sensatum visus est in pariete, ita hec habet
concedi: aer auditur, quia proprium sensatum auditus est in aere.

151. Preterea. De causa coloris non dicitur quod ipsa videtur, ut si
frigiditas sit causa albedinis non conceditur tamen quod frigiditas vide- 25
tur. Pari ratione cum pronuntiatio sit causa soni non debet dici quod
pronuntians sonum auditur; sed auditur, hoc constat. Ergo sonus est in
eo qui pronuntiat, quoniam similiter posset concedi frigiditatem videri
si albedo esset in frigiditate.

152. Solutio. Ad primum. Dicimus quod auditus non est species 30
tactus. Distinguitur enim a tactu per aptitudinem percipiendi sonum.

---

1 secundum *om.* P   animal *om.* P   2 tactu: tactum P; tacta V   auditu:
secundum aliud V   3 secundum *om.* PV   4 contactu: contractu V   apprehendat: apprehendit V   5 sensus *om.* V   7 Item: Amplius V   10 simili
modo *om.* P   est: erit V   11 quorum . . . sonus *om.* V   semper *om.* P
12 Item. In qua proportione *om.* V   12–13 lux . . . habet *om.* C   14 detectio:
deiectio V   autem *om.* P   15 illo: alio V   16 sit *om.* P   16–17 per
collisionem . . . nisi *om.* V   17 detectionem: destinacionem P   19 aere:
ere C   20 corporis: coloris CV   22 habet: potest CV   25 tamen *om.* C
27 sed: ut videtur *add.* P   constat *om.* V   28 posset concedi: credi V; contingit C   29 esset in frigiditate: sit effectus frigiditatis CV

7–8 Cf. Arist. *De An.* II. vi (418ª11–13); supra, n. 145.
30 Ad primum: cf. supra, n. 146.

147. Moreover, an animal perceives through an organ more easily by touch than by hearing. Also, sooner or later perception is not impeded whenever touch is there. Therefore, even though hearing grasps sound with ease by means of an organ in contact with air, and sound cannot be sensed through any other organ, there is no reason why that sense should not be touch. That sense is hearing, therefore hearing is touch.

148. Moreover, the proper sensible of sight is colour, and the proper sensible of hearing is sound. However, colour is said to be in the thing seen, similarly flavour is said to be in the thing tasted, and likewise concerning the other proper sensibles of senses. Therefore, in a similar way sound will be in the thing heard. Therefore, since bronze makes a sound, sound is in the bronze, and likewise regarding those things which always cause a sound when struck.

149. Again, the proportion by which light is constituted in respect of colour is the same proportion by which striking together or percussion is constituted in respect of sound. Indeed, light is the manifestation and revelation of colour; percussion, however, is the manifestation of sound: for in whatever manner sound comes from this body or from that, it cannot be observed except by means of striking together, just as in whatever manner colour is in this body or in that, it cannot be observed except through the revelation of light. Therefore, according to the way in which it is said that colour is in the body when there is no light, it should be said that sound is in the air when there is no striking; and so just as air is not the proper subject of the body, so neither is air the proper subject of sound.

150. Moreover, if air were the proper subject of sound, just as this is conceded: 'the wall is seen because the proper sensible of sight is in the wall', so this would have to be conceded: 'air is heard because the proper sensible of hearing is in the air'.

151. Again, it is not said that the cause of colour is seen, for if coldness were the cause of whiteness it would not be conceded that coldness is seen. For the same reason, since speech is the cause of sound it should not be said that the one uttering the sound is heard. Yet he is heard, and this is clear. Therefore, sound is in the one who speaks, since similarly it can be conceded that coldness can be seen if whiteness is in coldness.

152. Solution: With regard to the first ⟨question⟩, we state that hearing is not a kind of touch, since it differs from touch through its capacity to perceive sound. If it happens, however, that in contact

Si etiam in contactu aeris continget sentiri sonum a vi tangibili, hoc esset propter aeris levitatem vel asperitatem vel aliam consimilem formam inmutantem instrumentum tactus; non autem esset illud propter sonum.

153. Ad aliud. Dicendum quod hoc nomen 'sonus' inpositum est 5 passioni que generatur a collisione in aere et a concursu aliorum corporum; non autem imponitur passibili qualitati secundum aptitudinem cuius ex collisione fit sonus in aere, quia secundum diversas aptitudines diversarum passibilium qualitatum fiunt diversi modi soni in aere. Unde alterius modi sonus exit ab ere et alterius modi ut ab argento. Huius- 10 modi autem passibilis qualitas potest appellari sonoritas: hec autem est remotius principium auditus, sonus autem propinquius.

154. Ad ultimum. Dicimus quod sicut hec est vera per accidens: paries videtur, et hec est vera per se: lucidum videtur, ita cum sonus sit proprium sensatum auditus, et sic est in aere, hec est vera per accidens: 15 aer auditur, sed inusitata est; et hec est vera per se: sonus auditur.

155. Item. Experimento habemus quod aliquis sonus continuus est alii sono, et generat delectationem in audiente, aliquis autem propter eius absurditatem respuitur ab audiente. Queritur autem in quo sit delectatio ut in subiecto que generata est a soni concinnitate.　　　　20

156. Si est in anima; Contra. Corporale non agit in incorporale, quia omnis actio corporalis per contactum est. Sonus autem corporalis est, cum sit in re corporali; anima autem incorporalis est. Ergo sonus non agit in animam; ergo delectatio non fit in anima per sonum. Et delectatio est a sono, et non est in anima; et est in anima vel in corpore; ergo est 25 in corpore.

157. Preterea. Corporale habet agere in rem corporalem; ita si incorporale patiatur habet incorporale agere in ipsum. Si ergo in anima sit delectatio, illa fit ab alia re incorporali. Queritur a qua fiat immediate,

---

1 contactu: tactu PV　　contintet: contingeret V; contingit P　　sonum a vi tangibili: ipsum aerem vi tangibili P　　esset: contingeret V　　2 aeris om. P　　aliam: aliquam PV　　3 inmutantem: inmutentem P　　tactus: quod est nervus valde expansus add. V　　esset: esse P　　5 est om. V　　6 passioni: passivo C　　in aere om. P　　aliorum: aliquorum PV　　7 imponitur: inpediuntur P　　9 qualitatum: quantitatum V　　10 ut om. PV　　12 propinquius: propinquum C　　14 per se om. P　　ita om. C　　15 auditus: visus V　　et sic est: et sit CV　　16 sed . . . est om. P　　17 Item: Rursus V　　18 alii sono om. P　　delectationem: dilectionem V　　19 eius om. P　　absurditatem: surditatem P　　20 ut in subiecto om. C; sive in subiecto P　　subiecto: Queritur quid sit primum subiectum et proprium subiectum dilectionis add. V　　21 Si: Sed C　　in¹ om. P　　22 est² om. C　　29 illa fit: ipsa sit V　　fiat: fiant P　　immediate: in medietate V

5 Ad aliud: cf. supra, nn. 147–9.
13 Ad ultimum: cf. supra, n. 150.
13–16 Cf. Arist. De An. II. vi (418ª7–16).
17–19 Cf. Boeth. De Instit. Musica I. viii (195).

with air sound is heard through a tactile power, this will be because of the lightness or unevenness of the air or some other similar form affecting the organ of touch. It will not, however, be because of sound.

153. Regarding the next: It should be said that this word 'sound' is assigned to the action suffered which is produced by a striking together in the air and by the coming together of other bodies. A receptive quality, however, is not assigned according to the capacity ⟨of the bodies⟩ out of whose collision sound occurs in the air, because owing to the various capacities of different perceptible qualities, sounds occur in the air in diverse ways. Thus, sound comes from bronze in one way and from silver in a different way. For a receptive quality of this kind can be called sonority: this, however, is far from the source of hearing, whereas sound is close up.

154. Regarding the final question: We state that just as the following is true in an accidental manner: 'A wall is seen', and this is true in itself: 'A bright object is seen', even so, since sound is the proper sensible of hearing and so it is in the air, this is true in an accidental way: 'Air is heard', but it is unusual; and this is true in itself: 'Sound is heard'.

155. Again, we know through experience that some sounds which are continuous with others produce pleasure in someone hearing. Other sounds, however, are rejected by the hearer because of discordance. It is asked, however, in what, as in a subject, is the pleasure which has been produced by a harmonious sound.

156. What if it were taken to be in the soul? Against: The corporeal does not act upon the incorporeal, since all corporeal action happens through contact. But sound is corporeal, since it is in a corporeal thing; the soul, however, is incorporeal. Therefore, sound does not act upon the soul, and so pleasure does not arise in the soul by means of sound. But pleasure is from sound, and is not in the soul; and it is either in the soul or in the body, therefore it is in the body.

157. Moreover, the corporeal has to act upon a corporeal thing. Thus if the incorporeal is acted upon it has an incorporeal thing acting upon it. If, therefore, there is pleasure in the soul, that arises from another incorporeal thing. It is asked from what does it arise directly

et qualiter non est assignare, ut videtur; non est ergo delectatio in anima.

158. Si sit in corpore, Contra. Circumscripta anima nullus erit sensus corporis, et si nullus sit sensus corporis, a sono non fiet aliqua delectatio in corpore; ergo circumscripta anima in corpore non erit delectatio: 5 debetur ergo anime quod delectatio est. Et ad hoc quod sit delectatio preexigitur apprehensio eius a quo est delectatio. Hoc potest perpendi eo quod dormiens non apprehendens sonum non habet delectationem a sono. Et apprehensio est ab anima. Sic ergo cum delectatio sit post apprehensionem potius erit delectatio in anima quam in corpore.   10

159. Preterea. Corpus in se simpliciter consideratum insensibile est; unde, cum sine sensu preambulo non sit delectatio in corpore, simpliciter non est delectatio.

160. Item. Ab eodem in specie non exit nisi idem in specie; unde si agens sit hinc et inde eiusdem speciei et patiens similiter, et secundum 15 idem in specie sit agens hinc et inde, et secundum idem in specie sit patiens hinc et inde; res formata hinc et inde erit eiusdem speciei. Ergo cum quilibet aer sit eiusdem speciei cum alio aere, et aer sit subiectum soni patiens in generatione soni, omnes soni in aere formati ab agentibus eiusdem speciei, que sunt agentia secundum idem, sunt eiusdem speciei; 20 et in specie est adequatio. Ergo omnes huiusmodi soni erunt consoni. Nulla ergo erit dissonantia aliorum sonorum qui formantur ab eodem.

161. Preterea. Cum concinnitas et inconcinnitas sint dispositiones soni, ab eodem debent causari in esse a quo causatur sonus in esse. Cum igitur ex collisione corporum ad invicem et constrictione aeris intercepti 25 fiat sonus, ab eadem collisione ipsa dispositio soni causatur; propter quid ergo est quod quandoque causatur concinnitas in esse quandoque inconcinnitas. Similiter. Queritur propter quid unus sonus sit consonus alii sono et dissonus alii.

162. Item. Videtur posse ostendi quod concinnitas et inconcinnitas 30 non dicantur absolute sed in respectu, et quod unus et idem sonus sit concinnus et inconcinnus, et hac ratione: sicut aliquid dicitur esse

---

3 Circumscripta: inscripta P   nullus: nulla P   4 fiet: fiat V   5 in corpore . . . delectatio *om. per hom.* C   6–7 ad hoc . . . a quo: ad hoc quod delectatio est, et ad hoc quod sit dilectio V   9 post: per V   11 est *om.* P 14 exit: erit V   15 sit *om.* C   16 in specie *om.* P   hinc et inde *om.* P in specie *om.* P   18 alio: ergo C; aliquo V   aere *om.* CP   20 eiusdem . . . idem *om.* CV   22 aliorum: aliquorum PV   formantur: formant P 23–24 Preterea . . . ab eodem *om.* V   25 igitur: ergo V   ex: in P   intercepti: interempti PV   26 eadem *om.* P   27 ergo *om.* C   causatur: concausatur P   quandoque: quando C   29 alii[1]: aliter P   30 concinnitas et inconcinnitas: concinnum et inconcinnum CV   31 absolute sed: nisi CV 32 et[2]: in P   esse *om.* P

23 Cf. Boeth. ibid. I. iii (189–91); viii (195).

as it seems, and in what manner should it not be assigned. Therefore pleasure is not in the soul.

158. What if it were taken to be in the body? Against: If the soul were confined there, then there would be no sensation in the body, and if there is no sensation in the body, no pleasure would arise in the body from sound. Therefore, if the soul is confined by the body there will not be any pleasure, so it must be due to the soul that there is pleasure. And in order for there to be pleasure, a prerequisite is the perception of that from which there is pleasure. This can be observed since the sleeper who does not perceive sound does not have pleasure from sounds. And apprehension is from the soul. Thus since pleasure follows from perception, pleasure will be in the soul rather than in the body.

159. Moreover, the body considered simply in itself is without feeling, and so since there is no pleasure in the body without sensation leading the way, there is simply no pleasure in it at all.

160. Again, from one species comes forth only that which is of the same species. So it is that if an agent is both here and there of the same species, similarly the one being acted upon, and if by the same token it is acting both here and there, and by the same token is being acted upon here and there, then the thing which has been formed both here and there will be of the same species. Therefore, since any air is of the same species as other air, and air is the subject of sound, since it is acted upon in the production of sound, all sounds formed in the air by agents of the same species which are identical agents are of the same kind, and so there is a correspondence in kind. Therefore all sounds of this kind will be harmonious. Therefore, there will be no dissonance between one or other of the sounds which are formed by the same agent.

161. Again, since harmony and disharmony are modes of sound, they should be brought into being by the same thing from which the sound was brought into being. Since, therefore, sound arises from the collision of bodies and the striking of the intervening air, from the same collision the mode of the sound is caused. Therefore it is for this reason that sometimes harmony is brought into existence and sometimes disharmony. Similarly, it is asked for what reason one sound is in harmony with one sound and not in harmony with another.

162. Again, it seems that it can be shown that harmony and disharmony are not said in an absolute manner but relatively, and that one and the same sound may be harmonious and disharmonious. And for this reason, just as something is said to be large in comparison with

magnum in comparatione ad id quod est minus ipso, et illud idem dicitur
esse parvum in comparatione ad maius ipso, similiter aliquis sonus
alicui audienti est concinnus et delectabilis, et ille idem sonus alii
audienti est absurdus et inconcinnus. Sic ergo non debetur ipsi sono
inconcinnitas et concinnitas, immo audienti.　　　　　　　　　　　　　5

163. Item. Queritur ex qua natura sit quod ipse sonus faciat delectationem in audiente, et ex qua natura unus et idem sonus sit delectabilis
uni et indelectabilis et absurdus alii?

164. Solutio. Ad primam questionem dicendum est quod delectatio
proveniens ex sono est in anima. Et hic contingit quod in alio est motus 10
et in alio est species adquisita per motum; quoniam in corpore est
motus et passio, et in anima est delectatio adquisita per motum. Et est
simile: excutit hic dumos, occupat alter aves; et fit in ipsa anima
delectatio per apprehensionem inmutationis que exoritur in sensu
audientis a sono extra, et ita anima apprehendens in seipsa adquirit 15
delectationem per suam apprehensionem.

165. Ad hoc quod queritur quare unus sonus sit consonus uni sono
et dissonus alii, dicendum est quod hoc contingit propter commensurationem communem. Sunt enim omnes soni consoni qui communicantes
sunt, scilicet communem mensuram habentes; soni autem inconsoni 20
sunt in se illi quos non commensurat aliqua una et eadem aliquota.
Unde commensuratio est causa consonantie, incommensurabilitas est
causa inconsonantie.

166. Ad tertium. Dicimus quod secundum aliam intentionem nominis
dicitur esse sonus concinnus qui est homini delectabilis, et dicitur in 25
illa comparatione ad hominis dispositionem concinnus esse; et secundum

---

2 in comparatione: ad comparationem P　　　3 et² *om.* C　　　6 Item: Amplius V
12 est *om.* P　　　　　　13 occupat: accipit V　　　aves: oves P　　　　　15 audientis a:
auditis in C　　　　16 per suam: rei delectabilis V　　　　　17 queritur: dicitur V
18 propter: per CV　　18–19 commensurationem: mensurationem V　　19 qui: quia V
20 inconsoni: inconcina P　　　　　21 in: inter C; ad invicem V　　　illi *om.* P
22 consonantie: et tamen non quelibet superparticularis proportio in sonis facit con-
sonantiam *add.* V　　　　24 aliam: aliquam V　　　　25 concinnus: inconcinnus CV
et dicitur: dicitur enim sonus V　　26 illa: alia CV　　ad hominis dispositionem *om.* P

---

9 Ad primam: cf. supra, nn. 156–9. Cf. Arist. *De An.* ii. ii (413ᵇ22–24); iii (414ᵇ1–
6).
13 excutit hic dumos, occupat alter aves: anglice sic redditur a Georgio Wither:

　　　　　"Twas I that beat the bush,
　　　　　The bird to others flew.'
　　　　　(ed. F. Sidgwick, London, 1902, i, p. 151).

H. Walther, *Lateinische Sprichwörter u. Sentenzen des Mittelalters*, Göttingen, 1963,
i, n. 8409.
17 Ad hoc quod queritur: cf. supra, nn. 160–1.
19 Cf. Arist. *De An.* iii. ii (426ᵃ27–ᵇ7); Boeth. *De Instit. Musica* i. iii (189–91).
24 ad tertium: cf. supra, nn. 162–3.

what is smaller than it, and the same is said to be small in comparison with something bigger than it, similarly any sound to one hearer is harmonious and pleasing, and that very same sound to another hearer is harsh and unharmonious. So it is, therefore, that harmony and disharmony are not due to the sound itself but rather to the hearer.

163. Again, it is asked from what it arises that one sound gives rise to pleasure in the hearer, and from what it arises that the same sound is pleasing to someone and not pleasing and harsh to another.

164. Solution: It should be said in respect of the first question that the pleasure arising out of sound is in the soul. And here it is necessary that in one there is motion and in the other there is a species acquired through motion; because there is movement and the undergoing of action in the body, and in the soul there is pleasure acquired through motion. And there is the saying 'one man beats the bush, the other catches the birds'. And pleasure arises in the soul itself because of the perception of a change which arises in the sense of hearing from the sound outside. Thus the soul in itself perceiving derives pleasure by means of its perception.

165. With regard to the question as to why one sound is harmonious with one sound and unharmonious with another, it should be said that this happens because of a common measure. For all sounds are harmonious which share something in common, namely they have a common measure; dissonant sounds, however, in themselves are those which are not commensurate with one and the same aliquot part. Thus, a common measure is the cause of harmony, lack of it is the cause of disharmony.

166. Regarding the third objection, we state that according to another understanding of the word, a sound is said to be harmonious which is pleasing to a man, and it is said to be harmonious in relation to the disposition of a man. Again, according to another under-

aliam intentionem nominis dicitur sonus esse concinnus alii sono, et causa huius ultime concinnitatis superius assignata est. Secundum quod unus et idem sonus videtur esse concinnus uni audienti et absurdus alii, non est propter naturam ipsius soni, immo propter diversitatem dispositionum in audientibus; et hoc contingit sicut una et eadem lux, 5 nulla facta inmutatione in ipsa, est quibusdam oculis aspera, ut infirmis oculis, quibusdam autem est levis et non aspera, ut sanis oculis. Si autem audiens sit in debita consistentia existens, debite temperatus, debitam proportionem et commensurationem humorum habens, videbitur ei sonus concinnus, et erit ei delectabilis cum dispositione audientis 10 propter adequationem ipsius in commensuratione; et ille idem sonus in eadem equalitate existens erit inconcinnus homini intemperato et non habenti commensurationem humorum propter dissimilitudinem et inequalitatem que est inter suam consistentiam et sonum extra, sicut homini non debite temperato res, que dulcis est temperato, est amara. 15

167. Item. Supposito in *Physica* quod similia similibus gaudent et respuunt dissimilia. Anima autem incorporalis et spiritualis est; sonus autem est forma corporalis et a corporibus exiens, quia per collisionem eorum; corporale et incorporale sunt dissimilia; ergo a sono et propter sonum non est in anima delectatio. 20

168. Item. A sono qui auditur exoritur species in sensu audientis, et ab hac fit species in memoria, et ab ista in acie animi fit alia species. Utraque autem illarum que fiunt in sensu audientis et in memoria est corporalis, cum ipsa sit in corpore; illa autem que fit in anima, per hoc quod anima convertit aciem animi ad illam speciem que est in ipsa, est 25 incorporalis et spiritualis, et propter similitudinem fit delectatio, quia simile applaudit suo simili. Ergo anima habet delectationem propter

1 nominis *om.* C   2 ultime: intime V   3 esse *om.* P   concinnus: consonus V   6 ipsa: illa V   aspera: aspersa V   oculis *om.* V   7 Si: Similiter CV   autem: si *add.* V   8 consistentia: existentia P   10 sonus: quidam esse *add.* V   delectabilis: delectatio P; qui in consimili proportione se habet *add.* V   cum dispositione audientis *om.* P   12 equalitate: qualitate V   erit: exit C   13 et *om.* P   14 inequalitatem: equalitatem V   consistentiam: inconsistentiam V   15 est²: erit V   16 supposito: est *add.* CV   18 forma *om.* C   corporibus: corporalibus P   quia *om.* C   collisionem: collectionem P   21 auditur: audit P   22 ab ista: habita V   species *om.* P   23 et *om.* P   24 ipsa sit: ipse sint V   25 in *om.* V   ipsa: memoria P

2 superius: cf. supra, n. 165.
6–7 Cf. Aug. *Confess.* VII. xvi. 22: 'Et sensi expertus non esse mirum, quod palato non sano poena est et panis, qui sano suavis est, et oculis aegris odiosa lux, quae puris amabilis'.
14–15 Cf. supra, ll. 6–7.
16 in Physica: i.e. Arist. *De Gen. et Corr.* I. vii (323^{b}3 seq.), ad quem locum mittit Arist. *De An.* II. v (416^{b}35–417^{a}2). Cf. etiam *De An.* I. ii (405^{b}14–15); v (410^{a}23 seq.); II. iv (416^{a}30).

standing of the word, sound is said to be harmonious with another sound, and the cause of this latter way of being harmonious has been given above. Inasmuch as one and the same sound appears to be harmonious to one hearer and harsh to another, it is not because of the nature of the sound itself, but rather because of the diversity of dispositions in the hearers. And this happens just as one and the same light, without any change happening to it, is to some eyes harsh, such as to eyes which are sick. To others, however, the light is delicate and not harsh, as with healthy eyes. If, however, the hearer is in good health, of the right constitution, and having the right proportion and balance of humours, the sound will seem to him to be harmonious, and it will be pleasing to him because of the correspondence of the sound to a balance with the disposition of the hearer. And that very same sound existing in the same quality will be disharmonious to the unhealthy man who does not have the balance of humours because of the dissimilarity and imbalance which there is between his constitution and the sound outside, just as, to a man who is not correctly balanced, a thing which is sweet to one who is balanced is bitter.

167. Again, it is assumed in the *Physics* that similar things delight in similar things and reject dissimilar things. Now the soul is incorporeal and spiritual; sound, however, is a corporeal form and comes out from bodies by means of the collision between them. The corporeal and the incorporeal are dissimilar; therefore from sound and because of sound pleasure is not in the soul.

168. Again, a species arises in the sense of the listener from a sound which is heard, and from this a species enters the memory, and from this another species occurs in the eye of the mind. However, each of the species which is made in the sense of the listener and in his memory is corporeal since it is in the body; but the species which is made in the soul is incorporeal and spiritual, and because of its similarity pleasure occurs, because a similar thing welcomes what is similar to itself. Therefore, the soul has pleasure because of the experience

illam passionem que in ipsa fit, et non propter sonum qui est in acie animi vel propter inmutationem que fit in instrumento audientis vel in sede memoriali.

169. Si concedatur; Contra. Sicut in audiendo se habet auditus ad sonum extra, ita se habet anima ad speciem in memoria; cum voluntas 5 convertat aciem animi ad illam que in memoria est, et nullo sono existente, sed multo tempore sono prius perempto extra, adhuc species permanens est in memoria, que est ymago illius forme que prius fuit forma in sensu audientis, et anima, convertens se ad illam que est in memoria, apprehendit illam, et ex ea fit quidem in acie animi. Cum 10 igitur incorporale naturaliter habeat delectationem ab incorporali, quia simile applaudit suo simili, eque habebit anima delectationem in hac apprehensione et in illa, quando sonus est actu existens. Ergo sive sonus sit actu sive non sit actu, propter hoc non habebit anima minorem vel maiorem delectationem, quod experimento patet esse falsum; quon- 15 iam si sit recordatio in anima alicuius soni delectabilis et preteriti ipsa appetit iterum audire consimilem sonum.

170. Solutio. Dicimus quod in anima est delectatio a passione aurium que fit in sensu audientis a sono; et hec est ratio. Anima appetit perfectionem sicut et unaqueque res. Sed eius perfectio est a virtutibus 20 et a scientiis quibus sibi adquirit verum esse; scientias autem et virtutes adquirere non potest nisi dum ipsa est in corpore. Ut ergo ipsa adquirat sibi verum esse, ipsa appetit esse in corpore. Unde contingit quod cum ipsa sit perfecta scientiis et virtutibus non desiderat amplius esse in corpore, sed solum desiderat ut ipsa habeat verum esse, quod ipsa per 25 virtutes meruit. In corpore autem non potest esse anima nisi ipsum sit organicum et in debita consistentia existens; et ideo, ut conservetur esse eius in debita consistentia, anima delectationem habet cum aliquid conservat esse sui corporis; et, ut superius dictum est, inter sonum delectabilem et corpus organicum est similitudo, et simile conservat 30

---

1 ipsa: illa PV   1–2 sonum . . . propter *om.* CV   2 audientis: auditus P
3 memoriali: memorali C; memorata V   4 in audiendo *om.* CV   5–6 voluntas . . .
est *om.* V   6 convertat: convertit C   7 sono *om.* PV   extra: existit P
8 permanens est: permanet P   forme *om.* P   9 forma *om.* PV   10 apprehendit: apprehenderet C   quidem: que C   11 incorporale: in corpore P; cor-
poraliter V   delectationem: delectatio P   11–12 quia . . . simili *om.* P   12 suo
om. C   13 existens: consistens P   19 anima: enim *add.* P   20 eius:
hec P   21–23 scientias . . . esse[1] *om.* V   21 verum esse: virtutes et scientias
P   22 ipsa[1] *om.* P   24 sit: est P   scientiis et *om.* P   desiderat: sitit P
25 sed solum desiderat *om.* P; solum *om.* V   27 conservetur: servetur PV
28 eius *om.* C; corporis V   29 conservat: conservet P   ut: sic C   ut superius:
quoniam superius sicut V   30 similitudo: proportionalis *add.* V

6–7 nullo sono existente: cf. Alex. Nequam *De Nat. Rer.* I. xx (66).
29 ut superius: cf. supra, n. 166.

it undergoes and not because of the sound which is in the gaze of the
soul, nor because of the change which occurs in the organ of hearing
or in the seat of memory.

169. If this were conceded, then against this: Just as what is heard
in hearing is constituted in respect of the sound from outside, so also
is the soul constituted in respect of the species in memory. Since the
will turns the eye of the mind towards that which is in memory, and
without any sound existing, but with the sound from outside having been destroyed a long time before this, now the species persists
in memory, which is an image of that form which was beforehand a
form in the sense of hearing. And the soul, in turning its attention towards that which is in memory, grasps it, and out of that something
arises in the eye of the mind. Since, therefore, the incorporeal naturally has pleasure from the incorporeal, and since a similar thing welcomes its similar, the soul will equally have pleasure in this perception
and in the other, when the sound is actually existing. Therefore it is
not because of whether the sound actually is or actually is not that the
soul has a greater or lesser pleasure, as experience shows to be clearly
false; but because if there is the recollection in the soul of any pleasing
sound, the soul desires to hear an entirely similar sound again.

170. Solution: We state that there is pleasure in the soul from the
action upon the ears which occurs in the sense of hearing from sound.
This is the reason why. The soul like any other thing desires complete-
ness. However, its perfection is from virtue and knowledge, by means
of which it acquires true being for itself. It cannot, however, acquire
knowledge and virtue except for as long as it is in the body. Therefore,
so that it can acquire true being for itself, it desires to be in the body.
Thus, it happens that when it has become perfect by means of knowledge and virtue, it no longer desires to be in a body, but only desires
that it would have the true being which it has merited through virtue.
However, the soul cannot be in the body except when the body is
organized and existing in the correct constitution. Therefore, so that
the being of the soul may be maintained in the correct constitution,
the soul has pleasure when something maintains the being of its body.
Thus, as has been stated above, between a pleasurable sound and an
organized body there is a similarity, and a similar thing maintains its

suum simile in esse. Unde, cum corpus hominis conservatur in debita
temperantia, et quando revocatur ab indebita consistentia ad debitam
per sonum extra, habet anima delectationem a sono extra.

171. Dictum est superius quod sonus provenit ex violenta constrictione aeris inter percutiens et percussum; sed non est violenta aeris 5
constrictio nisi percussum sit resistens percutienti; sed resistentia est
a duritie. Ergo oportet quod durities sit in percusso ut sit sonus. Sed
in aere nulla est durities; ergo si nihil sit percussum nisi aer, non erit
ibi sonus, quod patet esse falsum; quoniam in tantum potest moveri
virga in aere quod erit ibi sonus, et corda cithare mota in aere facit 10
sonum.

172. Item. Corporis percussi resistentia facit ad esse soni; ergo
quanto maior est resistentia tanto maior est sonus, quod patet esse
falsum; quoniam si corde cithare resistat lapis vel aliquid durum non
tantus erit sonus, sicut si solus aer ei resistat in motu suo. 15

173. Item. Queritur que sit causa intensionis et acuminis in sono. Si
violenta constrictio, instantia est in proximo exemplo; si maior motus
et pluralitas motuum ut videtur in cordis cithare, instantia est in ere
percusso, quia non fiet maior sonus propter motum ipsius.

174. Solutio. Ad primum. Dicendum quod semper ubi fit sonus 20
audientibus ibi est resistentia; sed non omnis resistentia est a duritie,
et est ipsum resistere aeris ei loco duritiei. Sed queratur que sit causa
resistentie aeris cum ipse sit mollis. Dicendum est quod quidlibet
naturaliter appetit esse unum, et ideo naturaliter respuit divisionem; et
ideo aer de natura sui in quantum potest resistit divisioni. Et sicut con- 25
tingit duas nubes concurrere, et ex constrictione aeris intercepti inter
eas fieri sonum; ita possibile est duos aeres concurrere et fieri sonum
ex constrictione aeris constricti per concursum illorum.

175. Ad aliud. Dicendum est quod non est a maiori resistentia maior

---

1 hominis *om.* P   2 quando: quoniam C; quandoque P   ad debitam: in debita
temperantia P   5 percutiens: percusciens C; percustiens P   6 percutienti:
percuscienti C; percucienti P   7 ut: si C   8 ergo: igitur V   nihil: non P
9 in tantum *om.* P   10 ibi: ei C   12 ad *om.* P   soni: sonum V   14–15 non
tantus: tantus C; non tantum V   15 motu: m^{tl} P   16 Item: Similiter
V   acuminis: acucioris P   17 maior *om.* C   18 ut: non P   19 fiet: fit V
20 Solutio *om.* P   semper: similiter V   21 audientibus: quid P; *om.* V   est[1]:
sit P   22–23 causa resistentie: resistentia P   23 quidlibet: quilibet CV
25 resistit *om.* C   28 ex *om.* C   per concursum: percussum *add.* C   percur-
sum P   29 est[1] *om.* P

---

4 superius: cf. n. 145.
14–15 virga: cf. Macrobius *In Somnium Scipionis* II. iv. 3.
17 in proximo exemplo: cf. n. 172, ll. 14–15.
18–19 in ere percusso: cf. supra, n. 148, l. 10; n. 153, ll. 9–10.
20 Ad primum: cf. supra, n. 171. Vid. Avic. *De An.* II. vi (9^{vb}D).
29 Ad aliud: cf. supra, n. 172. Vid. Avic. loc. cit.

similar in being. Whence, if the body of a man is maintained in the right balance, and when it is brought back from an incorrect constitution to a correct one by means of a sound from without, the soul has pleasure from an external sound.

171. It has been stated above that sound comes from a violent striking of the air between something striking and something struck. However, there is no violent striking of the air unless the thing being struck resists the thing striking and resistance is due to hardness. Therefore, it is necessary that there is hardness in the thing being struck for there to be sound. But there is no hardness in air, and so if nothing but air is struck, there will be no sound. This is clearly false; for a stick has only to be moved in the air for there to be a sound there, and the strings of the lute when moved in the air make noise.

172. Again, the resistance of the body which has been struck causes sound to exist, therefore the greater the resistance the greater the sound. This is clearly false, for if the strings of the lute give resistance to a stone or something hard, the sound will not be as great as if only the air resisted it in its motion.

173. Again, it is asked what the cause is of increase and acuteness in sound. If it is the violent striking, there is an exception in the next example. If the cause is the greater movement and number of movements, as appears in the case of the strings of the lute, an exception is found in the bronze which has been struck, because a greater sound does not occur because of its movement.

174. Solution: Concerning the first question, it should be stated that where sound occurs to hearers, there is resistance there, but not all resistance is due to hardness as in the case of the resistance of air. However, it is asked what is the cause of the resistance of air, since air is soft? It should be stated that each and every thing naturally desires to be one, and so naturally rejects division. Thus air by its nature resists division in so far as it can. And just as it happens when two clouds run together, that out of the compression of the air which is struck between them sound occurs, so it is possible for two parts of air to run together and for sound to occur out of the compression of the air struck because of their running together.

175. Regarding the other objection, it should be stated that a greater sound does not arise from a greater resistance but from the

sonus, sed a velocitate et pluralitate motuum; et ubi velocior est motus ibi propter maiorem impetum est maior constrictio aeris, et ita maior est sonus. Contingit autem quod aliquis sonus magis sit acutus alio, tamen est minus audibilis eo; et hoc contingit quando ita est quod sonus magis acutus non est in materia tante quantitatis sicut reliquus; unde 5 fit quod sonus parve corde in cithara est magis acutus quam sonus maioris corde, sed est minus audibilis. Est autem magis acutus, quia plures sunt motus, et velociores sunt motus minoris corde quam maioris corde, minus tamen est audibilis minor corda; quia cum maior corda sit maioris quantitatis in aere maioris quantitatis movetur; et ideo est 10 eius sonus maior secundum distensionem in quantitate, et magis audibilis quam sonus minoris corde, qui subtilior est et maior in acumine. Et simile est in calore. Quandoque enim aliquis calor est maior in acuitate alio calore qui maior est eo in quantitate secundum distensionem. Similiter aliquis est magis albus alio, albedine magis accedente 15 ad puram albedinem, et tamen reliquus potest dici magis albus eo, si eius albedo distendatur in subiecto maioris quantitatis, ut habetur in *Sex Principiis*; et per hoc patet solutio tertie questionis.

XIII. *De reverberatione soni, scilicet de eccho*

176. Sequitur ut dicamus de reverberatione soni, ut de eccho. 20 Sciendum ergo quod aere informato per collisionem informatur successive aer motus secundum eandem figuram que recipitur in prima collisione, et aere sic moto et informato inveniente aliquam resistentiam in corpore duro et concavo reverberatur, et auditur sonus per illam resistentiam secundum figuram similem prime. Unde fit sonus consimilis 25 primo sono.

3 est *om.* P   aliquis *om.* CV   sit: est CV   4 tamen: cum P   audibilis eo: vel alio P   quando: quoniam P   5–6 unde fit: ut sit C   6–7 in cithara . . . corde *om.* V   8 sunt motus *om.* P   8–9 quam . . . tamen *om.* V   9 corde *om.* P   minor corda *om.* P   quia: quare C   10 in aere maioris quantitatis *om.* P   11 eius *om.* C   in quantitate: in aere P   12 sonus *om.* C qui: que CV   12–13 acumine: acuitate P   13 calore: color P *et in seq.* Quandoque enim: Quoniam P   14 acuitate: acumine CV   14–15 distensionem: distinctionem P   15 magis albus: maior P   albedine: ab *add.* P   17 albedo: color magis P   quantitatis *om.* V   21 ergo: igitur P   quod: in *add.* C 23 aere: in *add.* V   aliquam: quam P

1 Cf. Arist. *De An.* II. viii (420^{a}26–33); Boeth. *De Instit. Musica* I. iii (189–91); viii (195).

7 magis acutus: cf. Macrobius *In Somnium Scipionis* II. iv. 2: 'ut autem sonus ipse aut acutior aut gravior proferatur, ictus efficit qui dum ingens et celer incidit acutum sonum praestat, si tardior lentiorve, graviorem'.

17–18 in *Sex Principiis*: viii. 80–86.

18 tertie questionis: cf. supra, n. 173.

19 Cf. Arist. *De An.* II. viii (419^{b}25–33); Avic. *De An.* II. vi (9^{vb}DE); Alex. Nequam *De Nat. Rer.* I. xx (66).

speed and number of movements. And when the movement is quicker there is a greater compression of the air because of the greater force, and so the sound is greater. However, it happens that one sound will be more sharp than another, and yet is less audible than it. This happens when it is such that the sound which is sharper is not as great a quantity as the other in the matter. Thus it happens that the sound of a small string in a lute is more sharp than the sound of a bigger string, and yet is less audible. For it is sharper because there is a greater number of vibrations, and the movements of the smaller string are faster than the movements of the bigger string. Yet it is less audible than the bigger string because the bigger string is of a greater size and so it is moved with a greater quantity in the air. Thus its sound is greater according to the extent of its size, and it is more audible than the sound of the smaller string, which is thinner and higher in pitch. And it is the same with heat, for sometimes one heat is greater in sharpness than another heat which is greater than it in quantity according to its extent. Similarly, someone is whiter than another, coming closer in whiteness to pure whiteness, and yet the latter can be said to be more white than the former, if its whiteness is spread over a subject with a greater quantity, as is held in the *Six Principles*. Thus in this, the solution to the third question is clear.

XIII. *On the reverberation of sound, namely, on the echo*

176. We should next say something concerning the reverberation of sound, namely, regarding the echo. Therefore, you should know that when the air has been formed by means of a collision, the air which has been moved is successively formed according to the same form which it received in the first collision. When the air which has been moved and formed in this way encounters some resistance in a hard and concave body, it reverberates and a sound is heard by means of that resistance according to a form similar to the first. Thus a sound arises which is similar to the first sound.

177. Sed obicitur. Speculo opposito rei videnti apparet ei figura in speculo, et non oportet quod speculum sit multum distans ab oculo; et sicut per reverberationem soni fit eccho, ita per reflexionem radiorum apparet ymago in speculo, et per resistentiam corporis concavi fit reverberatio soni. Qua ratione ergo videtur forma in speculo non multum 5 distante sed prope posito, pari ratione, concavitate corporis resistentis non multum remota ab audiente, audietur sonus in reverberatione formatus. Ergo, cum in domo sint concavitates corporum resistentium motui aeris, homo existens in domo emittens sonum audiet alium sonum provenientem in eadem domo ex reverberatione.   10

178. Preterea. Quanto propinquior est sonus tanto melius auditur; ergo quanto propinquior est reverberatio soni tanto melius auditur; et ita potius auditur eccho in eadem domo in qua sumus per reflexionem factam in eadem domo quam si reflexio soni sit in concavitate remota: vel quare non.   15

179. Item. Sicut successiva est informatio soni in prima collisione, ita eodem ordine et consimili figura fit informatio soni in soni reflexione. Unde illud quod primo est informatum in prima collisione habet suum exemplum primo informatum in reverberatione: qua ratione ergo auditur finis et medium soni reverberati debet audiri principium eiusdem 20 soni. Sed experimento potest haberi hoc esse falsum. Principium enim soni reverberati non auditur.

180. Solutio. Ad primum. Dicendum est quod propter parvitatem spatii quod est inter primam informationem et reverberationem non contingit sonum reverberatum et sonum primo informatum in discretis 25 temporibus audiri; ita enim velox est motus aeris cum sui informatione quod ubi reverberatio est ita propinqua primo sono provenienti a prima collisione non potest discerni inter ipsum et primum sonum.

181. Ad secundum obiectum. Dicimus quod principium in eccho

1 opposito: apposito P   2 multum: nimium P; *om.* V   3 eccho: et ad auditum pervenit *add.* V   ita *om.* V   4 apparet: formatur P   6 posito: posita V   7 multum: iustum V   audiente: audito V   9 existens in domo *om.* P   sonum[2] *om.* P   10 in eadem domo *om.* P   12 ergo . . . auditur *om.* CP   13 ita: ipsa C   auditur: audietur C; audiretur V   13–14 in qua . . . domo *om. per hom.* V   14 in concavitate: magis P   16 sicut: si C   18 primo: primum P   19 ratione *om.* C   21 experimento: experimentum P   hoc esse: quod hoc est P   25 informatum: formatum P   26 cum sui informatione *om.* CV   27 ita *om.* P   sono: soni C   . 29 obiectum *om.* CV   dicimus: dicendum est P   in *om.* PV

1 Cf. Arist. loc. cit.
8 Cf. Avic. loc. cit.
21–22 Cf. Guill. de Conchis *Dragm.* VI (295): 'Echo solos verborum fines audimus'.
23 Ad primum: cf. supra, nn. 177–8. Vid. Avic. loc. cit.
29 Ad secundum obiectum: cf. supra, n. 179.

177. But it is objected: When a mirror is placed opposite the thing which is seen, its shape appears in the mirror, and it is not necessary for the mirror to be very far from the eye. And just as through the reverberation of sound an echo occurs, so in the same way by means of the reflection of the rays ⟨of light⟩ an image appears in the mirror, and by means of the resistance of a concave body the reverberation of sound happens. Therefore, just as a shape is seen in a mirror which is not very far away but which has been placed nearby, for the same reason, by means of the concavity of a resistant body which is not very far from the hearer, a sound formed by reverberation is heard. Therefore, since there are concavities of bodies resistant to the movement of air in a house, a man while in a house making a noise will hear another sound in the same house arising out of reverberation.

178. Moreover, the nearer a sound is, the better it is heard. Therefore, the nearer the reverberation of the sound, the better it is heard. Thus an echo will be better heard in the same house we are in by means of the reverberation made in the same house than if the reverberation of the sound were made in a concavity far away. But if not, why not?

179. Again, just as the forming of the sound in the first collision is successive, so the forming of the sound will occur in the same order and in the very same shape in the reverberation of the sound. Thus, that which is originally formed in the first collision has its copy formed first of all in the reverberation. For the same reason, therefore, that the end and the middle of the sound are heard, the beginning of the same sound should be heard. However, through experience we know that this is false, because the beginning of the sound which is bounced back is not heard.

180. Solution: Regarding the first objection, it should be said that because of the smallness of space that there is between the first forming and the reverberation, it does not happen that the reverberated sound and the sound which was first formed are heard in separate moments. For the movement of the air is so fast together with its shaping that where the reverberation is so near to the first sound coming from the first collision it is not possible to distinguish between it and the first sound.

181. Regarding the second objection, we state that the beginning

non auditur, quia propter velocitatem ipsius soni fit reditus ad audientem antequam formatur et totaliter fiat informatio et emissio soni; et ideo cum sonus rediens sit debilior primo sono propter eius debilitatem et fortiorem sonum audiri non potest. Et etiam principium soni eccho debilius est quam eius finis, quia primus aer rediens in reflexione ad 5 audientem invenit aera aliquantulum informatum a sono accedente ad reflexionem per quem aera est reditus soni. Sed finis eccho in reditu suo nullum habet impedimentum, et etiam cum maiori impetu percutit aer ipsum corpus resistens in fine reflexionis quam in principio; et ideo maior est eccho in fine quam in principio propter maiorem collisionem. 10

182. Dictum est prius quod a sono exoritur species in sensu audientis, et ad illam voluntas convertit aciem animi, et coniungit eam cum sono, et sic fit auditus. Sed cum due sint aures et in utraque aurium fiat passio que est ymago soni, cum ibi sint due passiones et anima convertit se ad utramque illarum, propter illas duas passiones debet ei apparere 15 duos esse sonos.

183. Preterea. Quanto sonus remotior est a suo ortu, tanto debilior est et in recedendo minoratur. Quantum ergo est ad aurem que propinquior est ortui ipsius soni maior apparebit esse sonus, quantum autem ad remotiorem minor; et ita est ibi distinctio secundum fortius et 20 debilius. Apparebit ergo anime quod ibi sint diversi soni, ubi non est nisi unus sonus.

184. Solutio. Dicendum est quod non apparent anime diversi soni propter illas diversas passiones que sunt in duabus auribus, quoniam ille due passiones sunt eiusdem soni ymagines, et acies animi eas 25 intuetur copulans eas cum uno et eodem sono. Ideo cum inveniat unum sonum iudicat ibi esse unum sonum.

185. Item. Contingat ita quod diversi soni sint quorum unus sit gravis et reliquus acutus, et eque cito perveniant ad aurem. Passio ergo

1 ipsius soni: inmutationis V   2 formatur et *om.* CV   ideo: ita P   6 aera: aerem V *corr.*   a *om.* CV   ad: apud CV   10 eccho *om.* V   collisionem: collectionem P   11 Dictum: Actum V   12 animi: anime P   14–15 convertit se: convertis P   15 ei apparere: appetere ibi P   17 Quanto: Quantum P est *om.* P   18 Quantum: Quanto P   19 est *om.* V   maior apparebit: magis appetabit P   quantum: quanto P   20 ibi *om.* CV   21–23 ubi . . . soni *om. per hom.* C   24 duabus: diversis P   25 eiusdem: ipsius P   eas: eos C   26 sono: qui sint causa illorum passionum *add.* V   27 sonum[2] *om.* P 28 contingat: contingit P   29 gravis: gravius C   et[1] *om.* P   perveniant: perveniunt P

11 prius: cf. supra, nn. 168–9.
17 Cf. Abaelard. *Glossae super Praedicamenta* (176–7): 'Saepe contingit, ut si quis longe a nobis remotus percutiendo aliquid cum ligno sonum faciat, nos qui remoti erimus, post ictum statim sonum non audiemus, quia non tam cito aer diffundi potest, ut sonum differat'.
28 Cf. Arist. *De An.* II. viii (420[a]26–33).

of an echo is not heard, since, because of the speed of the sound itself, it returns to the hearer before it is formed and before the shaping and emission of the sound have fully occurred. And so, since the
sound returning is weaker than the first sound, because of its weakness the stronger sound cannot be heard. Moreover, the beginning of
the sound in an echo is weaker than at its end, since the first air coming back in the reverberation to the hearer meets with air which has
been shaped only a very little by the sound happening at the reflection of the air through which the sound is returned. Yet the end of the
echo in its return does not have anything hindering it, and again the
air strikes the resistant body itself with a greater force at the end of
the reverberation than at the beginning, and so the echo is greater at
the end than at the beginning owing to the greater collision.

182. It has already been stated that a likeness arises from sound in
the sense of hearing, and the will turns the eye of the mind towards it
and links it with the sound, and thus hearing occurs. Yet since there
are two ears, and in each of the ears occurs the reception of an action
which is the likeness of the sound, since there are two actions then
and the soul directs itself towards both of them, because of these two
actions received it should appear to the soul that there are two sounds.

183. Moreover, the more distant a sound is from its source the
weaker it is, and it becomes less as it recedes. Therefore, the more
it is present to the ear which is nearer to the source of this sound the
greater the sound will appear, and in so far as it is more remote the less
it will appear. Thus there is a distinction there according to strength
and weakness. Therefore, it will appear to the soul that there are many
sounds where there is only one.

184. Solution: It should be said that because of the many modifications received in the two ears, many sounds do not appear to the
soul since those two modifications are likenesses of the same sounds,
and the eye of the mind intuits them by linking them with one and
the same sound. Thus, when it comes across one sound, it judges that
there is one sound there.

185. Again, it happens thus that there are various sounds of which
one is low and the other is high, and both equally quickly reach the
ear. The modification which is produced in the ear is made up out of

generata in aure est confecta ex duobus sonis oppositis. Sed si duo
sapores conmisceantur simul anima per gustum conmixtum apprehendet
conmixtionem que est in passione: non enim iudicat nisi secundum
representationem passionis invente per instrumentum sensus. Pari
ratione anima habet iudicare mixtum esse sonum a gravi et acuto propter 5
mixtam passionem in aure.

186. Item. Cum ymago gravis soni sit in aure, et ymago acuti soni
existat in alia aure, qua ergo ratione acies animi convertit se ad unam
illarum convertet se ad reliquam. Sed convertit se ad unam; ergo ad
utramque; et ita anima apprehendet simul et semel diversos sonos, et 10
ita plura simul.

187. Solutio. Ad primum. Dicendum est quod cum sensus audientis
sit compositum instrumentum in una sui parte potest esse ymago unius
soni et in alia sui parte ymago alterius soni; unde cum diverse sint
ymagines diversos iudicat anima esse sonos et distinctos: sicut in una 15
parte speculi apparet quandoque una ymago et in alia parte alia ymago.
Sed si ita esset quod unus et idem aer inmutatus simul a gravi sono et
acuto inmutaret aurem non distingueret unum sonum ab alio, immo ibi
esset sonus coniunctus. Sed tamen quia auditus fit secundum rectum,
sicut et visus, posset contingere quod anima discerneret gravem sonum 20
ab acuto; licet simul inmutaret aurem in contactu eiusdem aeris. Alia
enim linea recta ducenda esset ab aure ad ortum unius soni, et alia ad
ortum alterius soni.

188. Ad aliud. Dicendum est quod illum sonum apprehendit anima
prius ad quem intentio anime citius aciem animi convertit. Et quod 25
potius convertat se anima ad unum sonum quam ad alium hoc est
quandoque per casum et fortunam, quandoque autem per precedentem
apprehensionem alicuius rei habentis se magis ad unum sonum quam
ad alium.

189. Consequenter queritur propter quid sonus sit magis inmutatorius 30

2 conmixtum: permixtum P   apprehendet: apprehendit P   3 conmixtionem:
permixtionem P   nisi *om.* V   4 passionis invente per instrumentum: inven-
tam in instrumento C   7 Item: Preterea V   8 alia: aliqua P   ergo
om. C   convertit: convertat P   9 convertit: convertet P   10 anima *om.* V
15 iudicat: iudicas P   anima *om.* V   16 speculi: speciali P   apparet: appareret P
17 simul: insimul P   18 aurem: aerem V   sonum *om.* V   21 ab *om.* P   licet:
sed C; quamvis V   inmutaret: inmutarent C   in *om.* P   in contactu: cum tactu V
22 esset: est P   soni *om.* V   alia: linea *add.* V   24 *In* C, *fol.* 135ʳ–140ᵛ *habentur*
Sermones per distinctiones super Epistolis (vid. Introd.)   citius: cuius P   quod *om.* V
26 convertat: convertit V   anima: animam P   27 quandoque autem: quoniam P
30–p. 51, l. 1. inmutatorius . . . audientis: inmutatus sit auditus P

12 Ad primum: cf. supra, n. 185.
24 Ad aliud: cf. supra, n. 186.

the two opposite sounds. However, if two flavours are mixed together at the same time, the soul by means of a mixed taste perceives the mixture which is in the modification. Indeed, it only judges according to the representation of the modification which has been acquired by means of the organ of sense. For the same reason it happens that the soul judges that sound is mixed together from the deep and the high because of the mixed actions undergone in the ear.

186. Again, when there is a representation of a deep sound in one ear, and an image of a high sound in another ear, for the same reason that the eye of the mind turns itself towards one of these, it will turn itself towards the remainder. Yet it turns itself to one, therefore to both, and thus the soul grasps at one and the same time various sounds, and so many at the same time.

187. Solution: Regarding the first question, it should be stated that since the sense of hearing is a composite organ there can be a representation of a sound in one of its parts and another representation of a sound in another of its parts. Thus since the likenesses are different the soul judges that the sounds are also distinct, just as every now and again one image appears in one part of a mirror and another image appears in another part. However, if it were the case that one and the same air was modified at the same time by both a low and a high sound impressed upon the ear, it would not distinguish one sound from another, rather the sound would then be joined together. But since hearing takes place in a straight line just like sight, it can happen that the soul distinguishes the low sound from the high sound, namely, the ear is modified at the same time through contact with the same air. For one straight line can be led from the ear to the source of one sound and another to the source of another sound.

188. Regarding the next question, it should be stated that the soul perceives that sound first towards which the attention of the soul most quickly turns the eye of the mind. Moreover, that the soul turns itself more towards one sound rather than to another is sometimes due to chance and fortune. Sometimes, however, because of a previous perception of some thing, it is disposed more to one sound than to another.

189. Next it is asked for what reason does sound affect the sense of

sensus audientis et magis apprehenditur per auditum quam aliqua alia passio.

190. Quod sit ex natura instrumenti auditus sic videtur posse ostendi. Non est proprie subiectum soni nisi aer, et hoc non contingit nisi propter hoc quod sonus magis se habet ad naturam aeris quam ad naturam 5 alterius; provenit enim sonus ex constrictione aeris. Quod ergo instrumentum audientis magis inmutetur a sono quam ab alia passione, hoc non contingit nisi propter aeream naturam dominantem in ipso. Quia igitur auditus est aeree nature, ideo per auditum apprehenditur sonus.

191. Quod si concedatur erit obiectio de odoratu, qui secundum hanc 10 rationem premissam deberet apprehendere sonum. Nichil enim apprehendit anima per odoratum nisi id quod defertur ab aere ad instrumentum olfactus.

192. Ad hoc dicendum quod ita non est de olfactu sicut de auditu, quoniam auditus est aeree nature, et ideo inmutatur a sono qui provenit 15 ab aere colliso. Olfactus autem non inmutatur nisi a fumo resoluto a corpore odorifero, qui conmiscetur aeri ingredienti nares; unde non est simile quod inducitur pro simili.

XIV. i. *De olfactu*

193. Sequitur de olfactu. Primo ergo loco videndum est quid sit 20 olfactus. Habetur autem in commento super libro *de Anima* quod olfactus est vis ordinata in duabus carunculis dependentibus ab anteriore parte cerebri habentibus similitudinem cum mamillis. Potest autem queri quare odor potius sit cadens in illum sensum qui dicitur olfactus quam in alium sensum, et quare potius ille due caruncule sint recipientes 25 inmutationes ab odore quam ab alia passione.

194. Ad hoc dicendum quod odor est quasi proveniens a medio inter

---

1 et: quare *add.* V   apprehenditur: apprehendatur V   3 Quod: Et P
4 Non: Nihil P   proprie: proprium V   hoc *om.* V   5 habet: habeat P
6 alterius: rei *add.* V   8 contingit: contingat V   10 qui: quia V
11 premissam: premissum P   deberet: debet C   15 est: in *add.* C   ideo: non
add. C   inmutatur: mutatur V   16 colliso: et etiam nervus audibilis est expansus
nervus, et ideo in contactu aeris facilius inmutatur quam alia instrumenta sensuum *add.*
V   inmutatur: mutatur C   20 ergo *om.* P   est *om.* C   21–22 Habetur
... olfactus *om.* V   21 libro: librum P   27 dicendum: est *add.* V   a
om. P

---

9 aeree nature: cf. Arist. *De An.* II. viii (419^{b}19), et alibi passim; et infra, n. 196.
14 Ad hoc: cf. supra, nn. 190–1.
16 a fumo resoluto: cf. infra, n. 194.
19 *De olfactu*: cf. Arist. *De An.* II. ix (421^{a}7–422^{a}7).
21 in commento: i.e. Avic. *De An.* I. v (5^{ra}B); cf. Gundiss. *De An.* ix (68).
27–p. 52, l. 1 Cf. Arist. ibid. (421^{b}9).

hearing more and is grasped by hearing more than any other modification?

190. It seems that it can be shown to be due to the nature of the organ of hearing. The proper subject of sound is nothing other than air, and this only happens because of the fact that sound is constituted more like the nature of air than the nature of something else, for sound arises out of the contraction of air. Therefore, that the organ of hearing is more affected by sound than by any other modification can only happen because of the predominance of an airy nature in it. Thus, it is because hearing is of an airy nature that sound is grasped through hearing.

191. Now if this is conceded, then there will be an objection concerning smell, which, following the reason given above, should grasp sound. For the soul does not grasp anything through the sense of smell except that which is conveyed by the air to the organ of smell.

192. To this it should be stated that it is not the same in the case of smell as it is in the case of hearing, because hearing is of an airy nature and so is affected by the sound which arises from air which has been compressed. However, the sense of smell is only affected by the fumes given off by an odorous body which is mixed with the air going into the nostrils. Thus the situation is not the same with regard to that ⟨objection⟩ which was brought forward on the basis of a similarity.

XIV. i. *On the sense of smell*

193. Next regarding smell. And we must examine first of all what smell is. We know from the commentary on the book *On the Soul* that smell is a power located in two small pieces of flesh hanging down from the front part of the brain and having a similarity to nipples. However, it can be asked why odour comes under that sense which is called smell more than under another sense, and why these two small pieces of flesh are receptive of modification by odour rather than from other kinds of modification.

194. In this regard it should be stated that odour is something

aquam et aera. Dicitur autem ab auctoribus quod fumus resolutus a corpore odorifero est medium inter aquam et aera, et ille fumus habens in se proportionalem naturam aeris et aque est proprium subiectum odoris, et ille due caruncule que recipiunt inmutationes ab odore sunt consimilis complexionis. In eis enim equaliter dominantur aqua et aer; et 5 ideo sunt quasi medium inter aquam et aera, et propter hoc cum odor habeat esse in subiecto talis nature, ideo potius mediante tali instrumento apprehenditur odor quam alio.

195. Sed potest queri propter quid non habeamus similiter quemdam sensum secundum aliam quamlibet combinationem quorumlibet ele- 10 mentorum? Ut, sicut odoratus est quasi medium inter aera et aquam, similiter queritur quare non habeamus quemdam sensum medium inter aera et ignem, et sic de consimilibus; et si hoc, multo plures essent quam quinque.

196. Ad hoc dicendum est quod propter nimium fervorem cordis et 15 calorem cerebri necesse fuit advenire aera ad refrigerandum cerebrum et cor. Pulmo autem attrahit aera ad mitigationem fervoris utriusque; et ideo fit una via attractionis aeris per quam transit aer prius ad cerebrum et postea ad cor, et est illa via per nares et per guttur. Sed cum fumus sit quiddam quod sensu debeat apprehendi mediante odore, et 20 ille fumus cum odore conmisceatur partibus aeris qui ingreditur nares, et inveniat carunculas recipientes inmutationes ab odore, inmutat eas, et anima convertens se ad illam inmutationem apprehendit odorem. Et facte sunt ibi ille caruncule propter apprehensionem odoris transeuntis per illas carunculas mediante vi pulmonis attrahente aera. Cum autem 25 non sint alie forme sensibiles que non apprehendantur aliquo sensu, non oportuit esse plures sensus quam quinque. Per hanc eandem rationem habetur quare plures sint corporis sensus quam elementa, cum auditus

1 autem: tamen P   3 aque: aqua P   odoris: aeris P   6 inter . . . aera om. V   8 apprehenditur: comprehenditur V   alio: mediante add. V   10 secundum aliam: quantum est ad P   11 sicut om. P   est om. C   12 queritur om. C   habeamus: haberemus P   sensum om. P   16 advenire: invenire P   16–17 cerebrum et cor: aera C   ad . . . aera om. V per hom.   20 sit om. C   quod: in add. V   24 ibi om. P   25 per om. V   28 habetur om. V   quare: quod P

1 ab auctoribus: Arist. De Sensu ii (438^{b}24–25): 'Odor vero fumalis evaporatio est, fumalis autem evaporatio ab igne'; vid. v (443^{a}21–^{b}2). Cf. Plato Timaeus 66DE; Avic. De An. II. v (9^{ra}B).
15 ad hoc: cf. supra, n. 195.
18–19 Cf. Guill. de Conchis Dragm. VI (296).
28–p. 53, l. 1 auditus . . . aeree nature: cf. Arist. De An. II. viii (419^{b}19); III. i (425^{a}4); De Sensu ii (438^{b}20); etc. Vid. supra ad nn. 190–2.   visus ignee: Empedocles et Plato (Timaeus 45D) quos improbat Arist. De Sensu ii (437^{a}22–438^{a}4).

which arises from a midpoint, as it were, between water and air. However, it is said by various writers that the fume given off by an odorous
body is a midpoint between water and air, and that fume which has in
itself a proportionate nature of air and water is the proper subject of
odour, and those two small pieces of flesh which receive modifications
from an odour are of a similar constitution. For in them water and air
equally predominate, and thus they are a sort of midpoint between
water and air. Because of this, when smell exists in a subject of such
a nature, then by means of such an organ smell is grasped more than
by means of another organ.

195. However, it can be asked for what reason do we not have similarly some other sensation due to some other combination of some
other elements? Just as, if the sense of smell is a kind of midpoint
between air and water, it can similarly be asked why it is that we do
not have some sensation as a midpoint between air and fire, and so on
with similar cases. Again, if this were the case there would be more
senses than five.

196. To this it should be answered that because of the excessively
high temperature of the heart and the heat of the brain it was necessary
for air to be present to cool down the brain and heart. The lung draws
in air to lower the temperature of both. Thus there is one passage to
draw in air through which the air goes up to the brain and then to
the heart, and this is the passage through the nostrils and the throat.
However, since a fume is something which must be perceived by a
sense by means of an odour when that fume with the odour is mixed
together with the parts of the air which enters the nostrils, and arrives
at the pieces of flesh which receive the modifications from the odour, it
modifies them, and the soul in turning itself towards that impression
perceives the odour. And those little pieces of flesh are placed there
in order to grasp the odour passing through those pieces of flesh by
means of the power of the lungs drawing in air. Since, however, there
are not any other sensible forms which are not perceived by any sense,
it is not necessary for there to be more senses than five. For this same
reason we know why there are more senses of the body than elements,

ab auctoribus dicatur esse aeree nature, et visus ignee, tactus terree, gustus autem aquee.

197. Item. Queritur quare unusquisque sensus preter gustum plura habeat instrumenta, ut auditus duas aures et visus duos oculos.

Ad hoc dicendum quod illa questio est implicita falsi. Gustus enim 5 duo habet instrumenta, scilicet linguam et palatum; quoniam tam in lingua quam in palato potest exoriri species a sapore ad quam convertens voluntas aciem animi coniungit et copulat illam speciem cum sapore et eam apprehendit.

XIV. ii. *De odore, et quare potius inmutat olfactum quam alium* 10
sensum

198. Consequenter queritur de odore, cum odor sit proprium sensatum odoratus; utrum odor sit substantia vel accidens.

Si sit substantia, obiectio est in contrarium per hoc quod propria sensata aliorum sensuum sunt qualitates, ut color, qui est sensatum 15 visus, est qualitas, et sic de aliis. Pari ratione odor erit qualitas, cum sit proprium sensatum odoratus.

199. Preterea. Alia est obiectio per hoc quod corpus odoriferum adnihilatur fere omnino per resolutionem odoris ab ipso; et hoc esse non posset nisi odor esset substantia corporea. Quod si concedatur, 20 obicitur sic.

200. Omnis pars minor est suo toto; ergo illa pars que resolvitur a corpore odorifero est minoris quantitatis quam sit corpus odoriferum; ergo non apprehenditur mediante odoratu per tantam distantiam a corpore odorifero quantum est corpus odoriferum; quod patet esse falsum: 25 per hoc quod corpus odoriferum parve quantitatis afficiet circumquamque aera odore per magnam distantiam ab ipso. Unde sunt quedam animalia, ut vultures et tigrides, que per multa miliaria apprehendunt odorem distantem a corpore odorifero.

---

5 enim: autem PV   7 quam²: quem P   8 voluntas: anima P   9 eam: ita cum P   13 odor *om.* P   14 sit *om.* P   15 qui est *om.* P   23 sit *om.* V 26 afficiet: replebit P   28 multa miliaria: multas leucas P

---

1 tactus terree: cf. Arist. *De Sensu* ibid. (438ᵇ30).   gustus aquee: cf. Arist. *De An.* II. x (422ᵃ33–ᵇ10); *De Sensu* iv (441ᵃ3 seq.).

3–4 Nemesius *Premnon Physicon* viii (iuxta versionem Alfani 81); Iohan. Damasc. *De Fide Orthodoxa* (iuxta versionem Burgundionis) xxxii. 7 (128).

5–7 Cf. Nemesius ibid. ix (85); Iohan. Damasc. ibid. xxxii. 5 (127).

18–19 Cf. Avic. *De An.* II. v (8ᵛᵇA): 'cum diu odoraveris pomum, marcescit propter multum quod resolvitur ex illo'.

22 Cf. Avic. loc. cit.

27–28 quedam animalia: cf. Arist. *De An.* II. ix (421ᵇ12).   ut vultures et tigrides: cf. Avic. ibid. (9ʳᵃA).

since hearing is said by the experts to be of an airy nature, and sight is fiery, touch earthly, and taste watery.

197. Again, the question arises why each sense besides taste has more than one organ, since hearing has two ears and sight two eyes. To this it should be said that this question hides a falsehood. For taste has two organs, namely the tongue and the palate, since a species from a flavour can arise in both the tongue and the palate. The will turns the eye of the mind to this species and links and joins it with the flavour and perceives it.

XIV. ii. *Concerning odour and why the sense of smell is more affected than any other sense*

198. Next a question arises concerning odour. Since odour is the proper sensible of smell, is odour a substance or an accident?

If it is a substance, there is an objection against this in so far as the proper sensibles of other senses are qualities, such as colour, which is the sensible of sight and a quality, and the same is true of the others. For the same reason odour will be a quality since it is the proper sensible of smell.

199. Moreover, there is another objection inasmuch as an odour-bearing body is nearly completely annihilated through the emission of odour from it, and this can only be if the odour is a corporeal substance. Which if conceded can be objected to as follows.

200. Every part is less than its whole, therefore that part which is emitted by an odour-bearing body is less in quantity than the odour-bearing body. Therefore that part is not grasped by means of the sense of smell over a greater distance from the odour-bearing body than is the size of the odour-bearing body. This seems to be false, inasmuch as an odour-bearing body of a small quantity affects the surrounding air with an odour over a great distance from it. Thus, there are some animals such as vultures and tigers which perceive an odour over many miles distant from an odour-bearing body.

201. Solutio. Dicimus quod aliud est fumus et aliud est odor. Fumus est substantia corporea, odor autem est qualitas existens in fumo resoluto a corpore odorifero, et inmutatur aer ab illo odore, et ille aer inmutat alium; et sic est inmutatio successiva et continua quousque aer inmutatus attrahatur per inspirationem, et in meatu suo contingat 5 instrumentum odorandi. Et potest ita contingere quod ille fumus resolutus conmisceatur partibus aeris attracti, et tunc est vehementior inmutatio in odoratu propter vehementiorem odorem qui est in fumo resoluto, cum sit odor factus in aere ab odore fumi. — Et sciendum est quod nomen odoris quandoque sumitur abstractive, scilicet quando 10 sumitur pro qualitate sola; quandoque concretive, ut quando dat intelligere in concretione quadam simul fumum et odorem, scilicet qualitatem in ipso: et secundum hanc intentionem nominis est odor substantia corporea, et per evaporationem ipsius a corpore odorifero diminuitur ipsum corpus. Et hoc potest haberi per hoc quod si pori corporis 15 odoriferi constringantur non fiet evaporatio; et si fiat per confricationem calor, per calorem aperiantur pori, et fumus evaporans meatum habebit per poros apertos.

XIV. iii. *Utrum instrumentum odoratus sit alicuius odoris*

202. Sequitur ut videamus utrum instrumentum odoratus sit ali- 20 cuius odoris. Quod sit alicuius odoris, sic videtur posse ostendi.

Ab Aristotele habetur in libro *de Anima* quod apud sensum est delectatio in mediis, contristatio in extremis. Cum ergo apud odoratum sit delectatio, in medio odore erit delectatio. Redolens enim odor est medius inter duos olentes odores; sed propter simile est delectatio, 25 quoniam similia similibus gaudent. Cum igitur in odore sit delectatio apud odoratum instrumentum odoratus erit alicuius odoris.

203. Preterea. Ille due caruncule que sunt instrumenta odoratus, si ipse abscinderentur et tenderent ad corruptionem, ab ipsis egrederetur evaporatio fetidi odoris. Sic ergo, cum post huiusmodi permutationem 30

2 autem *om.* C   6 odorandi: ordinandi P   ille: ipse P   7 attracti:
contracti P   9 Et *om.* P   10 scilicet *om.* P   11 sumitur *om.* P
12 scilicet *om.* C   13 intentionem: interemptionem P   15 pori: porum P
16 constringantur: et *add.* V   fiat *om.* P   16–17 confricationem . . . calorem:
confirmationem et calorem P   17 aperiantur: aperientur C   et *om.* P   evaporans: evaporationis V   20 Sequitur: Restat V   21 alicuius . . . sic *om.* CV
24 enim: autem C   26 similia similibus gaudent: simile similibus gaudet P
in: ex V   sit: fit V   28 due *om.* V

1–9 Cf. Avic. *De An.* ii. v (9^{ra}B).
22 Arist. *De An.* iii. ii (426^{b}3–7).
26 similia similibus gaudent: cf. supra, n. 167.

201. Solution: We state that a fume is one thing and an odour is another. A fume is a corporeal substance; an odour, however, is a quality which exists in the fume given off by an odour-bearing body. The air is affected by that odour, and that air affects other air. Thus there is a continual and successive modification as far as the affected air which is drawn in through breathing, and in its passage reaches the organ of smell. And thus it can happen that that fume which is emitted is mixed together with the parts of the air which is drawn in, and then there is a stronger change in the sense of smell because of the stronger odour which is in the fume given off, since it is an odour made in the air from the odour of the fume. And it should be noted that the word 'odour' is sometimes understood abstractly, namely when it is understood as a quality only; and sometimes concretely, as when it gives one to understand that a fume and an odour are in some concrete case at the same time, namely a quality in it. And according to this understanding of the word an odour is a corporeal substance, and through its evaporation from the odour-bearing body the body itself is diminished. And this can be observed inasmuch as if the pores of the odour-bearing body are compressed no evaporation takes place; and if heat occurs through friction, the pores are opened owing to the heat, and the fumes evaporating will have an exit through the pores which have been opened.

XIV. iii. *Whether the organ of smell is a kind of odour*

202. Next we should examine whether the organ of smell is some kind of odour. That it is a kind of odour can be shown in the following way, it seems.

From Aristotle in the book *On the Soul* we know that with sensation pleasure is to be found in moderate things, pain in extreme things. When, therefore, there is pleasure with the sense of smell, there will be pleasure in a moderate odour. For the smell of an odour is midway between the smell of two extreme odours; however, pleasure is due to something similar since like things enjoy like. Since, therefore, there is pleasure in an odour in the sense of smell, the organ of smell will be of a certain odour.

203. Moreover, if those two little pieces of flesh which are the organs of smell are cut off and begin to rot, from them will come the evaporation of a stinking odour. And therefore, since after a change

in ipsis factam sint ipse fetentis odoris, et permutatio sit ab una qualitate in eius oppositam, ante illam permutationem fuerunt alterius
odoris, et ita ille caruncule, que sunt instrumenta olfactus, erunt alicuius odoris.

204. Contra. Due res eiusdem speciei in formis sensatis non simul 5
sunt in eodem subiecto, ut due albedines simul non sunt in eadem parte
eiusdem subiecti, quoniam si essent, tunc esset una albedo in alia albedine vel super eam. Pari ratione nec duo odores possunt esse simul in
eodem subiecto. Sed si instrumentum odoratus esset alicuius odoris,
cum ipsum inmutatur ab odore, tunc essent duo odores in eo simul; 10
sed hoc esse non potest. Relinquitur ergo quod instrumentum odoratus
non sit alicuius odoris.

205. Preterea. Ab Aristotele habetur in libro *de Generatione et Corruptione* quod simile a suo simili impassibile est. Ergo si instrumentum odoratus esset alicuius odoris non reciperet impressionem ab alio 15
habente odorem; sed recipit impressionem ab alio habente odorem;
ergo instrumentum odoratus non est alicuius odoris.

206. Solutio. Dicimus quod instrumentum odoratus non est alicuius
odoris. Ad hoc quod obicitur in contrarium. Dicendum est quod non
est delectatio ex odore quia recipiens impressionem ab odore sit alicuius 20
odoris, immo eo quod ipsum existens nudum ab odore appetit informari
ab odore; sicut materia, ut ipsa habeat esse actu in perfectione, respuens
nuditatem, appetit formam.

207. Ad aliud. Dicendum est quod illud quod recipit impressionem
ab alio ita se debet habere ad illud quod ipsa genere eadem sint et 25
similia, sed specie dissimilia. Unde cum albedo et linea non sint eadem
in genere, albedo non patitur a linea, nec e converso; sed contrarium
patitur a contrario, vel nudus ab informato: et contraria quidem eadem
sunt in genere, sed specie sunt diversa, et hoc habetur ab Aristotele in
libro *de Generatione et Corruptione.*   30

2 ante: ipse autem V   3 erunt: sunt V   7 esset *om.* V   una *om.* P
14-15 instrumentum . . . odoris: instrumentum esset alicuius odoris odoratus V
15, 16 alio: aliquo V   16 sed: licet V   20 ex: in P   21-22 appetit . . .
odore *om.* V   21 informari: inferiori P   22 actu: ut *add.* C; actum P   respuens: respiciens V   24 est *om.* P   25 alio: alia P   26 sed: et C
28 nudus: nudum P   29 hoc *om.* V   29-30 in libro *om.* P

13-14 Arist. *De Gen. et Corr.* I. vii (323^{b}4).
19 Ad hoc quod obicitur: cf. supra, n. 202.
24 Ad aliud: cf. supra, n. 205.
30 Arist. ibid. (323^{b}25-33).

occurs of this kind in them, they will have a fetid odour, and the change will be from one quality to its opposite, before that change they were of another odour, and thus those pieces of flesh, which are the organs of smell, will be of another odour.

204. Against: Two things of the same kind in sensed forms are not in the same subject at the same time. For example, two whitenesses are not in the same part of the same subject at the same time, for if they were, then there would be one whiteness in another whiteness on top of it. For the same reason two odours cannot be at the same time in the same subject. But if the organ of smell were of a certain odour, when it is affected by an odour, then there would be two odours in it at the same time. But this cannot be, therefore it follows that the organ of smell is not of a certain odour.

205. Moreover, we know from Aristotle in the book *On Generation and Corruption* that like cannot be acted upon by like. Therefore, if the organ of smell were of a certain odour it would not receive an impression from another thing which has an odour. However, it does receive an impression from another thing which has an odour, therefore the organ of smell is not of a certain odour.

206. Solution: We state that the organ of smell is not of a certain odour. To the objection to the contrary we say that there is no pleasure from an odour because the receiver of an impression from an odour is of a certain odour, rather because the recipient who exists free of odour seeks to be informed by the odour, just as matter, so that it will have actual being in perfection, rejects deprivation and seeks form.

207. Regarding the next question, it should be stated that that which receives an impression from another should be constituted in such a way in respect of it that these are of the same genus and similar, yet dissimilar in species. Thus since whiteness and a line are not the same as regards their genus, whiteness is not acted upon by the line, nor vice versa. However, one contrary is acted upon by another, or the deprived by the unformed. Indeed, contraries are the same as regards their genus but different in species, and this we know from Aristotle in the book *On Generation and Corruption*.

xv. *De gustu*

208. Dicto de olfactu dicendum est de gustu. Gustus autem sic describitur ab Avicenna. Gustus est vis ordinata in nervo expanso ad apprehendendum sapores ab extrinseco advenientes.

209. Sed obicitur. Nervus expansus est res corporea; sed rei cor- 5 poree non est apprehendere, sed solius anime; ergo falsa est superior discriptio, cum nec virtutis ipsius nervi sit apprehendere, nec etiam ipsius nervi.

210. Item. Videtur quod gustus sit species tactus, et hac ratione. Non est gustus in effectu nisi per contactum rei habentis saporem et ipsius 10 instrumenti. Sed ita fit tactus; et ita gustus est tactus. Forte dicet [aliquis] quod gustus distinguitur a tactu per hoc quod gustus habet apprehendere sapores, tactus habet apprehendere calidum, frigidum, humidum, siccum, grave, leve.

211. Sed Contra. Appelletur A tactus in quantum ipse est appre- 15 hensivus calidi et frigidi, et appelletur tactus B in quantum ipse est apprehensivus asperi et levis; secundum hoc tam A quam B habet esse in actu per tactum sicut gustus. Qua ratione ergo dicis B esse tactum particularem, eadem ratione debes dicere gustum esse tactum particularem; vel quare non. Sicut enim sapor non est una de primis 20 proprietatibus, similiter nec est asperitas, nec levitas, vel viscositas est de primis proprietatibus, immo de secundis.

212. Item. Instrumentum tactus est habens se in temperata complexione; unde cum nullum animal sit adeo temperate complexionis ut homo, in nullo animali viget tactus tantum ut in homine. Unde cum 25 tactus sit secundum complexionem receptam a primis elementis non deserit tactus animal; et est tactus apprehensivus qualitatum que sunt ab elementis primis, scilicet calidum, frigidum, siccum, humidum;

2 Dicto de olfactu dicendum est: Sequitur C; Consequenter dicendum est V   autem *om.* C   4 ab *om.* V   extrinseco: intrinseco P   5 res corporea: corporeus CV   sed *om.* P   9 Item: ut *add.* P   10 contactum: tactum P   11 ita²: videtur quod *add.* V   11–14 Forte . . . leve: Contra. Tactus habet apprehendere calidum et frigidum, siccum et humidum, grave et leve et huiusmodi; gustus vero nullum proprium sensatum tactus habet apprehendere. Ergo gustus non est tactus nec species tactus V   13 sapores: saporem P   14 grave: aridum P   15 Sed contra: Item V   appelletur: appellatur V   16 ipse est *om.* V   18 tactum: contactum P   19 debes *om.* P   21 asperitas: asperum P   levitas: leve P   23 est *om.* C   24 adeo *om.* V   25 tantum *om.* P   27 sunt *om.* V   28 siccum: et *add.* P

1 *De gustu*: cf. Arist. *De An.* II. x (422ᵃ8–ᵇ16).
2 Avic. *De An.* I. v (5ʳᵃB). Cf. Gundiss. *De An.* ix (68).
9 Arist. *De An.* III. xii (434ᵇ18–22): 'Gustus est sicut tactus quidam'; cf. II. ix (421ᵃ19); *De Sensu* iv (441ᵃ5); etc.
13–14 Cf. Arist. *De An.* II. iii (414ᵇ6–10).
24–28 Cf. Arist. ibid. II. ix (421ᵃ19–26); xi (423ᵇ27–29). Vid. p. xix.

XV. *Concerning taste*

208. Having spoken about the sense of smell, now we should talk about the sense of taste. Now taste is described as follows by Avicenna. Taste is a power located in a nerve which is spread out in order to grasp flavours arriving from outside.

209. However, it is objected: a nerve which is spread out is a corporeal thing, but it is not proper to a corporeal thing to perceive, but only to the soul. Therefore the above description is false, since it is neither proper to the power of the nerve itself to perceive, nor even to the nerve itself.

210. Again, it seems that taste is a kind of touch and for this reason: there is in effect no taste except through contact between the thing which has the taste and the organ itself. However, touch happens in this way; and so taste is touch. Perhaps someone might say that taste is distinguished from touch inasmuch as taste has to grasp flavours, whereas touch has to grasp hot, cold, moist, dry, heavy, and light.

211. However, against this: let touch in so far as it is perceptive of hot and cold be called A, and let touch in so far as it is perceptive of sharpness and lightness be called B. In this way both A and B have being in act through touch, just like taste. By what reason, therefore, do you say that B is a separate kind of touch? For the same reason you should say that taste is a particular touch; or if not, why not? For just as taste is not one of the primary qualities, similarly, neither sharpness nor lightness nor viscosity belong to the primary qualities, rather they belong to the secondary qualities.

212. Moreover, the organ of touch is made up of a temperate complexion. Thus, since no animal is of such a temperate complexion as a human being, touch does not thrive in any animal as much as it does in a human being. Thus, since touch exists according to the complexion received from the primary qualities, touch does not leave an animal. Again, touch is perceptive of qualities which are from the prime qualities, namely, hot, cold, dry, moist, and it happens in accordance with a

et contactus est secundum complexionem temperatam. Et tactus est in homine ad conservandum subiectum in esse, quia secundum eum percipimus nociva et apprehendimus expedientia. Sed cum omnis proprietas sit vel prima vel fluens a prima, cum tactus sit vis ordinata in nervo secundum temperatam complexionem, debet illud instrumen- 5 tum tactus eque recipere inmutationem ab omni qualitate fluente a primis qualitatibus; debet ergo secundum tactum fieri apprehensio saporis, cum sapor sit expediens vel nocivus et fluens a primis qualitatibus.

213. Item. Ubi est auditus in effectu? Est ad ipsum aer sonum 10 deferens. Similiter est aer medium deferens odorem ad odoratum. Queritur ergo quid sit medium deferens saporem ad gustum. Similiter quare ad hoc quod sit visus in effectu oportet quod ibi sit lux detegens et manifestans colorem visui. Similiter ad hoc quod sit auditus preexigitur collisio corporum que sonora sint, [et] per illam collisionem 15 detegitur et manifestatur eorum sonoritas auditui. Queritur ergo quid sit consimili modo detegens et manifestans saporem gustui?

214. Solutio. Dicendum est ad primum quod sensus descriptionis est: gustus est vis ordinata in nervo expanso, ut recepta inmutatione in instrumentum gustus per saporem, convertens sit voluntas aciem animi 20 ad illam inmutationem, et apprehendat anima saporem extra copulando passionem que est in sensu gustus cum sapore extra. Nec exigitur secundum sententiam descriptionis quod vel nervus vel vis nervi apprehendat, sed acies animi.

215. Ad aliud. Dicendum quod gustus non est species tactus. Quoniam 25 gustus est vis ordinata in instrumento quod est aquose nature. Unde, cum sapor habeat esse proprie in aquosa substantia, de aptitudine ipsius instrumenti gustus est ut ipsum recipiat impressiones a sapore. Saporis enim est esse in subiecto tali cum ipsum subiectum sit aquosum, ubi, scilicet aquea natura sit predominans; et ideo contingit quod ipsa 30

1 et contactus: et ideo contactus est magis vigens V   est[1]: fit P   Et om. P
2 in esse om. P   secundum: sicut C   3 percipimus: participamus P   cum om.
P   4 vel[1] om. C   5 illud: igitur V   7 ergo: igitur V; om. C   8 et om. C
11-12 deferens[1] . . . medium om. per hom. C   11 deferens[2]: differens P
13 quare: queritur P   quod[2]: ut V   15 sint: sunt V   17 consimili:
simili P   18 est om. V   19-20 in instrumentum gustus per saporem: et in instrumento gustus a sapore C   21 illam: animam C   apprehendat: apprehendit PV
copulando: concopulando P   22 gustus om. V   24 acies animi: aciem anime P
25 dicendum: est add. V   27 habeat esse: se habeat V   28-29 Saporis enim: Et
saporis V   29 subiectum om. C   30 aquea: aqua CV

1-3 Cf. infra, n. 221, ll. 18-19.
18 ad primum: cf. supra, n. 209. Cf. Guill. de Conchis Dragm. VI (296-7).
25 Ad aliud: cf. supra, nn. 210-12.
26-30 Cf. Arist. De An. II. x (422[a]33-[b]10); De Sensu iv (441[a]3 seq.).

balanced complexion. Moreover, touch is in a human being in order to keep the subject in existence since with it we perceive harmful things and perceive useful things. However, since all properties are primary or flow from primary qualities, and since touch is a power located in a nerve in accordance with a balanced complexion, that organ of touch should equally receive an impression from every quality flowing from the primary qualities. Therefore the perception of taste should also happen according to touch, since taste is either useful or harmful and it flows from the primary qualities.

213. Moreover, where is hearing effected? It is to hearing that the air brings sound. Similarly, air is a medium which brings odour to the sense of smell. Therefore it is asked what is the medium which conveys flavour to the sense of taste? Similarly, why in order for sight to be effected is it necessary for there to be light there to purge and manifest colour to vision? Similarly, for sound to exist it is required beforehand that there is a striking together of bodies which are re-sounding. Therefore, what is it which in a similar manner purges and manifests flavour to taste?

214. Solution: It should be stated regarding the first that the de-finition of the sense is: taste is a power in a nerve which is spread out, so that when an impression is received in the organ of taste by means of flavour, it is the will which turns the eye of the mind towards that modification, and the soul perceives the flavour which is outside, through linking the modification which is in the sense of taste with the flavour outside. Nor is it required according to the meaning of the definition that either the nerve or the power of the nerve perceives, but rather the eye of the mind.

215. Regarding the next objection, it should be said that taste is not a kind of touch, because taste is a power which has been located in an organ which is of a watery nature. Thus, since taste has its existence properly in a watery substance owing to the disposition of this organ of taste, it is such that it receives impressions from flavour. For it is of the nature of taste to be in such a subject since this subject is watery, namely, where a watery nature is predominant. Thus it happens that

lingua est spongiosa propter aquosam naturam, ut patens est in spongia
et spuma aque que raras habent partes, quia aquose sunt nature et
intermiscentur partes aeree. Unde contingit quod lingua est membrum
flexibile et mobile, quia ipsa est aquose nature que fluida est et habet
partes aereas sibi admixtas. Tactus autem est vis proveniens a temperata 5
complexione; unde alterius speciei est quam gustus: cum sit gustus vis
surgens maxime secundum aquosam naturam. Dicitur tamen a pluribus
auctoribus quod tactus est vis ordinata in nervo terree nature.

216. Ad tertium. Dicendum quod medium deferens saporem gustui
est saliva; et ideo dicit Aristoteles in libro *de Anima* quod sine saliva 10
non est sapor actu. Detectio saporis est resolutio corporis saporiferi per
subintrationem humiditatis.

XVI. *De tactu*

217. Sequitur de tactu. Unde primo loco videndum est quid sit
tactus. Tactus est vis ordinata in nervis cutis totius corporis extensis 15
fere per totam carnem animalis ad apprehendendum calidum, frigidum,
siccum, humidum.

Sed potest inprimis queri quare magis tactus distendatur in unamquamque partem animalis quam aliquis alius sensus.

218. Preterea. Dictum est superius quod instrumentum auditus 20
potius recipit inmutationem a sono quam ab alio; quoniam instrumentum auditus est aeree nature, et quod gustus recipit saporem quia
gustus est aquose nature. Sed instrumentum tactus in homine est
temperate complexionis, et ideo potest inmutari ab eis que temperatam
faciunt complexionem, ut a caliditate, et ab aliis primis qualitatibus. 25
Qua ratione ergo potest inmutari a quibusdam qualitatibus fluentibus
a primis, ut a viscositate, que est ab humiditate; et asperitate, que est

---

1 lingua *om.* V   patens est: patet V   3–5 Unde . . . admixtas *om.* P   5 proveniens: *scriptum sed corr. in* veniens C   10 sine *om.* C   11 non *om.* V
13 De tactu *om.* P; *in marg.* V   14 Sequitur: ut dicamus *add.* V   loco *om.* C
est *om.* C   15 totius corporis *om.* P   extensis: expansis PV   16–17 calidum
. . . humidum: caliditatem, frigiditatem, humiditatem, siccitatem P   22 gustus:
gustum P   23 aquose: aquee P   25 ab *om.* P   primis *om.* V
qualitatibus *om.* P   27 humiditate: humanitate P

---

1 lingua spongiosa: cf. Guill. de Conchis *Dragm.* VI (296).
7–8 Cf. supra, not. ad n. 196.
9 Ad tertium: cf. supra, n. 213.
9–11 Arist. *De An.* II. x (422ª17–19).
13 *De tactu*: cf. Arist. *De An.* II. xi.
14–17 Cf. Avic. *De An.* I. v (5ʳªʙ); Algazel *Metaphys.* II. iv. 3 (165).
20 superius: cf. supra, nn. 190–2, quoad auditum.
22 aeree nature: cf. supra, n. 196.
23 aquose nature: cf. supra, loc. prox. cit.

the tongue itself is spongy because of a watery nature, as is clear in sponges and foam that have fine parts, because they are of a watery nature and are mixed together with parts of air. So it happens that the tongue is an organ which is flexible and mobile because it is of a watery nature, and is fluid and has airy parts mixed with it. Touch, however, is a power which derives from a balanced complexion and thus it is of a different species from taste, since taste is a power arising mostly according to a watery nature. Moreover, it is stated by many writers that touch is a power placed in a nerve which is of an earthy nature.

216. In respect of the third objection, it should be said that the medium which carries flavour to taste is saliva; and so Aristotle says in the book *On the Soul* that without saliva there is in effect no flavour. The detection of flavour is the breaking down of a flavour-bearing body by means of the entering in of humidity.

XVI. *Concerning touch*

217. What follows concerns touch. Thus, in the first place we should look at what touch is. Touch is a power located in the nerves extended through all the skin of the entire body and throughout all the flesh of an animal in order to perceive the hot, cold, dry, and moist.

But one could ask first of all why touch is more extended in every part of an animal than any other sense.

218. Moreover, it has been stated above that the organ of hearing receives an impression better from sound than from anything else, for the organ of hearing is of an airy nature, and that the sense of taste receives flavour because taste is of a watery nature. However, the organ of touch in a human being is of a temperate constitution, and thus it can be affected by those things which affect a temperate constitution, such as by heat, and by the other primary qualities. Why, therefore, can it be affected by some qualities flowing from the primary qualities, such as by viscosity which is from humidity, and sharpness

a siccitate; et a gravitate, que est a frigiditate; et a levitate que causatur
in esse a caliditate simul et siccitate; pari ratione habet tactus recipere
impressionem ab aliis qualitatibus fluentibus in esse a quatuor primis,
ut albedine, nigredine, sapore; vel quare non. Et si posset ab eis inmutari mediante tactu erit apprehensio omnium qualitatum que fluunt 5
a primis qualitatibus.

219. Item. Queritur ex quo ita est quod inmutatio recepta in sensu
audientis non apprehenditur auditu, immo sonus extra; similiter impressio recepta in sensu videntis non apprehenditur mediante visu,
immo color extra; quare est ita quod mediante tactu apprehenditur 10
passio intra, ut in carne que sit, scilicet a qualitate extra?

220. Item. Potest queri quare non sit medium deferens apud tactum
sicut apud alios sensus, scilicet, auditum, visum, odoratum, ut aer
deferens apud gustum est saliva.

221. Solutio. Ad primum. Dicimus quod tactus distenditur per plura 15
membra quam alii sensus, quia tactus tribuitur animali a natura sue
complexionis; et ideo contingit quod tactus distenditur per plura
membra quam alii sensus. Est etiam tactus ad conservationem sui
subiecti, et ideo respuit tactus nociva. Si autem obicitur de cerebro et
osse et pulmone, que sunt membra insensibilia, dicendum est quod hoc 20
quidem contingit quia nervi tactus non distenduntur in illis membris.

222. Ad secundum. Dicendum est quod ille proprietates apprehenduntur mediante tactu que vel sunt prime vel propinque fluentes
ex primis; prime ut calidum, frigidum, siccum, humidum; propinque,
ut viscositas ex humiditate, ariditas ex siccitate, ponderositas ex frigidi- 25
tate, levitas ex caliditate et siccitate; ex caliditate enim et siccitate est
rarefactio et dissolutio, et ita levitas. Albedo et nigredo, et alii colores

---

1 causatur: curatur P   2 caliditate simul: causalitate similiter V   et: cum P
recipere: accipere P   4 posset: possit C   5 apprehensio: apprehensiva P
5–6 omnium . . . qualitatibus: substantialium qualitatum P   7 Queritur: Quare C
10 est: etiam C   11 scilicet: similiter V   12 Item: Similiter V   non: fieri P
13 scilicet: ut apud C   ut om. P   14 est: cum P; et V   16 tactus om. C   natura:
naturalia natura V   17 distenditur: distendatur P   18 Est: Et P
19 tactus om. V   20 que sunt om. P   insensibilia: intelligibilia P   21 quidem om. P   quia: quod P   nervi: ipsius add. V   distenduntur: diffunduntur V
23 tactu om. V   24 calidum . . . humidum: caliditas, frigiditas, siccitas, humiditas V   25 ex²: et P   26 levitas . . . siccitate: leve ex calido et humiditate P

---

14 apud gustum est saliva: cf. supra, n. 216.
15 Ad primum: cf. supra, n. 217.
15–18 Cf. Arist. De An. ibid. (423^{b}17–424^{a}2).
18–19 Cf. Arist. ibid. III. xii (434^{b}10–24); De Sensu i (436^{b}12–15); Avic. De An.
II. iii (8^{rb}CD).
20 osse: cf. Arist. De An. III. xiii (435^{a}24–25). Vid. p. xix.
22 Ad secundum: cf. supra, n. 218. Vid. Avic. ibid. (8^{ra}B); Arist. De Gen. et Corr.
II. ii, ad quem locum mittit Arist. De An. II. xi (423^{b}27).

which is from dryness, and by heaviness which is from coldness, and
by levity which is brought into existence by hotness and dryness at
the same time? For the same reason taste would be able to receive an
impression from the other qualities flowing into existence from the
four primary qualities, such as whiteness, blackness, taste; or if not,
why not? And if it can be affected by them, then by means of touch
there will be the perception of all of the qualities which flow from the
primary qualities.

219. Again, the question arises why it is the case that the change received in the sense organ of the hearer is not perceived through hearing but rather the sound outside is. Similarly the impression received
in the sense organ of sight is not perceived by means of vision but
rather the colour outside is. Why is it the case that in touch an affec-
tion is perceived within, namely in the flesh, yet it is from an external
quality?

220. Again, it can be asked why there is not a conveying medium
to be found together with touch just as with the other senses: namely,
with hearing, sight, and smell it is air, and with taste it is saliva which
is the conveying medium.

221. Solution: With regard to the first question we state that touch
is extended through more parts of the body than other senses, because touch is assigned to an animal by the nature of its constitution.
So it is that touch is extended through more parts of the body than
other senses, for the role of touch is to preserve its subject and so it
rejects harmful contact. If, however, an objection is raised concerning
the brain, bones, and lungs which are unfeeling parts of the body, it
should be stated that this in fact happens because the nerves of touch
are not extended throughout these parts of the body.

222. Regarding the second question, it should be said that those
properties which are perceived by means of touch are either primary
qualities or those flowing directly from them. The primary qualities
are hot, cold, dry, moist; the next are those qualities flowing directly
from them, such as viscosity from humidity, aridity from dryness,
heaviness from coldness, lightness from heat and dryness, since
rarefaction and separation are from dryness and heat. Whiteness and

et sapores remotiores proprietates sunt, et ideo a tactu non apprehenduntur.

223. Ad tertium. Dicendum est quod illud quod receptum est in
instrumento tactus non apprehenditur per tactum, quoniam nervus est
proprium instrumentum tactus; sed illud apprehenditur mediante tactu 5
quod recipitur in carne vel in pelle continguata nervo a qua fit passio
in nervo.

224. Ad quartum. Dicendum est quod vel caro vel pellis est medium
inmutationes deferens in tactu.

225. Item. Videtur posse ostendi quod caliditas et frigiditas non 10
sint propria sensata tactus nisi secundum viam accidentis, et hac
ratione. Caliditatis est dissolvere: unde cum caliditas advenit corpori
animalis ipsum dissolvit. Ipsa dissolutio aut erit nociva aut expediens,
aut medio modo se habens. Sed sive sic sive sic, cum tactus sit ad conservationem subiecti, tactus non apprehendet caliditatem nisi propter 15
dissolutionem provenientem a caliditate; secundum eandem viam
similitudinis non apprehenditur frigiditas nisi propter constrictionem;
et propter quod unumquodque est illud magis est, ut habetur in principio *Posteriorum Analyticorum*. Ergo dissolutio et constrictio per se
apprehenduntur mediante tactu, et caliditates et frigiditates per acci- 20
dens.

226. Item. Queritur utrum proprietates prime elementares sint sensibiles et elementa sensibilia per ipsas. Quod non, sic videtur posse
ostendi. Simplex non habet agere in compositum, immo compositum
in compositum. Sed elementa prima simplicia sunt; ergo non possunt 25
inmutare compositum. Sed unusquisque sensus est compositus; nervus
enim est res complexionata; ergo nervi non possunt inmutari a primis
qualitatibus elementaribus. Sed elementa non possunt sentiri nisi per
suas qualitates. Ergo elementa sensu apprehendi non possunt.

---

1 sapores: a primis *add.* V   sunt *om.* V   a *om.* V   1-2 apprehenduntur:
apprehendit P   3 est[1] *om.* P   est[2] *om.* C   in *om.* P   8 vel *om.* C
9 inmutationes *om.* P   tactu: gustu P   10 Item: Rursus V   non *om.* V
11 sint *om.* V   propria: prima P   nisi: sed P; ubi V   12 Caliditatis: caliditas PV
caliditas: calor P   13 expediens: expiciens P   14 sive[1] ... sic[2] *om.* V
15 tactus *om.* C   apprehendet: apprehenderet P   16 provenientem *om.* V
18 illud: ipsum PV   20 tactu: motu C   caliditates et frigiditates: caliditas et
frigiditas P   23 sic *om.* P   25 simplicia: sensibilia P   27 complexionata:
complexata P   nervi: sensibiles *add.* PV   non *om.* P   27-28 primis qualitatibus: proprietatibus P   29 apprehendi: comprehendi P

---

3 Ad tertium: cf. supra, n. 219.
8 Ad quartum; cf. supra, n. 220. Vid. Arist. *De An.* II. xi (423[b]26): 'medium
tactus est caro'; Avic. *De. An.* II. iii (8[rb]D).
14-15 cum tactus sit ad conservationem subiecti: cf. supra, n. 221.
19 Arist. *Anal. Post.* I. ii (72[a]29).

blackness, and the other colours and tastes, are more remote properties and so are not perceived by touch.

223. With regard to the third question, it should be stated that
that which has been received in the organ of touch is not grasped by
touch, because the nerve belongs to the organ of touch. However, it
is grasped by means of the touch which is received in the flesh or on
the skin which touches upon the nerve, and from the skin the action
upon the nerve arises.

224. Solution to the fourth point: It should be stated that either
flesh or skin is a medium carrying modifications to touch.

225. Again, it seems that it can be shown that heat and cold are not
proper sensibles of touch except in an accidental way, and with the following argument. It is in the nature of heat to distend: thus when heat
comes close to the body of an animal, it distends it. This distending is
either harmful or useful, or is somewhere in between. But whether it
is one or the other, since touch preserves the subject, touch does not
perceive heat except through the distending which flows from heat,
and in the same way cold is not perceived except through contraction; again, because something is greater when something else exists
because of it, as is held at the beginning of the *Posterior Analytics*.
Therefore, separation and contraction in themselves are grasped by
means of touch, and heat and cold by accident.

226. Again, it is asked whether the first elemental qualities can be
sensed and if the sensible elements can by means of them. That they
cannot can be proved as follows: A simple cannot act upon a composite, rather a composite acts upon a composite. But the first elements are simple, therefore they cannot act upon a composite. However, each sense is a composite, because a nerve is made up out of
simple things. Therefore nerves cannot be acted upon by the first ele-
mentary qualities. Yet elements cannot be sensed except through their
qualities. Therefore, the elements cannot be perceived by the senses.

227. Preterea. Sillium est herba complexionata, et non est adeo frigida ut aqua que summe frigida est, et tamen magis infrigidaret sillium hominem quam aqua. Sed hoc non contingit nisi quia sillium est res elementata, et ideo potest agere in rem elementatam; aqua autem simplex est nec potest agere in compositum. Si ergo concedatur quod 5 elementa sensu apprehendi non possunt.

228. Contra. Quanto aliquid est potentius alio tanto citius potest ipsum destruere; sed fortior est caliditas in igne quam sit in re elementata, ut in homine; ergo potius habet caliditas ignis destruere frigiditatem quam caliditas in elementato que debilior est. Ergo potius 10 elementatum frigidum habet destrui ab igne calido quam ab elementato calido. Sed tactus apprehendit ea que nociva sunt; ergo potius habet apprehendere calorem ignis quam calorem rei composite, et ita elementa sensibilia sunt.

229. Item. In libro *de Generatione et Corruptione* habetur quod 15 minus corrumpitur a pluri, ut minus calidum a magis calido; sed non potest corrumpi nisi per actionem; ergo in eo quod est minus calidum fiet passio a magis calido, ut ab igne. Potest ergo caro pati ab igne, et ita per tactum apprehendetur ignis; et ita potest convinci pura elementa posse sentiri. Quod bene concedimus, ut ostensum 20 est.

230. Multi tamen hoc concedere nolunt ratione preassignata in contrarium. Nos autem resistimus prime obiectioni, negantes hanc propositionem: simplex non potest agere in compositum, secundum quod elementa dicuntur esse simplicia; sed si sumatur simplex secundum 25 quod punctus est simplex, vel instans, vel anima, vera est illa propositio.

---

1 sillium: psillium V   complexionata: complexata P   2 ut: sicut P
3 sillium: psillium V   quia: quod P   4 potest: non potest V   7 Contra
om. C   est *om.* CV   8–9 elementata: scilicet *add.* P   9 homine: quoniam,
ut dicit Aristoteles in libro de Generatione et Corruptione, ignis est in tertio caliditatis
add. V   ignis: in igne P   `10 elementato: elemento V   11 elementatum:
quam *add.* P   elementato: elementa P; elemento V   16 pluri: plurali P   a
magis calido: et a maiore calitudo P   18 ut ab igne: sed non potest corrumpi V
19 apprehendetur: apprehenditur P   25 dicuntur: dicitur V   26 instans:
materia P   vel anima *om.* V

---

1 Sillium: Isaac, *Liber dietarum universalium*, c. x (xxxii^rb^). Textum citavi secundum
cod. Bodl. Auct. F.5.30 (2753), fol. 51^v^, saec. xiii: 'Psillium magis (plus *ed. Lugdun.*,
et ceteri codd. quos inspexi) refrigerat et extinguit sitim quam aqua, etsi aqua naturaliter
sit frigidior quam psillium, quia simplex est aqua'. Vid. p. xix.
12 nociva: cf. supra, n. 221.
15 Arist. *De Gen. et Corr.* I. vii (323^b^7–10).
22 Multi: cf. Avic. *De An.* II. iii (8^va^D).
22–23 in contrarium: cf. supra, n. 228.   prime obiectioni: cf. supra, n. 226.

227. Moreover, fleawort is a complex herb and it is not nearly as cold as water, which is cold to the highest degree, and yet fleawort will cool a man down more than water. Yet this should happen only if fleawort is composed of elements and thus can act upon something composed of elements; water, however, is simple, nor can it act upon a composite. Therefore, if it is conceded that elements cannot be grasped by the senses:

228. Against: The more something is stronger than another, the quicker it can destroy it. Yet the heat in fire is stronger than it is in a thing composed of elements, such as in a man, therefore the heat of fire is more able to destroy coldness than the heat in a thing made up of elements, which is weaker. Therefore, elemental coldness is more capable of being destroyed by the heat of fire than by elemental heat. Yet touch perceives those things which are harmful, therefore it is more able to perceive the heat of fire than the heat of a composite thing, and thus the elements can be sensed.

229. Again, in the book *On Generation and Corruption* it is stated that the lesser is corrupted by the more, such as the lesser heat by the greater heat. However, it cannot be corrupted except through an action, therefore in that which is less hot there occurs an action upon it from that which is more hot, such as from fire. Therefore flesh can be acted upon by fire, and thus fire is perceived through touch. Thus it is possible to be convinced that the pure elements can be sensed, which we readily concede, since it has already been demonstrated.

230. Many people, however, are unwilling to accept this because of the aforementioned argument against. We, however, oppose the first objection by denying this proposition: a simple thing cannot act on a composite, inasmuch as the elements are said to be simple things; yet if a thing is taken to be simple in the way that a point is simple, or an instant, or a soul, then that proposition is true.

231. Ad aliud dicendum est quod sillium magis infrigidat quam aqua,
quia aliquid de sillio incorporatur celebrata digestione, aqua pura nequaquam.

XVII. *De sensu communi*

232. Sequitur de sensu communi. Sed prius inquirendum est quare 5
intellectus sit abstrahens formas a materia et ab apendiciis materie, et
sensus nequaquam.

Quod sensus sit abstractivus sicut intellectus hac ratione videtur.

233. Qualitates, ut colores, sapores, caliditates, frigiditates, sunt propria sensata sensus. Subiecta autem illarum non apprehenduntur sensu 10
nisi per accidens. Quod ergo proprie et per se cadit in sensum est
qualitas. Sed propter nil aliud est intellectus abstractivus nisi quia
ipse apprehendit formam abstrahendo eam a materia et ab apendiciis
materie. Sed simile est de sensu, cum materia non apprehendatur per
sensum nisi per accidens. Ergo sensus est abstractivus.   15

234. Preterea. Si intelligeretur per impossibile quod caliditas esset
separata a subiecto, et color cum magnitudine similiter, adhuc inmutaret
caliditas tactum, et color cum magnitudine visum. Quod ergo sensus
apprehendat formas que sunt in materiis hoc non est nisi propter hoc
quod forme non possunt separari a materiebus, et non est de impotentia 20
sensus quod ipse non apprehendit formas separatas, sed est de impotentia forme secundum quod separari non potest a subiecto, nec apprehenditur sensu nisi secundum quod forma est ei presens. Est ergo
sensus quantum est in se abstractivus sicut intellectus.

235. Preterea. Visus apprehendendo colorem et magnitudinem 25
secundum quam distenditur color non apprehendit coaccidentia coloris
vel magnitudinis. Abstrahit ergo visus colorem a suis coaccidentibus;
est ergo visus abstractivus.

---

1 sillium: psillium V   2 aliquid de sillio: sillium P   pura *om.* P   5 Sequitur: Consequenter dicendum est P   inquirendum est: inquiratur V   6 abstrahens: abstractiores P   formas . . . materie *om.* P   apendiciis: appentiis V
9 Qualitates: scilicet *add.* P   caliditates, frigiditates: calor, frigus C   9–10 propria: proprie P   12 nil: nihil V   13 ipse *om.* V   materia: modo P   ab
om. C   15 Ergo . . . abstractivus *om.* P   17 color: corpori P   19 apprehendat: apprehendit P   est *om.* C   20 materiebus: materiabus V   21 est
om. C   22 forme: sensus P   secundum quod: que CV   23 forma *om.* P
24 quantum: inquantum V   abstractivus: abstractio P   26 secundum: ad P
coloris: corporis P

---

1 Ad aliud: cf. supra, n. 227.
4 *De sensu communi*: cf. Arist. *De An.* III. i (425^{a}14–^{b}3); ii (426^{b}8 seq.); Avic. *De
An.* IV. i (17$^{rb–vb}$A).
9–11 Cf. Arist. *De An.* II. vi (418^{a}7–25).

231. With regard to the next objection, it should be stated that fleawort cools more than water because something from the fleawort becomes part of the body after digestion has taken place; pure water never does.

XVII. *Concerning the central sense*

232. Next we should examine the central sense, but first of all we should investigate why the intellect abstracts forms from matter and from the appendages of matter, and the senses never do.

That sensation is abstractive like the intellect seems to be so by means of the following argument.

233. Qualities such as colours, tastes, heat, and cold are the proper sensibles of sensation. However, their subjects are not perceived through sensation except by accident. Therefore, that which properly and in itself falls under sensation is a quality. Yet the intellect is abstractive because of nothing other than this: it grasps the form by abstracting it from matter and from the appendages of matter. But it is similar in the case of sensation, since matter is not perceived by sensation except by accident. Therefore sensation is abstractive.

234. Moreover, if one were to take the case (albeit impossible) where heat was separated from its subject, and colour together with size similarly, then heat would affect touch, and colour together with size would affect sight. Therefore, that sensation grasps forms which are in material things is only because of the fact that forms cannot be separated from material things. It is not because of the inability of sensation that it does not grasp separated forms, but it is because of the inability of the form inasmuch as it cannot be separated from the subject, nor is something grasped by sensation except in so far as a form is present to it. Therefore, sensation is just as much abstractive in itself as the intellect is.

235. Moreover, sight, when it perceives colour and size inasmuch as colour is extended, does not perceive the accompanying accidents of colour or size. Therefore sight abstracts colour from its accompanying accidents and so sight is abstractive.

236. Solutio. Dicendum est quod sensus abstractivus non est, quia
ipse discurrit tantum circa singularia; intellectus quidem abstractivus
est, quia ipse discurrit tantum circa universalia abstrahendo ea a singularibus et ab accidentibus singularium.

237. Restat videre utrum sensus communis sit et quid sit. 5

Quod sit sensus communis potest haberi ab Augustino in libro *de
Libero Arbitrio*. Dicit enim sensum communem esse, vocans eum sensum
interiorem cui sensus exteriores communiter serviunt. Item, ab Avicenna habetur in commento *de Anima* quod sensus communis est vis
ordinata in anteriori parte cerebri recipiens impressiones fluentes in 10
ipsam a quinque sensibus exterioribus.

Quod autem huiusmodi sensus sit, sic videtur posse ostendi.

238. Visu non percipimus nos videre: quoniam quicquid videtur per
ymaginem videtur, et ymaginem sui ipsius non habet ipsa visio in sensu
videntis. Similiter tactu non percipimus nos tangere; et sic de aliis 15
sensibus exterioribus. Cum ergo percipimus nos videre, nos tangere,
hoc est per aliquam aliam vim quam sit aliqua vis sensuum exteriorum.
Sed experimento habetur, ut in anathomia patet, quod ab omnibus
sensibus exterioribus fluunt inmutationes in anteriorem partem cerebri,
et anima convertens se ad illas inpressiones coniungit et copulat eas 20
cum eis a quibus fiebant; sed cum ipse facte essent ab eis que sunt in
sensibus extra, acies animi intuendo ad eas extra percipit inpressionem
per eam quam reperit in instrumento tactus fieri visum. Sed illa vis que
est ordinata in anteriori parte cerebri est sensus communis, et mediante
illa vi percipimus nos videre, nos audire, nos tangere. Relinquitur ergo 25
quod sensus communis est.

239. Preterea. Ab Aristotele habetur in libro *de Anima*, quod motus,
et numerus, et magnitudo non sunt sensata alicuius sensus exterioris,
immo sunt sensata sensus interioris, qui est sensus communis.

1 est¹ *om.* P   2 ipse *om.* PV   discurrit: discurritur P   quidem: autem P
3-4 ea ... singularium: ei a singulis et ab accidentibus singularibus V   6-8 Quod
... serviunt *om.* P   8 Item: Sensus communis P   14 ymaginem sui ipsius:
suam ymaginationem P   15-16 et sic ... tangere *om. per hom.* C   17 hoc:
ergo *add.* C   aliquam *om.* P   18 anathomia: anothomia P   omnibus:
communibus P   21 cum² *om.* C   eis: ipsis V   sunt: fiunt V   22 extra¹
om. V   inpressionem *om.* P   ⌐23 reperit ... tactus: recipit in tactu P   in
om. C   visum: tactum V   25-26 Relinquitur ... est *om.* P   28 et¹: est C
et magnitudo *om.* P   exterioris: in exterioribus P   29 est: dicitur esse P

1 Dicendum: cf. supra, nn. 233-5.
2-4 Cf. Arist. *De An.* II. v (417ᵇ19-25); Avic. *De An.* II. ii (6ᵛᵇA).
6-7 Aug. *De Lib. Arb.* II. iv. 10-v. 12 (1246-7).
9 Avic. *De An.* I. v (5ʳᵇD).
13-17 Cf. Arist. *De An.* III. ii (425ᵇ12-25).
27 Arist. *De An.* II. vi (418ᵃ17-19); III. i (425ᵃ14-ᵇ4).

236. Solution: It should be stated that sensation is not abstractive, since it treats only of individual things; the intellect, however, is abstractive because it treats of universals by abstracting them from individual things and from the accidents of individual things only.

237. It remains to be seen whether there is a central sense and what it is.

That there is a central sense can be known from Augustine in the book *On Free Will*. For he states that there is a central sense, calling it the internal sense which the external senses together serve. Again we know from Avicenna in his commentary *On the Soul* that the central sense is a power located in the front part of the brain and which receives the impressions flowing into it from the five external senses.

That there is a sense of this kind, it seems, can be shown as follows.

238. Through sight we cannot perceive ourselves to see because whatever is seen is seen by means of an image, and sight does not have an image of itself in the sensation of seeing. Similarly, by touch we do not perceive ourselves to touch, and the same is the case with regard to the other external senses. Since, therefore, we do perceive ourselves to see and to touch, this is through some other power than any power of the external senses. But from experience, as is clear in anatomy, we know that from all of the external senses impressions flow into the front part of the brain, and the soul, in directing itself towards those impressions, joins and links those impressions with those things from which they arise. However, since these impressions have been caused by those which are in the external senses, the eye of the mind in turning outwards towards them perceives an impression through which that which it finds in the organ of touch is seen. Yet that power which is located in the front part of the brain is the central sense, and by means of that power we perceive ourselves to see, to hear, and to touch. Therefore it follows that there is a central sense.

239. Moreover, we know from Aristotle in the book *On the Soul* that movement, number, and size are not sensed by any external sense, but rather are sensed by the internal sense, which is the central sense.

240. Preterea. Ab Avicenna habetur quod nihil videtur nisi ubi ipsum
est. Ergo gutta pluvie descendens non videtur nisi ubi ipsa est. Sed ipsa
est rotunda; ergo secundum visum non videtur linearis, quia si videretur esse linearis tunc aliqua eius pars videretur esse ubi ipsa non est:
apparet tamen quod ipsa sit linearis. Est ergo hoc secundum aliam vim 5
quam secundum vim visibilem. Est autem illa vis, secundum quam
apparet illa gutta esse linearis, sensus communis; quoniam ab illa
passione que fit in oculo representante guttam extra fit inpressio in
anteriori parte cerebri; et sicut illa gutta est descendens linealiter, ita
inmutatio que fit in anteriori parte cerebri est linearis. Retinet enim 10
anterior pars cerebri inmutationem factam in eius parte superiori, et
inmutationem continue fluentem descendendo per ipsam sicut gutta
descendit, et acies animi convertens se ad illam inmutationem, quia
reperit illam linearem, ideo iudicat illud lineare esse, scilicet guttam
cuius ymago est illa inpressio que est in anteriori parte cerebri; et illa 15
vis dicitur esse sensus communis.

241. Si quis autem obiciat quod similiter erit apud visum, Dicendum
est quod non. Quoniam subiectum visus est inhabile ad retinendum
inpressionem propter cristallinum humorem aqueum, qui receptam
inpressionem non retinet.   20

242. Si concedatur quod sensus communis sit, obiectio erit in contrarium. Queritur ergo quare non habeamus intellectum communem
et ymaginationem communem sicut habemus sensum communem?

243. Item. Quod commune est potest predicari de hiis quibus ipsum
est commune: ut animal quod est commune suis speciebus predicatur 25
de qualibet suarum specierum; vere enim dicitur: homo est animal,
asinus est animal. Ergo cum sensus communis sit sic communis quinque sensibus exterioribus habet predicari de unoquoque illorum. Erit
ergo quelibet harum propositionum vera: visus est sensus communis;
auditus est sensus communis; tactus est sensus communis, et odoratus. 30
Sed constat quod hoc est falsum. Videtur ergo quod nihil sit sensus
communis.

---

1 habetur *om.* V   2 ipsa: ipsum P   3 videretur: videre P   4 aliqua
om. P   5 quod ipsa sit *om.* C   Est *om.* V   hoc *om.* C   6 visibilem:
sensibilem C   7 esse *om.* C   9–10 et sicut . . . cerebri *om. per hom.* C
9 est: que est C   14 reperit: recipit V   ideo: non P   18 Quoniam: Quia
P   inhabile: habile P   19 inpressionem: inpressiones P   aqueum, qui: aqua P
21 Si: ergo *add.* C   21–22 obiectio . . . contrarium *om.* C   22 ergo *om.* C
24 potest: habet P   25 animal *om.* C   26 enim: autem C   animal: et
add. V   27 sic *om.* P   29 harum: illarum P   propositionum *om.* CV
30 et odoratus *om.* P; et sic de aliis sensibus *add.* V   31 ergo: enim P

1–2 ubi ipsum est: Avic. *De An.* I. v (5^{rb}D).
2–5 Avic. loc. cit.; III. vii (15^{rb}A).

240. Moreover, from Avicenna we know that nothing is seen except where it is. Therefore a falling drop of rain is only seen where it is. Yet a raindrop is round and therefore according to sight it will not be seen to be like a line, since if it were seen to be like a line then some part of it would be seen to be where it is not. Yet it appears that it is like a line. Therefore, this happens according to a power other than the power of seeing. That power, according to which that drop appears to be like a line, is, however, the central sense. Since from that modification which happens in the eye, and which represents the raindrop outside, an impression arises in the front part of the brain, and just as the raindrop falls down like a line, so the impression which occurs in the front part of the brain is like a line. For the front part of the brain retains the impression made in its upper part, and the impression continually flowing falls through it just as the raindrop falls. Moreover, the eye of the mind in turning itself towards that impression, because it finds the impression to be like a line, therefore judges the thing to be like a line—that is, the raindrop whose image is the impression which is in the front part of the brain. That power is said to be the central sense.

241. If, however, someone objects that the case of sight is similar, it should be stated that it is not. This is because the subject of sight is incapable of retaining the impression because of the watery crystalline humour, which having received the impression does not retain it.

242. If it is conceded that there is a central sense, there will be an objection to the contrary. Therefore, it is asked why we do not have a common intellect and a common imagination just as we have a common or central sense.

243. What is common can be predicated of those of which it is common. For example, 'animal', which is common to its species, is predicated of any of its species. Thus, it can be truly said: 'A man is an animal', 'An ass is an animal'. Therefore, since the central sense is common in this way to the five external senses, it can be predicated of each of them. Therefore each of these propositions is true: 'sight is a common sense', 'hearing is a common sense', 'touch is a common sense', and 'smell is a common sense'. Yet it is clear that this is false. Therefore, it seems that the central sense is nothing.

244. Solutio. Dicendum est secundum Aristotelem in libro *de Anima,*
et secundum alios philosophos, sensum communem esse. Et dicitur
esse communis eo quod ipse communiter recipit inpressionem a quinque
sensibus extra. Non autem dicitur esse communis secundum illam viam
communitatis secundum quam genera et species dicuntur esse com- 5
munia, sed secundum illam viam communitatis secundum quam aliquis
alveus recipit aquam fluentem a diversis fontibus per diversos rivulos.
Nec est ponendum propter hoc esse communem intellectum vel communem ymaginationem. Non enim sunt diversa instrumenta subser-
vientia intellectui, nec sunt diverse vires secundum quas fit receptio 10
formarum intelligibilium, nec sunt diverse vires secundum quas est
receptio inpressionum a rebus ymaginabilibus; quia, cum unica sit
cellula ymaginativa, in ea est unica vis ordinata secundum quam est
receptio inpressionum fluentium in ipsam ab unico, hoc est, a sensu
communi. Similiter intellectus ab unico suam recipit inpressionem. Sed 15
ille sensus qui dicitur esse communis a pluribus recipit inpressiones, tum
a visu tum a tactu tum a gustu, et ita sensus exteriores diversi sunt et
subservientes communiter isti sensui qui dicitur esse communis.

245. Item. Obicitur sic. In qua proportione se habet species illa que
fit in sensu cernentis a specie corporis extra ad illam speciem extra, in 20
eadem proportione se habet passio que exoritur in sensu communi ad
illam speciem que est in sensu cernentis: quoniam sicut illa que est in
visu est quasi proles, et eius parens est species corporis extra, ita et illa
que est in sensu communi est quasi proles, et illa que est in sensu
cernentis est quasi eius parens. Sed ita se habet illa passio que est in 25
visu, quod voluntas, convertens aciem animi ad passionem que est in
sensu videntis, coniungit et copulat illam ymaginem cum sua parente
extra, scilicet cum specie corporis mediante eius ymagine existente in
oculo, et sic fit apprehensio per visum. Ergo se ita debet habere illa
impressio que est in sensu communi ad illam formam que est in sensu 30

2 secundum *om.* C   4 autem *om.* V   illam *om.* V   5 secundum quam:
quod P   6–7 secundum² . . . alveus *om.* V   7 alveus: communis qui *add.* C
8 hoc *om.* V   9 sunt *om.* P   10 intellectui: intellectum C   10–11 fit
. . . quas *om. per hom.* C   12 ymaginabilibus: ymaginibus P   unica: una C
15 unico: uno V   16 dicitur esse: est C   a *om.* C   inpressiones: inpressionem P
17 ita: isti P   diversi: diversa C   18 isti: illi PV   esse *om.* P   19 sic
om. C   20 ad . . . extra: ad formam corporis extra V; *om. per hom.* C
22 speciem *om.* P   in *om.* C   23 quasi: sicut C   extra: erit
P   25 Sed: si C   26 voluntas: anime *add.* V   27 videntis: et *add.* V
28–29 mediante . . . oculo *om.* CP   29 se: si P   30 communi: que est *add.* C

1–2 secundum Aristotelem et alios philosophos: vid. supra, nn. 232, 237.
4–6 Cf. supra, n. 243.
8–9 Cf. supra, n. 242.
17–18 Cf. Avic. *De An.* I. v (6ʳᵃG).
23 proles, parens: cf. Aug. *De Trin.* XI. vii. 11–12.

244. Solution: It should be stated in accordance with Aristotle in the book *On the Soul* and according to other philosophers that there is a central sense. And it is said to be 'common' because it receives impressions in common from the five external senses. It is not, however, said to be common according to that kind of commonality according to which genera and species are said to be common, but according to that kind of commonality according to which a trough receives water flowing from different springs through different streams. Nor because of this should a common intellect or a common imagination be posited. For there are not different organs serving the intellect, nor are there different powers according to which the reception of the intelligible forms is received. Nor are these different powers according to which there is the reception of impressions from imaginable things. Since there is only one seat of imagination, there is only one power located there, according to which there is the reception of impressions flowing into it from one thing only, and this is from the central sense. Similarly the intellect receives its impression from one thing only. Yet that sense which is said to be 'common' receives its impressions from many, both from sight and from touch and taste. Thus, the external senses are diverse and together serve that sense which is said to be common.

245. Again, it is objected as follows: to the extent to which a species is constituted which arises in the sense of distinguishing from the species of an external body, to the same extent is the modification constituted which arises in the central sense in respect of that species which is in the sense of distinguishing. For just as that which is in sight is, as it were, an offspring, and its parent is the species of the external body, so also that which is in the central sense is like an offspring and that which is in the sense of distinguishing is like its parent. But that modification which is in sight is constituted in the same way because the will, in turning the eye of the mind towards the impression which is in the sense of seeing, joins and links that image with its parent outside, namely with the bodily species by means of its image existing in the eye, and in this way perception through sight occurs. Therefore, that impression should be constituted likewise which is in the central sense in respect of that form which is in the sense of seeing because

videntis, quod voluntas convertens aciem animi ad illam ymaginem que
est in sensu communi coniungendo et copulando eam cum ea que est
eius parens in oculo existens, habet apprehendere illam que est in oculo,
et secundum eandem viam similitudinis currit res inter sensum communem et inter alios sensus exteriores. Ergo mediante sensu communi 5
non fit apprehensio nisi illarum passionum que sunt in sensibus exterioribus; quod patet esse falsum: quoniam per sensum communem fit
apprehensio formarum extra que generant suas passiones in sensibus
exterioribus.

246. Item. Ut habetur ab Avicenna in commento, alia est vis secun- 10
dum quam est species impressionis receptio, et alia est vis secundum
quam est impressionis recepte retentio, ut patet in aqua est vis receptiva
impressionis, sed non vis impressionis recepte retentiva. Similiter in
oculo est vis receptiva ymaginis, sed non retentiva; in ymaginatione
autem est vis receptiva impressionis fluentis ab ea que est in sensu 15
communi, et licet pereat illa que est in sensu communi, tamen adhuc
remanet illa que est in ymaginatione et formata a sensu communi. Ipsa
enim constituta in ymaginatione stabilimentum sue permanentie ibi
habet a vi retentiva. Est ergo in ymaginatione duplex vis corporalis:
una que est receptiva similitudinis corporis extra, et alia vis que est 20
formate similitudinis retentiva, et utraque istarum virium subservit apprehensioni facte per vim ymaginativam. Qua ratione ergo distinguitur
ymaginatio secundum vim receptivam debet ipsa distingui secundum
vim retentivam.

247. Similiter potest ostendi quod duplex est vis ipsius memorie, et 25
ita plures essent virtutes quam distinguantur, et non esset ymaginatio
una sola vis sed plures. Similiter et memoria.

248. Solutio. Ad primum. Dicendum est quod voluntas anime convertens aciem animi ad similitudinem corporis, que est in sensu communi, et coniungens eam cum parente sua, que est in sensu exteriori, 30
non ibi sistit suam apprehensionem, immo cum illa parens sit proles
speciei extra, per illam convertit aciem animi ad speciem extra, et propter

---

1 convertens: convertans P   ymaginem: ymaginationem V   3 parens: que
est *add*. V   illam: eam V   4 currit: erit C   res: ens V   11 impressionis
receptio: receptio impressio C   15 autem *om*. P   18 constituta: instituta CV
19 corporalis: et *add*. V   20–21 corporis . . . retentiva *om*. V   20 extra *om*. C
21 istarum: illarum P   21–22 apprehensioni: apprehensionem V   24 retentivam: receptivam CV   26 distinguantur: distinguuntur V   esset: est C
27 una *om*. P   et *om*. C   30 sensu: communi *add*. V   31 apprehensionem:
impressionem P

---

10 in commento: i.e. Avic. *De An.* I. v (5^{rb}D).
28 Ad primum: cf. supra, n. 245.

the will, which directs the eye of the mind towards that image which is in the central sense, by joining it and linking it with that which is its parent existing in the eye, has the role of perceiving that likeness which is in the eye, and according to that way of similarity the thing runs between the central sense and between the other external senses. Therefore, by means of the central sense a perception does not take place except of those impressions which are in the external senses. This is clearly false, because by means of the central sense the perception of the forms outside occurs which give rise to its impressions in the external senses.

246. Again, as we know from Avicenna in the Commentary, the power according to which there is the reception of the species of the impression is different from the power according to which there is a retention of the species of the impression, as clearly there is a receptive power of the impression in water, but there is not a retentive power of the impression received. Similarly in the eye there is a receptive power of the image, but not a retentive power. In the imagination, however, there is a receptive power of the impression flowing from that which is in the central sense, and even if that which is in the central sense perishes, yet that which is in the imagination and which has been formed by the central sense still remains. For this likeness which is constituted in the imagination has the basis of its permanence there from the retentive power. Therefore, there is a twofold corporeal power in the imagination: one which is receptive of a likeness of the external body and another power which is retentive of the likeness formed, and both of these powers serve the perception made by means of the imaginative power. Therefore, because the imagination is subdivided according to the receptive power, it should also be subdivided according to the retentive power.

247. In a similar way it can be shown that the power of memory itself is twofold. Thus there would be many powers that are subdivided and the imagination would not be a single power but many, and the same is true of memory also.

248. Solution: With regard to the first question, it should be stated that the will of the soul which directs the eye of the mind towards the likeness of a body which is in the central sense, and which it joins with its begetter which is in the external sense, does not cease its apprehension there. Rather since it is the begetter of an offspring of the external species, by means of it it turns the eye of the mind to the ex-

vehementem similitudinem que est inter ymaginem et ymaginatum non
distinguit anima inter ea; et cum ymaginatio discurrat non tantum circa
presentia, immo circa absentia, anima inveniens similitudinem corporis
in ymagine copulat eam cum illa quam prius habuit in sensu, et mediante
illa cum corpore quod prius vidit.   5

249. Ad aliud. Dicendum est quod, ut ostensum est, diverse sunt
vires, vis receptiva et vis retentiva, et utraque illarum virium subservit
ymaginationi. Et sunt ille due virtutes corporales; nec ponendum est
illas duas virtutes esse de ymaginatione, nisi dicatur quod ymaginatio sit
vis corporis et non vis anime.   10

XVIII. *De ymaginatione*

250. Sequitur de ymaginatione. Unde primo loco videndum est quid
sit ymaginatio et qualiter distinguatur ymaginatio ab aliis viribus.
Ymaginatio est vis ordinata in anterioris partis cerebri extremitate,
scilicet in prima concavitate, retinens impressiones receptas a sensu 15
communi. Unde quia vis ymaginativa retentiva est passionum receptarum a sensu communi, que sunt similitudines rerum singularium, per
ymaginationem rerum absentium potest fieri apprehensio; sensus autem
communis, cum de sua aptitudine non possit retinere ymagines singularium receptas a quinque sensibus exterioribus, non potest per eum 20
fieri apprehensio rerum absentium, immo presentium tantum.

251. Sed videtur secundum hoc posse ostendi quod ymaginatio sit
idem quod memoria, vel quod frustra sit ymaginatio animali data.
Quoniam per memoriam potest idem apprehendi, et eodem modo quod
apprehenditur per ymaginationem. Quoniam ymaginatio apprehendit 25
res prius existentes in sensu, et memorie est memorari preteritorum;
et ita memorie est apprehendere res secundum quod ipse prius fuerunt
existentes in sensu. Vel ergo ymaginatio est idem quod memoria, vel

2 et *om.* V   4 ymagine: ymaginem P; ymaginatione V   6 est *om.* P
9 ymaginatione: ymagine P   nisi: sibi V   dicatur: dicitur V   12 Sequitur:
Consequenter agendum est V   primo loco: primum P   14 Ymaginatio: dupliciter
accipitur apud auctores: uno modo sic: ymaginatio est vis apprehensiva rerum significatarum per singulas dictiones; alio modo, prout hic accipitur ymaginatio *add.* V
15 retinens: primas *add.* P   16–17 Unde ... communi *om. per hom.* V
17 singularium: extra *add.* P   18 ymaginationem: vim ymaginativam P   rerum:
singularium *add.* P   absentium: apud eum. Iterum absentium *add.* V   19 sua
om. V   22 sit: est P   23 ymaginatio: memoria C; ymago P   animali
om. V   24–25 quod apprehenditur: sicut P   27 ipse *om.* P   28 existentes *om.* C

6 Ad aliud, ut ostensum est: cf. supra, n. 246.
14–17 Cf. Avic. *De An.* i. v (5^rbD).
26–27 Cf. Arist. *De Memoria* i (449^b15–30). Cf. infra, cap. XX.

ternal species and, because of the strength of the likeness between the image and the thing imagined, the soul does not distinguish between the two of them. Since the imagination deals with things which are not present as such but, on the contrary, are absent, when the soul comes across a likeness of the body in an image, it links it with that which it previously had in sensation, and by means of this with the body which it had previously seen.

249. Regarding the other objection, it should be stated, as has been shown, that there are distinct powers, a receptive and a retentive power, and both of these powers serve the imagination. And these two powers are corporeal; nor should it be posited that these two powers are of the imagination, unless it is stated that the imagination is a power of the body and not a power of the soul.

XVIII. *Concerning imagination*

250. What follows next concerns the imagination, and first of all we should examine what the imaging faculty is, and how the imagination is distinguished from the other powers. The imagination is a power located in the furthest extremity of the front part of the brain, namely in the first ventricle, which retains the impressions received from the central sense. So, because the imaging power is retentive of the impressions received from the central sense, which are the likenesses of individual things, by means of the imagination it is possible for the perception of absent things to occur. The central sense, however, because of its disposition cannot retain the images of singular things received from the five external senses; it cannot apprehend absent things but only things which are present.

251. However, it seems that according to this it can be shown that the imagination is the same as memory, or that the imagination is given to an animal for nothing because through memory the same thing can be perceived, and in the same way that it is perceived through imagination. For the imagination perceives a thing which previously existed in sensation, and what memory does is to remember things of the past. Thus what memory does is to perceive things inasmuch as these existed previously in sensation. Therefore either imagination is the same as memory or imagination is worthless. This

inanis est ymaginatio, quod constat esse falsum: nihil enim sine causa quod est a creatore rebus est insitum.

252. Solutio. Dicendum est quod virium apprehendentium quedam apprehendunt et operantur simul, ut habetur in commento *de Anima*, quedam apprehendunt et non operantur simul. Apprehendunt et ope- 5 rantur simul, ut vis estimativa et vis memorialis, que apprehendunt res et componunt eas ad invicem vel dividunt. Per memoriam enim fit apprehensio rerum et recordatio quod prius eedem res fuerunt apprehense; unde in ipsa recordatione est compositio; per vim autem ymaginativam apprehenditur solummodo res cuius ymago describitur in 10 subiecto illius virtutis, sed non componendo vel dividendo aliquam passionem in aliquo subiecto fit per eam apprehensio. Unde vis ymaginativa apprehendit simpliciter rem non simul operando.

253. Si quis autem obiciat de hoc quod ymaginatio intuetur rem absentem et sensatam prius, et ita apprehendit rem ut videtur percipi- 15 endo quod ipsa fuit prius in sensu; ergo simul operatur cum sua apprehensione, quoniam rem ymaginatam prius fuisse in sensu quedam compositio est. Dicendum est ad hoc quod licet ymaginatio intueatur rem secundum quod ipsa fuit prius sensata, non tamen est per ipsam perceptio quod res illa prius fuerit in sensu. Illud enim prius in ymagi- 20 nationem non cadit, sed in memoriam.

XIX. *De estimatione*

254. Sequitur de estimatione. Unde videndum est quid sit estimatio et ad quid sit. Ab Avicenna habetur quod estimatio est vis ordinata in media concavitate cerebri ad apprehendendum intentiones non sensatas 25 que sunt in rebus singularibus et sensibus, diiudicans utrum res sit fugienda propter intentionem si ipsa intentio sit nocitiva, vel appetenda propter intentionem si ipsa sit expediens: ut vis que est in ove diiudicans

---

1 sine causa *om.* V   2 quod: que C   quod est *om.* P   3 apprehendentium: apprehensivum P   quedam: que V   4 simul *om.* CV   5 apprehendunt ... simul: non P   6 estimativa: existimativa P   7 componunt: opponunt CV   vel: et P   8 eedem: heedem C   12 in: ab CV   14 de hoc *om.* P 15 sensatam: sensitiva P   17 fuisse: esse P   18 est *om.* P   ymaginatio *om.* P 19 fuit: est P   sensata: suscepta P   20 perceptio: recordatio V   fuerit: fuit C 21 in memoriam: memoria P   23 Sequitur: Restat ut dicamus V   24 sit *om.* P   habetur: habemus P   25 ad: et P   sensatas: sensitivas PV   27 si: sed P   nocitiva: nosciva P   appetenda: appetitiva V

1-2 Arist. *De Caelo* I. iv (271ᵃ34): 'Deus et natura nihil frustra faciunt'; *De An.* III. ix (432ᵇ21); xii (434ᵃ31); Avic. *De An.* v. iv (24ᵛᵇA).
3 Dicendum: cf. supra, n. 251.
4 in commento *de Anima*: i.e. Avicennae, I. v (5ʳᵃC).
20-21 Cf. supra, ad n. 251.
24 Avic. *De An.* ibid. (5ʳᵇE).

can be seen to be false, for nothing which has been placed in things by the Creator is without a reason.

252. Solution: It should be stated that of the perceiving powers some perceive and act at the same time, as is held in the commentary *On the Soul*, and some perceive and do not act at the same time. Those which perceive and operate at the same time, such as the estimative power and the memorizing power, are those which perceive things and either link them to each other or separate them from each other. For by means of memory the perception of things occurs and the recollection that these same things were perceived. Thus in this recollection there is a linking together. However, by means of the imaginative power only the thing is perceived whose image is delineated in the subject of that power, but no linking together or separating of any property in any subject arises by means of this perception. Thus, the imaginative power simply perceives the thing but does not operate at the same time.

253. If, however, someone were to object regarding this that the imagination intuits an absent thing which was previously sensed, and thus it apprehends the thing, it would appear, by perceiving that it was first of all in sensation, that it therefore operates at the same time with its apprehension, because a thing imagined to have been in sensation first of all is a kind of composition. To this it should be stated that, although the imagination intuits a thing inasmuch as it was previously sensed, the perception that that thing was first in sensation does not happen through the imagination. For 'before' does not fall under the imagination but rather under memory.

XIX. *Concerning estimation*

254. Next concerning estimation, and so we should examine what estimation is and what it is for. From Avicenna we know that estimation is a power placed in the middle ventricle of the brain in order to perceive non-sensed intentions which are in individual and sensed things. It judges whether a thing should be avoided because of an intention if this intention is harmful, or is to be desired because of an intention if this is useful. An example is the power which is in a sheep

quod ab hoc lupo est fugiendum, et quod huius agni, qui est agnus ipsius ovis, est miserendum. Intentionem appellat Commentator qualitatem singularem non cadentem in sensum, que est vel rei nocitiva vel expediens. Nocitiva, ut illa proprietas que est in lupo propter quam ovis fugit lupum; expediens, ut illa proprietas que est in ove propter quam 5 eam appetit agnus.

255. Sed obicitur. Ymaginatio nihil apprehendit nisi secundum ymaginem quam recipit a sensu communi; similiter sensus communis nihil intuetur nisi secundum similitudinem rei extra formatam in ipso a sensu exteriori. Pari ratione ergo cum estimatio sit vis ordinata post 10 ymaginationem nihil formabitur in instrumento estimationis nisi ab impressione prius formata in ymaginatione, et quicquid est in ymaginatione fit a sensu. Ergo nihil accipitur ab estimatione nisi illud quod prius apprehendebat sensus. Ergo estimatio non intuetur intentionem nisi intuitionis similitudo prius fuerit constituta in sensu. Ergo intentio 15 est res cadens in sensum.

256. Preterea. Cum lupus sit res separata ab ove, qualiter constituetur in estimatione similitudo intentionis existentis in lupo, nisi prius fuerit inmutatio formata ab intentione existente in lupo in sensu ovis, cum sensus sit medium inter sensatum et estimationem? Qualiter 20 enim posset ignis remotus ab homine calefacere hominem nisi aer in medio reciperet calorem a caliditate ignis?

257. Solutio. Dicendum est quod intentio est res accepta ab estimatione non cadens in sensum ut mediante sensu apprehendatur ab anima, ita quod non exigatur vis alia ad intentionis apprehensionem, nec est 25 eius ymago in sensu vel in ymaginatione; sed fit in estimatione ymago intentionis [per] apprehensionem, nec est eius ymago in sensu vel ymaginatione, sed fit in estimatione ymago intentionis non existente aliqua similitudine intentionis in aliquo eorum que sunt inter instrumentum estimationis et subiectum intentionis, sicut superius dictum est de visu, 30

---

2 ipsius: huius CV   miserendum: miserandum P   3 que: qui P   nocitiva: nosciva *et deinceps* P   5 lupum: eum P   est: inest C   7 apprehendit: apprehenditur P   8 recipit *om.* P   9 extra *om.* C   11 formabitur: formatur V   in *om.* P   ab *om.* V   12 impressione: estimatione P   13-15 Ergo . . . sensu *om. per hom.* CV   17 res *om.* P   18 estimatione: ovis *add.* V   similitudo: sumendo P   19 intentione: intuitione *et sic deinceps* P   existente in lupo *om.* P   20 cum . . . estimationem: et eius ymagine sicut P   21 enim *om.* P   posset: possit V   ab homine *om.* P   25 ita quod non exigatur: quia nec exigitur C   ad intentionis apprehensionem *om.* P   26 in² *om.* C   27-29 [per] apprehensionem . . . intentionis *om.* PV   29 inter: in V

---

2 Commentator: i.e. Avicenna ibid. (5^{ra}c).
7 Cf. supra, n. 250.
8-10 Cf. supra, n. 237.
30 superius dictum est de visu: cf. supra, n. 56.

which judges that it should flee from this wolf, and that this lamb, which is the lamb of this sheep, should be looked after. The Commentator (Avicenna) calls an intention an individual quality which is not picked up by sensation, which is either harmful or useful to a thing. Harmful, such as that quality which is in a wolf and because of which the sheep flees from it; useful, such as that property which is in the sheep and because of which the lamb approaches it.

255. But it is objected that imagination perceives nothing except according to the image which it receives from the central sense. Similarly the central sense intuits nothing except according to the likeness formed in itself of the external thing by the external sense. Therefore, for the same reason, since estimation is a power which comes after imagination, nothing will be formed in the organ of estimation except from the impression formed first of all in the imagination, and whatever is in the imagination has arisen from sensation. Therefore estimation does not intuit an intention unless a likeness of the intention was first of all constituted in sensation. Therefore, an intention is a thing which comes under sensation.

256. Moreover, since the wolf is a separate thing from the sheep, in what way is a likeness of the intention which exists in the wolf constituted in the estimation unless there was first of all an impression formed in the sensation of the sheep by the intention which exists in the wolf, since sensation is a medium between the thing sensed and the estimation? For in what way can a fire which is far off from a man heat the man, unless the air in between receives heat from the heat of the fire?

257. Solution: It should be stated that an intention is a thing received by the estimation and which does not fall under sensation so that it would be perceived by the soul by means of sensation. So it is that it does not require another power in order to perceive the intention, nor is its image in sensation or in imagination. However, an image of the intention arises in estimation by means of perception, nor is its image in sensation or imagination, but an image of the intention occurs in the estimation without there existing any likeness of the intention in any of those which are between the organ of estimation and the subject of the intention, as was said above concerning sight, that

quod ymago rei vise sit in visu, non tamen in aere intermedio est consimilis ymago. Sed quia illud alicui videbitur difficile ad intelligendum
potest dici quod similitudo intentionis fit in sensu et in ymaginatione, sed
anima secundum eas non apprehendit, quoniam sensus et ymaginatio non
sunt nature concordantis cum proprio subiecto intentionis. Sed instru- 5
mentum estimationis est consimilis nature cum eo quod est per se et
proprie subiectum intentionis, et ideo secundum vim estimativam fit
apprehensio intentionis.

258. Item. Ut dictum est superius, per estimationem tum fit compositio tum fit divisio. Sed non est secundum estimationem componere 10
aliqua nisi prius apprehendantur extremitates illius compositionis. Ergo
cum ovis secundum estimationem componit in eius anima hoc, scilicet
lupum esse fugiendum, prius apprehendit hoc ipsum 'fugere' et rem
designatam per hunc terminum 'lupum'. Sed hoc ipsum 'fugere' est
universale, et ille terminus 'lupum' significat universale. Sic ergo per 15
estimationem apprehenduntur universalia. Possunt ergo bruta animalia
apprehendere universalia.

259. Preterea. Circa compositionem vel divisionem estimationis est
veritas vel falsitas. Ergo cum verum sit lupum esse fugiendum ab ove,
secundum vim estimationis potest percipi illud esse verum; pari ratione 20
et eius compositum esse falsum. Sic ergo cum per vim estimativam
possit perpendi quid sit verum et quid sit falsum, bruta animalia possunt discernere verum a falso; possunt ergo bruta animalia uti mutuo
disputationibus.

260. Solutio. Ad primum. Dicendum est quod universalia non pos- 25
sunt apprehendi a brutis animalibus. Estimatio autem non apprehendit
nisi singularia; unde secundum vim estimativam non apprehenditur
quod a lupo sit fugiendum, sed apprehenditur quod ab hoc lupo sit
fugiendum, qui est in sensu, vel prius fuit in sensu; et cum per hunc
terminum, 'hoc lupo', significetur singulare, illud quod significatur 30
ulterius per hunc terminum 'fugiendum' trahitur ad singulare per hanc
determinationem 'ab hoc lupo'.

---

1 quod... in visu: fit ymago rei vise C   9 per: secundum P   10 fit *om.* P
14 ipsum: nomen P   15 ille: iste V   significat: fugit P   Sic *om.* C   18 vel:
et V   estimationis *om.* P   19 vel: et P   20–21 estimationis... vim *om. per hom.* P
22 sit² *om.* CV   23 ergo *om.* V   26 animalibus *om.* P   autem: etiam P
27 apprehenditur: apprehendit visio V   28 quod¹ ... apprehenditur *om.* V
apprehenditur *om.* P   29 vel ... sensu *om.* C   30 illud: id P   significatur:
significamus C   31 terminum *om.* C   fugiendum: fugitur P

---

9 superius: cf. supra, n. 252.
25 Ad primum: cf. supra, n. 258.
27 Cf. Avic. *De An.* IV. i (17^{va-b}B).

the image of the thing seen is in sight, but there is not something similar to the image in the intervening air. However, because this might
seem to someone to be difficult to understand, it can be said that a
likeness of the intention occurs in sensation and in imagination, but
the soul does not perceive according to them, because sensation and
imagination are not by nature in agreement with the proper subject
of the intention. Yet the organ of estimation is of a similar nature to
that which is in itself and properly the subject of an intention, and
thus according to the estimative power the perception of an intention
occurs.

258. Again, as has been said above, both composition and division
happen by means of estimation. However, it is not in keeping with
estimation to link some things together unless the extremities of that
composition have already been perceived. Therefore, when the sheep
according to estimation links this together in its soul, namely to flee
from the wolf, it first of all perceives this 'to flee from' and then the
thing designated by this term 'wolf'. Yet this same 'to flee from' is a
universal, and that term 'wolf' signifies a universal. Therefore in this
way universals are perceived by means of estimation. Therefore brute
animals perceive universals.

259. Moreover, regarding the composition or division of estimation
there is either truth or falsity. Therefore, since it is true that the sheep
should flee from the wolf, according to the power of estimation it can
perceive that it is true and, for the same reason, that its composition
could be perceived to be false. Therefore, since by means of the estimative power what is true and what is false can be inferred, brute
animals can discern the true from the false; therefore brute animals
can reciprocally make use of reasonings.

260. Solution: With regard to the first question, it should be stated
that universals cannot be apprehended by brute animals. For estimation only perceives individuals. Thus, according to the estimative
power it is not perceived that a wolf should be fled from, but rather
it is perceived that this wolf should be fled from, which is a sensation, or which was first of all in a sensation. And since by means of
this term 'this wolf' an individual is signified, that which is later signified by this term 'to be fled from' is linked to the individual by this
determination 'from this wolf'.

261. Ad aliud. Dicendum quod licet in vi estimativa sit compositio vel divisio et circa illam sit veritas vel falsitas, non tamen per vim estimativam percipitur ibi esse veritas vel falsitas, sed per intellectum tantum et per rationem. Unde a brutis, etsi percipitur id in quo est veritas vel falsitas, non tamen percipitur ab hiis verum secundum quod verum, 5 vel falsum secundum quod falsum. Non enim apprehendunt veritatem vel falsitatem, cum intellectu careant et ratione.

xx. *De memoria*

262. Sequitur de memoria. Unde primo videndum est quid sit memoria. Memoria est vis ordinata in posteriori parte concavitatis cerebri reci- 10 piens impressiones ab estimatione et eas retinens. Et quia in memoria reponuntur ymagines, voluntas convertens aciem animi ad memoriam invenit ibi ymagines per quas intuetur res prius apprehensas: unde memoria est ad memorandum prius apprehensa.

263. Sed obicitur. Memoria recolit per ymaginem repositam in suo 15 instrumento res quas prius intuebatur et recolit se prius eas inspexisse. Sed huiusmodi memorari non potest nisi ita sit quod ipsa apprehendat hoc ipsum prius: sed quicquid apprehendit per ymaginem apprehendit. Ergo in memoria est ymago prioritatis.

264. Preterea. Audito nomine prioritatis non apprehenderetur prio- 20 ritas nisi agnosceretur res que illo nomine significatur; sed non agnosceretur nisi memoria eam retineret. Relinquitur ergo quod ymago prioritatis est in memoria, cum ipsa meminit se illud prius percepisse. Sed nihil imprimit ymaginem in memoria nisi quando ipsa res presens adest. Si ergo in memoria est ymago prioritatis aliquando affuit prioritas 25 in presentia sui ei imponens ymaginem. Ergo tunc potuit vere dici prioritas est, quod est falsum.

265. Preterea. In memoria habet exoriri species ab estimatione, cum estimatio sit proxima vis eius ante. Oportet ergo, cum similitudo

---

1 dicendum: est *add.* V   2 vim: viam P   3 ibi *om.* V   4 per *om.*
P   etsi *om.* P   in quo: quod P   5 hiis: ipso P; ipsis V   5–6 secundum
quod falsum *om.* V   6 enim: tamen P   7 et ratione *om.* V   9 Sequitur:
Dicto de estimatione, dicendum est *add.* V   primo: primum P   10 concavitatis:
concavitate P   12 animi *om.* C   13–14 unde . . . apprehensa *om.* P
17 huiusmodi memorari: hoc reminisci P   ipsa *om.* PV   21 agnosceretur:
acgnoscentur V , que: in *add.* V   23 illud *om.* P; rem V   prius: re *add.* P
26 ymaginem: ymaginationem P; suam *add.* V   potuit: ponit P   28 exoriri:
exori C   29 eius: ei V   cum similitudo: quod in similitudo P

---

1 Ad aliud: cf. supra, n. 259.
10 Avic. *De An.* I. v (5^{rb}E).
12–13 Cf. Arist. *De Memoria* i (449^{b}10–29).
15 Cf. supra, n. 262: Avic. *De An.* IV. i (17^{vb}C).
29 proxima vis eius ante: cf. Avic. *De An.* I. v (5^{rb}E).

261. With regard to the other question, it should be said that although there is composition or division in the estimative power, and regarding that there is truth and falsity, yet it is not by means of the estimative power that truth or falsity is perceived to be there, but only by means of the intellect and by means of reason. Thus, even if brute animals perceive that in which there is either truth or falsity, yet the true inasmuch as it is true is not perceived by them, nor the false inasmuch as it is false, for they do not perceive truth or falsity since they lack understanding and reason.

xx. *Concerning memory*

262. Next concerning memory, and firstly we should examine what memory is. Memory is a power located in the rear ventricle of the brain which receives impressions from estimation and retains them. And because images are deposited in memory, when the will turns the eye of the mind towards memory it finds images there by means of which it intuits things which have previously been perceived. Thus memory serves in order to remember things previously perceived.

263. But it is objected that memory recalls things which it has previously intuited by means of an image preserved in its organ, and recalls that it has examined them before. Yet it cannot remember these kinds of thing unless it is the case that it perceives this 'before'. However, whatever it perceives it perceives by means of an image. Therefore, there is an image of 'before' in memory.

264. Moreover, when the word 'before' is heard, previousness will not be apprehended unless the thing which is signified by that name is recognized. However, it will not be recognized unless memory retains it. It follows, therefore, that there is an image of 'previousness' in memory, since it remembers itself to have perceived it previously. However, nothing impresses an image in memory except when this thing is present. If, therefore, there is an image of 'previousness' in memory, at some time there was 'previousness' in its presence and impressed an image on it. Therefore, it can then be truly said that 'previousness' exists—which is false.

265. Moreover, a species in memory arises from estimation, since estimation is the nearest power which comes before it. Therefore it is

prioritatis sit in memoria, quod eius ymago prius fuerit in estimatione. Potest ergo estimatio intueri hoc ipsum prius et ita memorari. Ad quid ergo erit memoria, cum estimatio vel ymaginatio posset facere quicquid facit memoria?

266. Item. Videtur quod cognitio rei antecedit memoriam eiusdem 5 rei, quoniam non est memoria nisi eorum que prius ceciderunt in animam et quorum ymagines reposite sunt in anima. Sed si reposite sunt in anima, quando ibi reponebantur, novit anima res illas per suas ymagines ibi receptas, et ita si quis meminit rem aliquam prius novit illam.   10

267. Item. Si quis rem agnoscit, meminit se prius inspexisse eam, quoniam non est noscere rem in prima visione rei; et ita sequitur: si agnoscit, meminit, et si meminit, prius novit, ut superius ostensum est. Ergo a primo si agnoscit, prius novit; et ita nulla erit cognitio prima. Qualiter ergo veniet aliquis in cognitione alicuius rei?   15

268. Preterea. Secundum hoc occurret Menonis ambiguitas, qui dicebat neminem aliquid addiscere nisi id quod ipse prius novit; quoniam si discit, novit in cognitionem eius quod discit; et si cognoscit, meminit illud, et ita si discit, meminit. Et ita a primo non discit aliquid nisi id quod prius novit.   20

269. Solutio. Ad primum. Dicendum est quod prioritas potest accipi a memoria, quoniam per memoriam recolimus quod prius vidimus res; et ymago prioritatis relinquitur in memoria a presentialitate, cum res presens imprimit memorie suam ymaginem, et illa ymago que tunc fuit presentie similitudo semper post illud tempus impressionis 25 erit ymago prioritatis et preteritionis. Illud idem enim quod nunc est

---

1 eius *om.* C   prius *om.* V   2 memorari: reminisci P   3 ymaginatio: ymago P   posset: possit P   5 rei *om.* P   eiusdem: huius V   5-6 eiusdem rei *om.* P   6 memoria: memori V   ceciderunt: cecidere P   7 reposite²: recepte C   8 illas: quia *add.* P   9 ibi: in P   receptas: repositas P   9-10 prius novit illam: eam prius novit P   11 agnoscit: noscit P   se prius inspexisse *om.* P   eam: illam V   12 noscere: cognoscere V   15 veniet: meminit P   in cognitione *om.* P   16 occurret: occurrit V   Menonis: mentionis C   Zenonis PV   17 ipse *om.* P   18 quoniam: quia V   novit: venit V   eius *om.* V   novit . . . discit: meminit et si meminit, novit C   et *om.* P   20 id *om.* P   23 presentialitate: pre facilitate P   25 fuit: est P   similitudo: sumendo P   semper: similiter V   26 enim *om.* P   nunc: modo PV

5-7 Cf. supra, n. 262.
7-10 Cf. Aug. *De Trin.* XI. vii. 11; *Confess.* X. xv. 23.
11 Cf. Plato *Meno* 81A seq. (iuxta versionem Henrici Aristippi 21 seq.); Aug. *De Trin.* XII. xv. 24.
13 ut superius ostensum est: cf. n. 266.
16 Menonis ambiguitas: cf. Arist. *Anal. Prior.* II. xxi (67ª21); *Anal. Post.* I. i (71ª29); Cicero *Tusc. Disp.* I. 24. 57.
21 Ad primum: cf. supra, nn. 263-6.

necessary, since there is a likeness of 'previousness' in memory, that its image was previously in estimation. Therefore, estimation can intuit this same 'previously' and thus remember. What therefore will memory be for, when estimation or imagination can do anything that memory can?

266. Again, it seems that the knowledge of a thing comes before the memory of the same thing, for there is no memory except of those things which first came into the soul and whose images are preserved in the soul. However, if they are preserved in the soul, when they are deposited there, the soul knows those things by means of their images which have been received there, and thus if someone remembers any thing, he has previously known it.

267. Again, if someone recognizes a thing, he remembers that he previously examined it because one does not know a thing upon the first sight of a thing. Thus it follows that if one recognizes something one remembers it, and if one remembers it one has known it previously, as has been shown above. Therefore, if one recognizes something from the start, one has known it previously, and thus there will be no knowledge. In what way, therefore, does someone come to a knowledge of any thing?

268. Moreover, in this way Meno's uncertainty comes about. Meno said that no one could learn something unless he knew it previously. For if he learns something, what he learns comes into his knowledge,[6] and if he knows it, he remembers it, and so if he learns something, he remembers it. And so he does not learn anything new, but only that which he previously knew.

269. Solution: With regard to the first question, it should be stated that 'previousness' can be grasped by memory because by means of memory we can recall that we have previously seen a thing. Thus, the image of 'previousness' is left in memory by presentness, since a thing which is present impresses its image upon memory, and that image which was then a likeness of the present will after the time of the impression always be an image of previousness and pastness. For

---

[6] Reading (as suggested by Werner) 'venit in cognitionem' with V rather than 'novit in cognitionem' as in the edition.

presens post hoc erit preteritum, et dicetur fuisse prius in quolibet
tempore quod modo est futurum.

270. Ad aliud. Dicendum est quod hec hypotetica vera est: si agnoscit, meminit; sed hec non tenet: si meminit, prius novit. Contingit enim
reponere multas ymagines in memoria rebus tamen ymaginum non 5
notis. Non enim cognoscitur res per unicam apprehensionem, immo per
multas.

271. Item. Quicquid intuetur memoria, intuetur per ymaginem repositam in subiecto memorie; sed contingit aliquem recolere quod illa
ymago est in aliquo subiecto; sic ergo intuetur memoria suum subiectum. 10
Sed aliud est ymago et aliud subiectum ymaginis. Ergo in ipso subiecto
memorie, quando illud memoria intuetur, est ymago representans illud
subiectum. Queritur ergo a quo causetur illa ymago in esse; non nisi
a re cuius est illa ymago, et illa res est subiectum in quo ipsa est ymago.
Ergo ipsum subiectum constituit in esse suam ymaginem que in ipso 15
est; ergo idem in seipsum agit, quod esse non potest.

272. Item. Memoria meminit quandoque se meminisse, et ita memoria
meminit memoriam meminisse. Intuetur ergo memoria seipsam; intuetur etiam hoc ipsum meminisse. Invenit ergo duas ymagines per quas
apprehendit illa duo: aliter enim audito hoc, scilicet memoriam memi- 20
nisse, non acciperet illud quod significatur per illas voces. Queritur ergo
qualiter ille due ymagines formentur in subiecto memorie, cum estimatio eas prius non apprehenderit.

273. Ad primum. Dicendum est quod memoria apprehendit subiectum illud in quo reponuntur ymagines sine ymagine representante illud 25
subiectum. Quoniam illud subiectum semper presto est ipsi memorie;
se meminisse autem meminit per ymaginationem illius rei cuius rei
per illud meminisse fuit recordatio; quoniam per illud meminisse fuit
recordatio in effectum; quoniam per illud meminisse fuit firmius insita

1–2 et . . . futurum *om.* V   3 hec *om.* C   5 reponere: imponere V
6 notis: usitatis C   9–10 quod . . . est *om.* C   10 sic ergo: et sic C   11 et
om. C   12 memorie: memoria P   14 illa[1] *om.* C   res est *om.* V   in: vel
P   ipsa: illa V   15 ipsum *om.* C   17 quandoque: quoniam P   19 etiam:
ergo C   hoc ipsum: se ipsam P   invenit: inveniat V   20 enim: autem
V   audito: audit P   scilicet: sed P   21 acciperet: accipiet P   significatur:
signamus P   ergo *om.* P   22 qualiter: quare C   ille *om.* CP   due: duas
C   formentur: orirentur P   23 eas: ea P   apprehenderit: apprehendit C
25–26 illud[1] . . . subiectum *om.* V   25 reponuntur: reponit C   26 Quoniam:
Quia V   Quoniam illud subiectum *om.* P   27 ymaginationem: ymaginem
V   illius rei *om.* P   28 quoniam: cum *add.* V   fuit[2]: fuerit V   28–29 fuit[2]
. . . meminisse *om.* P   29 effectum: effectu V   quoniam *om.* V

3 Ad aliud: cf. supra, nn. 267–8.
17–18 Cf. Aug. *Confess.* x. xiii. 20.
24 Ad primum: cf. supra, nn. 271–2.
24—p. 74, l. 4 Cf. Aug. ibid. xv. 23.

that same thing which is now present will after this be past, and will be said to have been in any time which will soon be.

270. With regard to the next question, it should be stated that this hypothetical statement is true: 'if one recognizes, one remembers'. However, this does not hold: 'if one remembers, one knew previously', because it happens that many images are deposited in memory but without one knowing the things of which these are the images. For a thing is not known through a single perception but rather through many.

271. Moreover, whatever is intuited in memory is intuited by means of an image deposited in the subject of memory, but it happens that someone recollects that the image is in some subject; thus, memory intuits its subject. Yet an image is one thing and the subject of the image is another. Therefore, in this subject of memory, when memory intuits, there is an image representing that subject. Therefore, the question arises from what is that image brought into existence? From nothing other than the thing which that is an image of, and that thing is the subject in which the image is. Therefore, this subject brings into existence its image which is in itself, and so the same thing acts on itself, which cannot be the case.

272. Moreover, memory remembers whenever it has remembered, and so memory remembers memory to have remembered. Therefore, memory intuits itself, and it even intuits this same 'to have remembered'. Therefore it comes across two images by means of which it perceives these two. Otherwise once this is heard, namely 'to have remembered memory', it would not grasp what is meant by those words. Therefore, the question arises of how these two images are formed in the subject of memory, when estimation does not perceive them previously.

273. Regarding the first, it should be stated that memory grasps that subject in which the images are stored without an image representing that subject, for that subject is always present to memory itself. In remembering itself, however, it remembers by means of the imagining of that thing, which was recalled by means of that 'to have remembered', because by means of that 'to have remembered' there was actually a recall, and because by means of that 'to have re-

illa ymago in memoria; et cum illud meminisse fuerit a memoria per apprehensionem eius quod est meminisse, intuetur memoria seipsam per effectum procedens ad suam causam: causata enim sunt suarum causarum vestigia.

274. Si autem queritur quare ita sit quod memoria meminit se 5 meminisse et visus non videre possit se vidisse, Dicendum est quod hoc contingit quia in memoria reponuntur ymagines, et manent rebus absentibus. Ideo memoria est eorum que prius intuebatur anima, visus autem est non retinens ymagines rerum visarum. Immo absentibus rebus visibilibus absunt earum ymagines a visu; et ideo visus est presentium 10 tantum, et non potest visus videre se vidisse, sicut memoria se meminisse.

275. Sciendum autem quod Aristoteles distinguit in fine libri *de Anima* inter memorari et reminisci dicens: 'Differt autem ab ipso memorari reminisci non solum secundum tempus, sed quoniam ipso memorari et homines et multorum aliorum animalium participant 15 multa. Sed reminisci nullum animalium que cognoscuntur nisi homo. Causa autem est quia reminisci est ut syllogismus quidam, quod aliquis prius vidit aut audivit aut aliquid huiusmodi passus fuit.' Hinc habetur quod reminisci est universalium tantum, memorari est et universalium et singularium, et ideo Aristoteles in eodem libro *de Anima* dicit quod 20 'meditationes salvant memoriam in reminiscendo'.

XXI. *Utrum sit reminisci oblivionis*

Sequitur videre utrum quis possit reminisci oblivionis et memoria accipere oblivionem quod contingit per memoriam.

276. Quod possit, videtur posse haberi per hoc quod quis meminit 25 se scivisse aliquid et oblitum esse eiusdem, sed recolendo se oblitum esse eiusdem meminit oblivionem.

277. Sed obicitur. Oblivio est reposite ymaginis in memoria deletio: inventum enim in memoria sui presentia delet. Sic ergo cum ipsa sit destructio eiusdem quod memoria habet reminisci, memoria recolere 30

---

2 quod: que C   8 Ideo: Item V   intuebatur anima: intuebantur animam P
9 ymagines: ymãg P; mãg C   12–21 Sciendum . . . reminiscendo *om.* P
23 videre: ut videamus V   utrum: qualiter P   23 memoria *om.* C   24 quod
contingit *om.* C   25 possit, videtur posse: potest P   26 sed: et C
27 eiusdem *om.* V   28 est: et C   deletio: delectio C   29 inventum:
notatum P   sui: sua P   delet: delit P   30 eiusdem quod: eius que P   memoria habet *om.* V

9–10 Cf. Algazel *Metaphys.* II. iv. 5 (173).
12–18 Aristoteles . . . de Anima: i.e. *De Memoria* ii (453ᵃ6–11).
20–21 Arist. ibid. i (451ᵃ12–13).
25–27 Cf. Aug. *Confess.* x. xvi. 24.
28–p. 75, l. 2 Cf. Aug. loc. cit.

membered' that image was more firmly embedded in memory. Since that 'to have remembered' was by memory through the perception of that which it is 'to have remembered', memory intuits itself by means of an effect which proceeds towards its cause, for things which are caused are traces of their causes.

274. If, however, it is asked why it is that memory remembers that it has remembered and sight cannot see that it has seen, it should be stated that this happens because images are stored in memory, and remain when the things are absent. Therefore, memory is of these things which the soul intuited previously. Sight, however, is not retentive of images of things seen. Rather, when the visible things are absent the images of them are absent from sight. Thus, vision is of present things only, and sight cannot see that it has seen in the way that memory can remember that it has remembered.

275. It should be known that Aristotle distinguishes between remembering and reminiscing at the end of the book *On The Soul*, saying: 'For to remember differs from to reminisce not only according to time but because both human beings and many other animals share in remembering many things. But except for man none of the animals that are known reminisce. The reason for this is that to reminisce is like a certain syllogism, which someone has previously seen or heard, or has experienced something like this.' Hence we know that reminiscing is of universals only, remembering is of both universals and individuals, and thus Aristotle says in the same book *On The Soul* that 'meditations in reminiscing preserve memory'.

XXI. *Whether there is reminiscing of forgetfulness*

Next we should see whether someone can recollect forgetfulness and with memory grasp forgetfulness because this happens by means of memory.

276. That someone can it seems can be known as follows: someone recollects that he knew something and that he has forgotten it, but in remembering that he has forgotten he remembers forgetfulness of the same.

277. But it is objected: Forgetfulness is the wiping out of an image stored in memory; from its presence it destroys what has been found in memory. So it is that, since forgetfulness is the destruction of the same thing that memory is a recollecting of, memory cannot recollect

oblivionem non potest. Ipsa enim oblivio est ymaginis privatio per quam habet fieri recordatio. Ergo eius non erit ymago in memoria; ergo memoria non potest intueri oblivionem, cum non inveniat ymaginem que sit signum et vestigium oblivionis.

278. Preterea. Quando oblivio presto est, oblivio est privatio memorie; 5 sed si memoria aliquid accipit recolendo non privatur memoria. Ergo, cum oblivio sit privatio memorie, memoria reminisci non potest oblivionem, quod experimento percipimus esse falsum.

279. Solutio. Dicendum est quod si aliqua res sit totaliter oblita, non potest memoria recolere se oblitam esse illius rei. Sed si memoria par- 10 tem illius rei teneat, et eam non perfecte teneat, propter illum defectum recolit se oblitam fuisse illius rei, et per illam partem que retinetur a memoria, meminit memoria se prius scivisse illam. Ut si aliquis sciverit aliquam propositionem per aliquod medium quod fuerit ratio illius scientie potest retinere memoria ymagines illius propositionis, 15 deleta ymagine representante medium illi propositioni appropriatum; retinebit tamen ymaginem medii in universali, et erit deletio ymaginis ipsius specialis medii in particulari. Unde per ymaginem medii in universali recolet memoria quod animus per aliquod medium habuit scientiam ipsius propositionis; sed propter defectum illius similitudinis 20 medii particularis appropriati illi propositioni recolit memoria se oblitam fuisse per quod medium animus venit in scientiam propositionis.

280. Item. Animus potest convertere se ad ymaginem repositam in memoria intuendo ibi esse ymaginem, eo quod aliter non esset cogitare de rebus absentibus et nimis distantibus nisi per ymagines presentes que 25 sunt similitudines earum rerum, quas intuendo animus cogitet de rebus quarum ille ymagines sunt signa et vestigia; sicut ymago in speculo est signum rei cuius ipsa est signum et vestigium. Sed sicut anima percipit ymaginem presentem ei representantem rem absentem, ita potest memoria postea recolere animum cogitasse de illa ymagine, et ita 30 memoria recolit ibi esse ymaginem. Sed illud quod intuetur memoria,

1–2 per . . . recordatio om. P   2 erit: est C   8 quod om. P   10 illius: ipsius P   11 illius om. P   13 Ut: Nec V   aliquis: quis P   16 representante: representate P   18–19 in universali: nihil C   20 illius similitudinis om. C   21 recolit: recolet P   22 venit: venerit P   23 repositam: receptam P   24 esse om. V   cogitare: cognoscere P   26 sunt om. P   cogitet: cognoscit P; cogitat V   27 quarum: quare P   sicut: sed cum P   29 ei om. CV   30 cogitasse: cognovisse P

9 Dicendum: cf. supra, nn. 277–8.
9–10 Cf. Aug. De Trin. XI. vii. 12.
10–13 Cf. Avic. De An. IV. iii (19ᵛᵃB); Aug. De Trin. loc. prox. cit.; Confess. x. xix. 28.
23 Cf. Aug. De Trin. XI. viii. 13 (994–5).

forgetfulness. This is because forgetfulness itself is the privation of an image by means of which remembering takes place. Therefore, its image will not be in memory, and so memory cannot intuit forgetfulness, since it does not come across an image which is a sign and trace of forgetfulness.

278. Moreover, when forgetfulness is present, forgetfulness is the privation of memory. However, if memory grasps something by recollecting, memory is not lacking. Therefore, since forgetfulness is the privation of memory, memory cannot recollect forgetfulness. However, through experience we perceive this to be false.

279. Solution: It should be stated that if something is completely forgotten, memory cannot recollect that it has forgotten that thing. However, if memory holds onto part of that thing, and does not completely hold onto it, then because of that incompleteness it recalls itself to have forgotten about that thing, and by means of that part which is retained by memory, memory remembers that it had previously known it. Just as if someone had known a certain proposition through some means which was the reason for that knowledge, memory can retain the images of that proposition, even when the image representing the middle term appropriate to that proposition has been deleted. It will, however, retain the image of the middle term as a universal and the deletion will be of that specific middle term in particular. Thus, by means of an image of that middle term as a universal, memory recalls that the mind has knowledge of this proposition through a certain middle term. However, because of a defect of that likeness of the particular middle term proper to that proposition, memory recalls that it has forgotten the middle term through which the mind arrived at the knowledge of the proposition.

280. Again, the mind can direct its attention towards the image stored in memory by intuiting that there is an image there. Otherwise there would not be a thinking about absent things and not even of very distant things, except by means of images which are present. These images are likenesses of those things, which when the mind intuits them it knows about the things of which these images are signs and traces, just as an image in a mirror is a sign of a thing of which it is a sign and trace. However, just as the soul perceives an image present to it representing an absent thing, so memory afterwards can recall that the mind thought about that image, and thus memory recalls that there is an image there. However, that which memory intuits

intuetur per ymaginem; ergo, cum memoria ipsam intueatur ymaginem, per illam ymaginem ipsam inspiciet. Sic ergo ymaginis existentis in memoria erit aliqua ymago in memoria, et iterum pari ratione illius secunde ymaginis erit tertia ymago, et sic deinceps in infinitum.

281. Item. Si ymago habeat ymaginem, aut prima constituet secun- 5 dam per se aut per accidens. Si per se, pari ratione et secunda tertiam, et tertia quartam per se, et sic in infinitum. Sed si aliquis per se et proprie est causa alicuius secundum quod ipsum est, ipso existente existit eius effectus. Cum eo ergo erunt simul constitute in effectum infinite ymagines in memoria; quod esse non potest, quoniam ubi est infinitas 10 ibi non est terminus. Si per accidens, queritur per quod accidens, et per quid aliud ab ipsa ymagine constituetur ymago ymaginis in memoria non est assignare.

282. Item. Queritur qualiter Deus, qui est prima causa omnium, ab anima accipiatur in intellectu eius et memoria. Quod non per ymaginem 15 sic ostenditur.

283. Finitum et infinitum sunt dissimilia; quelibet ymago est finita, prima causa omnium est infinita; ergo prima causa nullam habet ymaginem sui representantem seipsam; ergo primam causam non intuetur per ymaginem anime intellectus nec memoria.   20

284. Item. Quicquid intelligitur, vel quod est intelligibile, est intelligibile vel per materiam vel per formam. Sed prima causa omnium, hoc est Deus, nullam habet materiam nec formam. Ergo prima causa omnium est incomprehensibilis et supra omnem rationem et intellectum. Sed nihil significamus nisi id quod intelligimus et preconcipimus ante- 25 quam idem significemus. Ergo cum prima causa non sit intelligibilis, prima causa non est significabilis; ergo prima causa significari non potest. Quid ergo significamus per hoc nomen 'Deus'?

285. Solutio. Ad primum. Dicimus quod ymaginis existentis in memoria non est alia ymago. Cum enim illa ymago sit presto illi 30

---

1 memoria: ipsa P   ipsam *om.* P   intueatur: inspiciat CV   2 illam: aliam C   ipsam: eam P   inspiciet: inspicit C   4 ymaginis: ymagines V   et sic *om.* V   5 Item: Similiter V   constituet: constituit C   7 et sic *om.* C 9 effectum: effectu C   11 queritur ... accidens *om.* V   12 constituetur: constituatur P   14 Item: Similiter V   qualiter: quare V   15 eius *om.* V ymaginem: ymagines P   18 infinita: finita V   19 seipsam: ipsam V   20 anime intellectus: anima P; anime *om.* V   21 vel quod: quod etiam P   21–22 est intelligibile *om.* C   23 hoc: que C   nec: vel P   formam: quia ipse est supra omnem formam et supra omnem materiam *add.* V   24 rationem: propositionem P   25 et: quod *add.* P   preconcipimus: preconcepimus P   26 idem significemus: illud significamus P   prima: ipsa V   29 dicimus: dicendum P

7–9 Cf. Arist. *De Gen. et Corr.* II. x (336ª27–28).
29 Ad primum: cf. supra, nn. 280–1.

is intuited by means of an image; therefore, since memory intuits the image itself, by means of another[7] image it looks into itself. Therefore, in this way there will be some image in memory of an image existing in memory, and again for the same reason there will be a third image of this second image, and so on to infinity.

281. Again, if an image has an image, either the first sets up the second as such or accidentally. If as such, then in the same way the second will set up the third, and the third a fourth as such, and so to infinity. However, if something is the cause of another as such and properly inasmuch as it is, as long as it exists, so will its effect. Therefore, together with it there will be infinite images brought into effect in memory at the same time. This cannot be the cause, because where there is infinity there is no end. If in an accidental manner, it is asked by means of what accident; and one cannot assign anything other than the image itself by means of which an image is constituted of an image in memory.

282. Again, it is asked in what way is God, who is the first cause of all, grasped by the soul in its understanding and memory? That it is not by means of an image can be shown as follows.

283. The finite and infinite are dissimilar; any image is finite and the First Cause of everything is infinite; therefore, the first cause does not have an image of itself representing itself; therefore neither the intellect nor the memory of the soul intuits the First Cause by means of an image.

284. Again, whatever is understood, or what is intelligible, is intelligible either by means of matter or by form. But the first cause of everything, which is God, has no matter or form. Therefore the first cause of everything is incomprehensible and above all reason and understanding. However, we do not express anything except what we have understood and preconceived before we express the same. Therefore, since the First Cause cannot be understood, the first cause cannot be signified; therefore the first cause cannot be expressed. What, therefore, do we mean by this word 'God'?

285. Solution: With regard to the first question we state that there is not another image of the image existing in memory. For since that

---

[7] Following Werner and the reading given by C, 'aliam' rather than 'illam'.

memorie, ipsa memoria seipsa eam intuetur; sed res extra per ymagines inspicit.

286. Ad aliud. Dicendum est quod primam causam non intuetur anime acies tantum per ymagines intus habitas, sed per res apprehensas extra que sunt effectus cause prime: creature enim sunt signa et vestigia 5 creatoris. Sic ergo anima intelligendo creaturam venit in intelligentiam creatoris, et per cognitionem boni quod hic est venimus in cognitionem summi boni, et per cognitionem vite que apud nos est, et gaudii quod hic est, venimus in intelligentiam vere vite.

287. Item. Contrariarum causarum contrarii sunt effectus: ut caliditas 10 et frigiditas; sed duo contraria simul non possunt esse in eodem; ergo duo contrarii effectus simul non possunt esse in eodem. Sed letitia et tristitia contrariantur ad invicem; ergo earum effectus simul non possunt esse in eodem.

288. Sed contra. Ut prius ostensum est, in memoria reponuntur 15 ymaginationes ad quas voluntas convertens aciem animi intuetur res quarum sunt ille ymagines; unde memoria est quasi venter animi et sinus. Sed huiusmodi ymaginationes exoriuntur in memoria a rebus quarum sunt ymagines. Ergo sicut res sunt contrarie et ita earum ymagines sunt contrarie. Ergo in memoria non sunt simul et semel 20 reposite ymagines letitie et tristitie, albi et nigri, quod falsum est.

289. Preterea. Aliquis existens in letitia potest recordari tristitie, et e contrario; sed letitia et tristitia sunt contraria; sunt ergo affectiones contrarie in anima, quod esse non potest.

290. Item. Per memoriam fit intuitio rerum prius apprehensarum; 25 sed sicut contingit aliquem recolere se prius apprehendisse res sensatas, ita potest aliquis recolere se apprehendisse universalia, ut genera et species que sensibilia non sunt, sed solum intelligibilia. Accipit ergo memoria universalia ea intuendo sicut sensibilia. Sed, ut habetur

1 seipsa eam: ipsam P   3 est om. P   4 anime . . . tantum: anima P
5 cause: esse P   6 Sic: Et sic V   creaturam: creatorem P   7 cognitionem²: cognitione V   9 vere: beate P   10–11 caliditas et frigiditas: caliditatis et frigiditatis C   11 contraria om. C   13 simul om. C   15 prius: superius P   16, 18 ymaginationes: ymagines V   19 Ergo sicut res: Sed si cause CV   et ita om. CV   22 aliquis: aliquid C   26 sicut: si V   27 se: prius add. V

3 Ad aliud: cf. supra, nn. 282–4.
5–9 Rom. i. 20; Sap. xiii. 1–5; cf. Aug. De Trin. xv. ii. 2–3.
10 Arist. De Gen. et Corr. II. x (336ᵃ31).
11 Arist. Physic. v. vi (230ᵇ30–31).
15 Ut prius ostensum est: cf. supra, n. 262.
17–18 quasi venter animi et sinus: cf. Aug. De Trin. XII. xiv. 23: 'Quasi glutiens in ventrem ita in memoria reposuerit, poterit recordando quodam modo ruminare, et in disciplinam quod sic didicerit traiicere'. Vid. etiam Confess. x. xiv. 21: 'memoria quasi venter est animi'.

image is present to that memory, memory itself intuits it, but it examines things outside by means of images.

286. Regarding the next question, it should be stated that the eye of the mind does not intuit the first cause only by means of images which are held within, but by means of things perceived externally which are effects of the first cause, for creatures are signs and traces of the Creator. Therefore, in this way the soul in understanding a creature comes to an understanding of the Creator, and through a knowledge of a good which is here we come to a knowledge of the Highest Good, and through a knowledge of the life which is ours, and of the happiness which is here, we come to a knowledge of the true life.

287. Again, the effects of contrary causes are contrary, such as heat or coldness; but two contraries cannot be in the same thing at the same time; therefore two contrary effects cannot be in the same thing at the same time. However, happiness and sadness are opposed to each other, therefore their effects cannot be in the same thing at the same time.

288. But against this, as has already been shown, images are stored in memory. When the will turns the eye of the mind towards them, it intuits the things of which these are the images; whence memory is as it were 'the bowels and bosom' of the spirit. However, these kinds of mental image arise in memory from the things of which these are images. Therefore, just as things are contrary, so in the same way these images are contrary. Therefore, the images of happiness and sadness, of white and black, are not stored at one and the same time in memory, which is false.

289. Moreover, someone while being happy can remember sadness, and the other way round; therefore there are contrary emotions in the soul, which cannot be the case.

290. Again, by means of memory the intuition occurs of things which have been perceived previously. However, just as it happens that someone remembers that they had previously perceived things which were sensed, in the same way someone can recollect that they have perceived universals, such as genera and species which cannot be sensed but can only be understood. Therefore, memory grasps universals by intuiting them just like sensible things. Yet, as is held by

ab Augustino et ab Aristotele, bruta animalia habent memoriam qua
contingit ipsa memorari, ut per memoriam redeunt apes et aves ad suas
mansiones; et in memoria convenimus cum eis. Possunt ergo bruta
animalia per memoriam intueri universalia, quod esse non potest cum
intellectu careant et ratione.   5

291. Solutio. Ad primum. Dicendum est quod bene possunt in
memoria simul reponi ymagines duorum contrariorum; nec tamen ille
ymagines contrarie sunt, quoniam ipsi non sunt immediati effectus
contrariorum, immo valde remoti effectus.

292. Et si quis obiciat: immediati effectus contrariorum sunt con- 10
trarii, et illi effectus habent iterum alios effectus immediatos, et illi
erunt contrarii; et sic erit percurrere usque perveniatur ad ymagines
repositas in memoria, et ita iterum erunt ille ymagines contrarie.

293. Ad hoc. Dicendum est quod contrarietas est in eis que sunt
activa et passiva ad invicem, ut in calido et in frigido; sed tantus potest 15
fieri cursus in causatis causantibus alia ex se quod tam debilis esse
erunt aliqua causata quod non possunt causare aliqua contraria, ut
patens est in grosso exemplo. Ignis calefacit aerem sibi proximum, et
ille aer calidus sibi proximum, et ita percurrit inmutatio caloris in aere
quod in tantum est debilis esse ipse calor quod non potens est ex se 20
generare calorem, sed tepiditatem; et tepiditas semper secundum distantiam a suo ortu diminuetur: unde in tantum potest debilitari quod
ex se non generat similem inmutationem; quoniam illud causatum quod
est posterioris esse habet esse debilius et instabilius: ut patet in hominibus qui sunt secundum temporis successionem posterioris esse quam 25
sui antecessores, et ideo sunt debilioris esse, quia magis recedunt a suo
primo ortu. Et cum debilior sit eorum natura in posterioribus semper

---

1 ab² *om.* P   qua: contra V   6 est *om.* V   7 simul *om.* V   8 effectus
om. V   9 remoti: rememorati C   10 Et *om.* PV   11 iterum *om.*
CV   et² *om.* P   12 usque: donec V   13 iterum *om.* P   ille: due *add.* V
15 in² *om.* P   16 causatis causantibus: creatis creantibus V   17 aliqua¹
om. CV   causata: creata *codd.*   possunt: potuerunt P   18 aerem: aera P
19 inmutatio: mutatio P   aere: aera P   20 in *om.* P   23 ex se: ex in se
V   similem: consimilem V   24 habet esse *om.* C   instabilius: quam res antecedens *add.* V   25–26 qui . . . antecessores: secundum posterioris esse secundum
temporis successionem quam homines anterioris temporis V   26 debilioris: debi-
liores V   27 ortu: omnis enim causa prima magis est influens super creatum
primum quam super secundum *add.* V

---

1 ab Augustino: *Confess.* x. xvii. 26: 'Habent enim memoriam et pecora et aves,
alioquin non cubilia nidosve repeterent, non alia multa, quibus assuescunt; neque enim
et assuescere valerent ullis rebus nisi per memoriam'. Cf. *Contra Epistolam Fundamenti*
xvii (185).   ab Aristotele: *De Memoria* i (450ª15–19); ii (453ª8).
4 universalia: cf. supra, n. 258.
6 Ad primum: cf. supra, n. 287.
14–15 Cf. Arist. *De Gen. et Corr.* II. ii (329ᵇ24).

Augustine and by Aristotle, brute animals have memory by means of which it happens that they remember, so that by means of memory bees and birds return to their homes. Thus as regards memory we are the same as they are. Therefore, brute animals can intuit universals by means of memory. This, however, cannot be since they lack understanding and reason.

291. Solution: With regard to the first question, it should be stated that the images of two contraries are well able to be stored at the same time in memory, nor indeed are these images contrary, because these are not immediate effects of contraries; rather they are really remote effects.

292. Moreover, if someone were to object: the immediate effects of contraries are contrary, and these effects also have other immediate effects, which will be contrary. So they will run on until they reach the images stored in memory, and so again those images will be contrary.

293. Regarding this, it should be stated that contrariety lies in those things which are naturally active and passive, such as a hot thing and a cold thing. However, the process in caused things causing others out of themselves can go so far that some caused things will become so weak that they cannot cause further contraries, as is clear in the following rough example. Fire heats the air which is next to it, and that hot air heats the air next to it, and so the modification due to heat runs through the air to a point where the heat is so weak that it is not able in itself to generate heat but lukewarmness. Moreover, lukewarmness always diminishes according to the distance from its source: so it can become so weak that it does not produce a similar modification. For something caused which comes into existence later has a weaker and more unstable existence, as is clear in the case of men who, according to the succession of time, exist later than their ancestors, and thus have a weaker existence, because they are further away from their original source. And since the nature of those who come after is always weaker than those who come before, inasmuch as

quam in prioribus secundum quod alia fluunt ex aliis, contingit quod
facilius corrumpitur res huius speciei 'homo' in hoc tempore quam in
tempore propinquo post creationem hominum. Sed ipsius nature est
conservare suam speciem in esse, ut nature hominis est conservare
hominem in esse. Et ideo ne species deficiat propter facilitatem cor- 5
ruptionis singularium, eo quod minoris durationis sunt modo singularia
quam prius fuerunt, natura plus et plus multiplicat singularia. Et ideo
contingit quod secundum temporis successionem semper in posteriori
etate, quantum est ad viam nature et intentionem, debent plures esse
homines quam in precedenti etate ut salvetur species in illa multi- 10
tudine. Sed si aliquis homo esset semper durans et immortalis, cum in
eo posset hec species 'homo' salvari, natura hominis non multiplicaret
singularia, sicut natura solis non producit plures soles, quia hec species
'sol', in sole qui nunc est, salvatur propter eius perpetuitatem et incor-
ruptibilitatem. 15

294. Ad aliud. Dicendum est quod affectio et forma cogitata non sunt
contrarie; ut quando aliquis letatur, tunc est letitia eius animi affectio,
et si in eodem tempore recolat tristitiam prius habitam est illa tristitia
cogitata et eum non afficiens.

295. Ad ultimum. Dicendum est quod fortior et subtilior est memoria 20
in homine quam in brutis animalibus propter intellectum, qui est in
homine, qui accipit formas universales; et memoria potest postea easdem accipere, cum quelibet vis apprehensiva in homine sit vis illius
anime que est rationalis. In homine enim illud efficit solum anima
rationalis, que est et sensibilis, et etiam plus quam in brutis efficiat 25
anima sensibilis. Bruta autem animalia carent intellectu, et ideo bruta
animalia debilioris sunt memorie quam rationalia. Simile contingit in

---

2 facilius *om.* V   homo *om.* C   tempore: quod nunc est *add.* V   2–3 in
tempore *om.* P   7 fuerunt *om.* CV   multiplicat: multiplicent P   11 esset
om. P   14 est *om.* P   16 affectio: mentis affirmacio P   cogitata: ex-
cogitata P   18 illa: ipsa V   20 memoria *om.* V   23 sit: sed C
24 illud efficit: operatur C   25 que . . . sensibilis *om.* P   sensibilis: et vegetabilis *add.* V   25–26 et etiam . . . sensibilis *om. per hom.* V   25 quam: quod
P   efficiat: efficit P   26 intellectu: et cellula memorialis in homine est maioris
siccitatis quam in brutis animalibus, et ideo magis viget retentio apud memoriam hominum quam apud memoriam brutorum animalium *add.* V

3–11 Cf. Arist. *De An.* II. iv (415^{a}29–^{b}7).
16 Ad aliud: cf. supra, nn. 288–9.
20 Ad ultimum: cf. supra, n. 290.
22 formas universales: cf. supra, n. 260.
26–27 Cf. supra, n. 275.
27–p. 80, l. 2 Arist. *De An.* II. ix (421^{a}9–13): 'Sensum hunc non habemus certum,
sed peiorem multis animalibus. Prave enim odorat homo, et nihil odorat odorabilium
sine letitia et tristitia, sicut non existente certo eo quo sentimus'; cf. *De Sensu* v (444^{b}6–
13); Avic. *De An.* II. v (8^{vb}A–9^{ra}B).

others flow from others, it happens that the stuff of this species 'man' is destroyed more easily in this time than in the time soon after the creation of men. However, it is proper to each nature to preserve its species in existence, so that it is of the nature of man to preserve man in existence. And thus, lest a species were to fail because of the ease with which individuals are destroyed and because new individuals are of a lesser duration than they were previously, nature multiplies individuals more and more. Thus it happens that as time goes by, in so far as this is according to the way and intention of nature, in the latter age there have to be more men than in the previous age so that the species is preserved in such a great number. However, if one man were to be immortal and everlasting, since the species 'man' could be preserved in him, the nature of man would not multiply individuals, just as the nature of the sun does not produce many suns, because this species 'sun' is preserved in the sun which exists now because of its perpetuity and incorruptibility.

294. As regards the next question, it should be stated that an emotion and the form thought about are not contraries. For example, when someone is gladdened, then the happiness is an emotion of his spirit, and if at the same time he recalls the sadness which happened previously, that sadness is thought about and does not affect him.

295. With regard to the last question, it should be stated that memory in man is stronger and more acute than in brute animals because of the understanding which is in a man who grasps universal forms. Then afterwards memory can grasp them, since whatever perceptive power is in man is a power of that soul which is rational. For in man it is the rational soul alone which achieves that, and it is also sensible, and the sensitive soul achieves more than that which is in brute animals. This is because brute animals are lacking in understanding,[8] and so brute animals have a weaker memory than rational animals. A similar thing happens in the senses, for the sense

[8] V adds here: . . . and the part of the brain concerned with memory in man is drier than that in brute animals, and thus the power of retention in human memory is greater than that of brute animals . . .

sensibus, quoniam odoratus fortior est in uno animali quam in alio, ut in cane quam in homine.

XXII. *De anima rationali*

296. Sequitur de anima rationali et de eius viribus. Primo autem loco querendum est utrum anima rationalis sit corporea vel incorporea. 5 Quod sit incorporea sic ostenditur. Ab anima rationali habemus quod ratiocinamur et intelligimus. Si ergo anima rationalis sit corpus, a corpore habemus quod ratiocinamur et intelligimus. Sed hoc non est a natura corporis in quantum est corpus, quoniam secundum hoc omne corpus ratiocinaretur et intelligeret. Ergo quod ratiocinamur et intel- 10 ligimus non debetur nature corporis ergo debetur alii, et illud, quicquid sit, est anima rationalis; ergo anima rationalis non est corpus, nec natura corporis.

297. Item. Si anima rationalis est corpus. Sed omne corpus est animatum vel inanimatum. Ergo anima rationalis, si ipsa sit corpus, 15 ipsum est animatum vel inanimatum. Si animatum, ergo habet animam, et illa anima iterum aliam animam, et sic in infinitum. Si inanimatum, hoc esse non potest, quia secundum hoc nunquam conferret vitam.

298. Preterea. Aristoteles in libro *de Generatione et Corruptione* pro inconvenienti habet aliquod corpus esse non sensatum. Sed si anima 20 esset corpus, aliquod corpus esset non sensatum; nulla enim anima sensu apprehenditur. Relinquitur ergo quod anima non est corpus.

299. Preterea. Unumquodque corpus aliquibus formis sensibilibus afficitur, ut vel tangibilibus vel visibilibus, et sic de aliis. Sed anima nullis formis sensibilibus afficitur; ergo nulla anima est corpus.   25

300. Quod ipsa sit corpus sic ostenditur. Contingit hominem apprehendere spatia terrarum que ipse prius videbat et de illis cogitare. Sed non posset aliquis ita intueri nisi per ymagines in memoria repositas,

---

1 sensibus: sensibilibus P   quam in alio *om.* P   4 de² *om.* C   4–5 autem . . . est: ergo queritur P   6 sit incorporea: non sit corporea C   quod: quidem V   8 a: de V   10 ratiocinamur: ratiocinemur P   12 sit: illud *add.* V   ergo anima rationalis *om.* C   14 est¹: sit C   15–16 Ergo . . . Si animatum *om.* C   16–18 Si animatum . . . vitam *intervertit* P: Si inanimatum . . . vitam. Si animatum . . . infinitum.   17 iterum: habet *add.* V   Si: est *add.* V   19 *De Gen. et Corr.*: Generationis et Corruptionis P   21 anima *om.* C   24 ut vel . . . aliis *om.* C   24–25 Sed anima nullis: nulla anima aliquibus C   27 cogitare: cognoscere P   28 ita: ista V

4–9 Cf. Avic. *De An.* v. ii (22ᵛᵇA), et supra, n. 30.
14–18 Cf. Costa ben Luca *De Differentia Animae et Spiritus* iii (132); Gundiss. *De An.* ii (40); Alanus de Insulis *Contra Haereticos* 1. xxvii (329B).
19–20 Arist. *De Gen. et Corr.* 1. v.
23–25 Cf. Arist. loc. cit.; Costa ben Luca ibid. (131); Gundiss. loc. cit.; Alanus loc. cit.

of smell is stronger in one animal than in another, such as in a dog rather than in a man.

XXII. *Concerning the rational soul*

296. What follows next concerns the rational soul and its powers. Firstly we should enquire into whether the rational soul is corporeal or incorporeal. It can be shown as follows that it is incorporeal. We have reasoning and understanding from the rational soul; if therefore the rational soul were a body, then reasoning and understanding would be had from a body. However, this does not arise from the nature of a body inasmuch as it is a body, because in this way every body would reason and understand. Therefore, that we reason and understand is not due to the nature of the body, therefore it is due to another, and that, whatever it is, is the rational soul. Therefore, the rational soul is not a body, nor of the nature of a body.

297. Again, is the rational soul a body? Yet every body is either animate or inanimate. Therefore the rational soul, if it is a body, is either animate or inanimate. If it is animate, then it has a soul, and that soul again would have another soul, and so on to infinity. If it were inanimate, this could not be, because as such it would never confer life.

298. Moreover, Aristotle in the book *On Generation and Corruption* holds that it is contradictory for there to be a body and it not to be sensed. However, if the soul were a body, then one body would not be sensed, since no soul is grasped by a sense. Therefore, it only remains that the soul is not a body.

299. Moreover, every body whatsoever is affected by some sensible forms, such as by tangible or visible forms, and similarly regarding others. But the soul is not affected by any sensible forms, therefore no soul is a body.

300. That the soul is a body can be shown as follows. It happens that someone perceives some areas of land which he saw previously and then he thinks about them. However, no one can think about things in this way except by means of images stored in memory since

cum illa que ipse intuetur memorando sint ab eo remota et absentia;
ut aliquis iacens in lecto et cogitans metitur terre quantitatem per
spatium et radium firmamenti et spatium signorum; sed non contingit
metiri nisi per dimensionem. Ergo oportet quod ipse ymagines rerum
extra per mutationem quarum anima metitur ymaginata habeant ali- 5
quam dimensionem, ut si mensuraret in longum, est ibi porrectio; si in
latum, est ibi in duas partes distentio. Sed ille ymagines sunt in anima
ut in subiecto; ergo cum ille ymagines dimensiones habeant in longum
et in latum et in spissum, et ipsa anima easdem habebit dimensiones;
et anima est substantia et habens secundum hoc triplicem dimensionem; 10
ergo anima est corpus.

301. Item. Anima est in corpore: aut igitur est intrinsecus aut
extrinsecus, aut partim intra, et partim extra. Si totaliter forinsecus,
non esset vivibilis vel vegetabilis intra. Si totaliter intrinsecus, non
sentiret stimulum pungentem vel calorem in extremitatibus corporis. 15
Ergo ipsa anima secundum quid est intra et secundum quid in extremitatibus. Sed simul et semel non potest esse in eodem instanti et hic et
ibi totaliter secundum sui essentiam; ergo secundum aliquam sui partem
est hic, et secundum aliquam sui partem est ibi. Ergo tanta est anima
quantum est ipsum corpus in quo est vita et tactus. Ergo habet triplicem 20
dimensionem; ergo ipsa est corpus, quod esse non potest, quoniam
secundum hoc duo corpora essent simul in eodem loco, vel unum corpus
divideretur omnino in infinitum, ut illud corpus in quo est anima divideretur in infinitum per subintrationem anime, cum non esset sumere
aliquam ita parvam partem corporis animati in qua non esset pars anime. 25

302. Preterea. Si anima sit corpus et est in corpore, aut illa duo corpora sunt continua, aut contigua; sed sive sic sive sic, erit secundum
hoc sumere aliquam partem ubi erit aliqua pars anime contiguata vel
continuata parti corporis quod non est anima. Ergo cum ibi non sit
anima ubi ulla pars corporis est, in illa parte corporis non est anima, et 30
ita illud corpus non est vegetatum.

1 eo: ea P   remota: rememorata C   2 et cogitans: cogitando V   metitur:
metiri P   3 et radium *om.* C   4 rerum: que sunt *add.* V   6 mensuraret: mensuretur P; mensuret V   in: ibi P   porrectio: inporrectio V   7 ibi
om. V   in *om.* P   distentio: distensio P   8 dimensiones: dimensionem P
9 et³ *om.* V   11 ergo *om.* C   12 aut: vel V   igitur *om.* P   13 intra
. . . extra: intrinsecus et partim extrinsecus P   et: aut C   14 vivibilis vel
vegetabilis: utile P; vivibile vel vegetabile V; vivibilis vel vegetabile C   15 extremitatibus: extremitatem V   16 intra: intrinsecus P   et *om.* CV   17 et²
om. C   19 hic . . . est² *om.* V   tanta: tantum V   23 in¹ *om.* P   24 in:
per V   subintrationem: subtractionem P   sumere *om.* V   25 aliquam ita
parvam partem: aliqua parva pars V   esset: esse P   29 quod: ubi V   30 ubi
ulla: ut V   ulla: est ibi P   est² *om.* P   31 vegetatum: vegetativum P

12–19 Cf. Guill. de S. Theod. *De Natura Corporis et Animae* II (712).

those things which he thinks about by remembering are distant from him. Just as when someone lying in bed and thinking measures a quantity of land by means of the extension and radius of the heavens and the extension of the constellations. However, measuring only happens by means of spatial dimensions. Therefore, it is necessary that these images of external things, through the changing of which the soul measures things imagined, have some spatial dimension, such that if it measures lengthwise there is a straight line there and if in width, there is extension in two directions. However, these images are in the soul as in a subject, therefore since those images have dimensions in length, breadth, and thickness, the soul itself will also have those dimensions. Thus, the soul is a substance and in this way has three dimensions, therefore the soul is a body.

301. Moreover, the soul is in the body, therefore either it is outside or inside the body, or partly within and partly without. If it were totally outside, it would not be capable of life or of growth within. If it were totally inside, it would not feel a sharp sting or heat at the extremities of the body. Therefore, the soul itself in a certain respect is inside and in a certain respect at the extremities. However, at one and the same time it cannot be completely in the same situation both here and there according to its essence, therefore according to some part of it it is here, and according to some part of it it is there. Therefore, the soul is as a body itself is in which there is life and touch. Therefore, the soul has three dimensions; and so the soul is a body. This cannot be because then two bodies would exist at the same time in the same place, or one body would be completely divided into infinity, so that that body in which the soul is would be divided into infinity when the soul enters into it, since one cannot find a part of an animated body which is so small that there is no part of the soul in it.

302. Moreover, if the soul is a body and is in a body, either those two bodies are continuous or they are contiguous. However, whether they are or are not, one could take some part where there is a part of the soul contiguous to or continuous with a part of the body which is not the soul. Therefore, since there is not a soul there where any part of the body is, in that part of the body there is not a soul, and thus that body is not living and growing.

303. Solutio. Dicimus quod anima est substantia incorporea. Et ad primam obiectionem, Dicendum est quod quando anima metitur spatia terrarum vel spatia firmamenti per ymagines eorum ad quas ipsa se convertit, tunc sunt ipse ymagines reposite in cellula memoriali, que corporea est, et non in ipsa anima. Sole enim ymagines intelligibiles 5 reponuntur in anima.

304. Ad aliud. Dicendum est quod cum anima habeat esse in corpore ut in eo adquirat suam perfectionem per adeptionem virtutum et scientiarum semper convertit suas intentiones ad ymagines repositas in corpore que assimilantur rebus extra, ut per intuitionem earum 10 ratione mediante sibi adquirat scientias et virtutes. Nec oportet quod anima sit ibidem sita ubi illa sita est ymago quam ipsa intuetur, sicut si videam parietem, propter hoc non sum in pariete. Unde quamvis anima sentiat in extremitate corporis stimulum pungentem, propter hoc non est anima ibidem existens secundum essentiam in extremitate 15 illius corporis.

305. Sed obicitur. Illa potentia percipiendi qua anima percipit et intuetur passiones in extremitate corporis constitutas et in cellula cerebri, aut est in anima ut in subiecto, aut in corpore ut in subiecto. Si in corpore ut in subiecto. Contra. In corpore est potentia recipiendi 20 illas impressiones et in corpore est potentia easdem percipiendi. Ergo qua ratione dicitur de corpore vere quod ipsum recipit illas passiones, de corpore potest dici vere quod ipsum percipit easdem. Ergo vel ipsum corpus est ipsa anima, vel frustra est ipsa, cum corpus posset exercere easdem operationes quas anima exerceret. Si potentia percipiendi sit in 25 anima ut in subiecto, et potentia recipiendi passiones extra sit in extremitatibus corporis, queritur qualiter anima possit percipere passiones illas cum in illa non sit ibi presens.

306. Similiter queritur, cum corpus omnino vegetetur ab anima,

3 spatia *om.* P   4 tunc *om.* P   5 enim *om.* C   intelligibiles: intelligibilium P   7 habeat: habet V   8 eo: ea C   adeptionem: adoptionem C
10 intuitionem: intentionem V   earum: eorum P   11 adquirat: adquiratur C
12 intuetur: eam *add.* C   15 secundum essentiam *om.* V   essentiam: existentiam P   17 et *om.* P   18 extremitate: illius *add.* V   cellula: cellulis P
19 aut²: est *add.* V   20 recipiendi: percipiendi P   21 easdem percipiendi: percipiendi eandem P   22 recipit: percipit V   23 percipit: percipitur P   24 ipsa¹: ipsum P   27 anima: illa V   28 in *om.* C   illa: ipsa P   ibi presens: in presens ibi P

1–2 ad primam obiectionem: cf. supra, n. 300.
4–6 cellula memoriali: cf. Avic. *Canon* i. fen i, doct. 6, summa 1, cap. 5 (76rb–va): '. . . que est virtus conservativa et memorialis. Et est thesaurus eius . . .'.
7 Ad aliud: cf. supra, n. 301.
7–11 Cf. Avic. *De An.* v. i (22^vbD); Algazel *Metaphys.* ii. v. 3 (185). Cf. infra, nn. 324, 336.

303. Solution: We state that the soul is an incorporeal substance. And, regarding the first objection, it should be stated that when the soul measures spaces of land or distances of the heavens by means of their images towards which it turns, then these images are stored in the seat of memory which is corporeal, and not in the soul itself. For only intelligible images are stored in the soul.

304. Regarding the next objection, it should be stated that since the soul has its being in the body so that it will acquire its perfection in it through the obtaining of virtue and knowledge, it always directs its attention to the images stored in the body which are assimilated to things outside, so that through the intuition of them by means of reason it acquires knowledge and virtue for itself. Nor is it necessary that the soul be located in the same place where the image which the soul intuits is located, so that if I see a wall, I am not in the wall because of this. Thus although the soul feels a sharp sting in the extremity of the body, because of this the soul does not exist in its essence in the extremity of that body.

305. But it is objected that that power of perceiving by which the soul perceives and intuits the actions undergone which are located in the extremity of the body and in the seat of the brain, is either in the soul as in a subject or in the body as in a subject. If it is in the body as in a subject, ⟨it can be objected⟩ against this that there is in the body the potential to receive those impressions and in the body there is the potential to perceive them. Therefore, because it is truly said of the body that it receives those sensations, it can be truly said of the body that it perceives the same. Therefore, either the body is the soul itself, or the soul is of no use, since the body can carry out the same operations which the soul can. If the potential to perceive is in the soul as in a subject, and if the potential to receive external actions is in the extremities of the body, it is asked in what way the soul can perceive these actions since the soul is not present there.[9]

306. Similarly it is asked, since the body in every way is vivified by

---

[9] Following C and omitting 'in'; cf. no. 308, ll. 13–14 'ipsa . . . presens'.

qualiter totum corpus habeat vegetationem ab ipsa nisi ipsa distendatur
per illud totum.

307. Ad hoc diceret forte aliquis quod hoc est per transmissionem
spirituum vitalium. Contra. Illi spiritus aut sunt corporei, aut incorporei; aut animati, aut inanimati. Si animati: ergo habent animam; et 5
ita occurrit precedens dubitatio de ipsis. Si inanimati: ergo vitam conferre non possunt, vel vegetationem. Si sunt non corporei, aut sunt
substantie, aut non. Si sunt substantie, qualiter transmittuntur nisi
vitam habeant, et etiam locum obtineant. Si non sunt substantie: ergo
accidentia sunt: ergo transmitti non possunt, quia accidentia non 10
possunt permutare suum subiectum.

308. Solutio. Dicimus quod potentia percipiendi est in anima ut in
subiecto, et potest intueri passiones corporis, licet ipsa ibidem non sit
cum ipsis passionibus presens secundum sui essentiam; quia, ut assignata est ratio superius, ipsa semper, ut suam habeat perfectionem, habet 15
convertere suas intentiones ad inmutationes formatas in corpore, et
omnes illas potest intueri inclusa in corpore habens se ad similitudinem
centri. Sicut enim si centrum esset oculus videret totam circumferentiam, et si esset illa circumferentia speculum, ipse oculus posset
videre ymagines usquequaque parte circumferentie constitutas in 20
speculo, et tamen non esset ipse oculus actu existens secundum sui
essentiam in unaquaque parte ipsius circumferentie. Similiter anima
existens in corpore potest intueri ymagines formatas in ipso corpore, non tamen erit presens secundum sui essentiam cum ipsis
ymaginibus.   25

309. Ad aliud. Dicendum est quod, licet vegetatio per totum corpus
sit, non tamen oportet quod anima distendatur per totum corpus. Sicut
si centrum circuli esset igneum, ipsum illuminaret totum circulum, et
illuminatio procedens ab ipso distenderetur per totum; et tamen ipsum

---

1 corpus *om.* P   3 aliquis: quis P   hoc *om.* C   transmissionem: transmu-
tationem P   4 corporei: corpora P   4–5 incorporei: non corpora P
aut incorporei *om.* V   5 animati[1, 2]: animata P   inanimati: inanimata P   6 de:
ex C   inanimati: inanimata P   8 sunt *om.* P   transmittuntur nisi: transmutantur in P   9 obtineant: habeant C   sunt *om.* P   10 transmitti: trans-
mutari P   11 permutare: transmutare V   13 ipsa *om.* P   16 convertere:
convertens V   17 potest: ipsa *add.* V   similitudinem: dissimilitudinem P
19 si *om.* P   20 parte *om.* V   20–22 constitutas . . . circumferentie *om. per
hom.* C   26 aliud: hoc P   27 tamen *om.* V   28 si centrum: intelligere quod tōtū P

3–4 per transmissionem spirituum vitalium: cf. Costa ben Luca ibid. i–ii (121–30);
Guill. de S. Theod. ibid. I (700B–701C); Guill. de Conchis *Dragm.* VI (268–9); Alfredus
Anglicus *De Motu Cordis* x (37–45).
6 precedens dubitatio: cf. supra, n. 305.
15 superius: cf. n. 304.
26 Ad aliud: cf. supra, n. 306.

the soul, in what way can all of the body have growth and flourishing from the soul unless the soul is extended throughout all of the body?

307. Regarding this, perhaps someone might say that this is by means of the sending out of vital impulses. Against: those impulses are either corporeal or incorporeal; either animate or inanimate. If they are animate, therefore they have a soul, and thus the preceding doubt arises concerning them. If they are inanimate, they cannot therefore bestow life, nor growth and flourishing. If they are incorporeal, either they are substances or not. If they are substances, in what way are they sent out unless they have life, and also have a location? If they are not substances, therefore they are accidents, therefore they cannot be sent out, because accidents cannot transform their subject.

308. Solution: We state that the potential to perceive is in the soul as in a subject, and it can intuit the affections of the body, even if it is not present in the same place as its affections, according to its essence. Thus, for the reason given above, in order for it to achieve its perfection, it always has to turn its intentions towards the impressions formed in the body, and it can intuit all of these while enclosed in a body, being constituted in the manner of a centre. For just as if the centre were an eye, it would see all of the circumference, and if that circumference were a mirror, the same eye would be able to see images displayed in the mirror as far as any part of the circumference, and yet the eye itself would not exist according to its essence at any part of the circumference itself, similarly a soul existing in a body can intuit images formed in the body itself, but it will not however be present according to its essence with the images themselves.

309. Regarding the next objection, it should be stated that, even if life and growth is throughout all of the body, it is not however necessary that the soul is extended throughout all of the body. It is the same as if the centre of a circle were fiery it would illuminate all of the circle, and the illumination coming from it would be extended throughout all of it, and yet the centre itself is simple and lacks quantity and is not

centrum simplex est carens quantitate et non distensum. Similiter ab anima est vegetatio in totum corpus procedens et distensa per totum, non tamen ipsa anima distensionem habet. Sol autem totum mundum illuminat, non tamen est secundum sui essentiam ubique presens ubi est eius illuminatio.   5

Ostensum est autem per premissa quod anima est incorporea substantia.

310. Sed forte dicet quis quod in homine tres sunt anime diverse, scilicet anima vegetabilis, qua convenit homo cum arboribus et plantis; et anima sensibilis in qua convenit homo cum brutis animalibus; et 10 anima rationalis in qua convenit homo cum angelis et ceteris intelligentiis. Dicet etiam quod anima rationalis est incorporea, sed anima sensibilis et anima vegetabilis sunt corporee et distenduntur per totum corpus vegetatum quod tactu utitur.

311. Contra. Demonstretur aliqua pars extrema corporis vegetati et 15 sentiens, ut digitus. Inde sic. Hoc est corpus animatum et sensibile, quia est vegetatum et utens sensu, quia tactu, et corpus animatum sensibile convertitur cum animali. Ergo cum illud corpus sit animatum, sensibile, ipsum erit animal, quia hec est definitio animalis: animal est corpus animatum sensibile. Sed illud est digitus hominis; ergo digitus hominis 20 est animal. Sed eodem modo potest ostendi de infinitis partibus hominis, quod ipse sunt animalia. Sic ergo essent de integritate hominis infinita animalia contiguata vel continuata ad invicem.

312. Item. Anima sensibilis est substantia corporea, ut dicit Respondens, et distenditur per totum corpus hominis. Sed aliud est corpus 25 quod est vegetatum, et aliud anima sensibilis, que est dans sensum; et anima vegetabilis que est vegetans est aliud. Ergo cum tam anima sensibilis quam anima vegetabilis sit substantia corporea, que est in corpore sensato, vel sunt ibi diversa corpora simul in eodem loco circumscripto, vel unum corpus aliud corpus subintrat. Sed duo corpora in eodem loco 30

---

1 simplex *om.* C   3 habet: non habebit P   4 est *om.* C   6 est²: sit C
8 Sed: Si C   dicet: dicat C   quis: aliquis P   9 convenit homo: conveniunt
homines P   10 in *om.* V   convenit homo: conveniunt P   11 convenit homo:
conveniunt P   13 anima *om.* P   16 sentiens: sensuati P   et: est *add.* V
17 est *om.* C   19 ipsum *om.* P   erit: est V   19–20 quia ... sensibile: ergo
illud est animal P   20 illud: istud P   21 Sed eodem: Simili P   21–22 de
infinitis . . . animalia: de qualibet parte hominis integrali P   22 essent: erunt P
23 contiguata: contiguentia P   26 aliud: est *add.* V   27 est² *om.* P   cum
m. V   29 sensato: vegetato C   ibi: in P   simul *om.* C   loco *om.* V
o corpus² *om.* P   30–p. 85, l. 2 Sed duo . . . subintrat *om. per hom.* C

---

6 Ostensum est autem per premissa: cf. supra, nn. 296–9.
8–12 Cf. Gundiss. *De An.* iv (44). Cf. supra, cap. IV.
16–20 Cf. Arist. *De An.* ii. ii (413ᵇ2–9).
24–25 ut dicit Respondens: cf. supra, n. 310.

XXII. CONCERNING THE RATIONAL SOUL

extended. In the same way life and growth comes from the soul, goes forth into all of the body, and is diffused throughout, and yet the soul itself does not have extension. For the sun illuminates all of the world, and yet it is not present according to its essence everywhere where its illumination is.

Moreover, through what has been put forward[10] it has been shown that the soul is an incorporeal substance.

310. Yet perhaps someone will say that there are three different souls in man, namely the vegetative soul which man shares in with trees and plants, the sensitive soul which man shares in with brute animals, and the rational soul which man shares in with the angels and the rest of the intelligences. Someone might also say that the rational soul is incorporeal, but that the sensitive soul and the vegetative soul are corporeal and are extended throughout the vivified body which makes use of touch.

311. Against: any extremity of a vivified body may be shown to be also sentient, such as a finger. From which it follows: this is an animated and sensible body, because it is living and flourishing and uses sensation, because it uses touch, and 'animated and sensible body' is interchangeable with 'animal'. Therefore, when that body is animated and capable of sensation that will be an animal, because this is the definition of an animal: an animal is an animated sensible body. However, this is what the finger of a man is; therefore the finger of a man is an animal. In the same way, however, it can be shown regarding infinite parts of man that these are animals. Therefore, there will be infinite animals contiguous or continuous with each other making up a complete man.

312. Moreover, the sensitive soul is a corporeal substance, as the Respondent[11] says, and is extended throughout the human body. However, the body which is vivified is one thing, the sensitive soul which gives sensation is another, and the vegetative soul from which there is growth is another. Therefore, since both the sensitive soul and the vegetative soul are a corporeal substance which is 'in the sensible body', either there are various bodies enclosed there at the same time in the same space or one body enters into another body. However, two bodies do not exist in the same place at the same time

---

[10] In nos. 296–99.   [11] This refers to no. 310.

simul non sunt: hoc est notum in principio intellectus in naturalibus;
ergo unum corpus aliud subintrat.

313. Quod dicet forte. Dicet enim quod anima secundum quod est
corporea et vegetabilis subintrat corpus in quo est; sed illud corpus in
quo illa anima subintrat non omnino dividitur in illa subintratione, 5
quoniam secundum hoc esset divisio in infinita facta, et ita illud corpus
omnino divisum periret. Relinquitur ergo quod ibi non est omnino divisio
facta. Ergo est ibi aliqua pars corporis subintrati in qua particula nulla
est particula anime subintrans, et illa pars corporis subintrati habet
longum, latum, spissum, et ita in eius medio nulla erit vegetatio nec 10
aliquis sensus propter defectum anime vegetabilis et anime sensibilis,
et ita illud corpus erit purum corpus, non vegetatum; erit ergo secundum
hoc solummodo vegetatio et sensus in contiguatione et in continuatione
anime cum corpore.

314. Item. Si anima sensibilis et vegetabilis sint corporee, illa anima 15
et corpus in quo sunt, aut contiguantur, aut continuantur. Si continuantur: tunc est diversarum rerum continuatio, quod est contra
Aristotelem in *Physicis*. Si contiguantur: tunc contingit quod ex una
parte erit corpus purum et ex alia erit anima, quia contigua sunt quorum
termini simul sunt. Non erit ergo in illo corpore puro vegetatio vel 20
sensualitas nisi in contiguatione solum, quia qualiter sentiret ipsum
corpus in medio puncto sui, cum anima sit contiguata ei extra.

315. Forte dicet aliquis, quod ab anima transmittuntur spiritus vitales
et spirituales, et quod a vitalibus est vita, a spiritualibus est sensus.

316. Contra. Illi spiritus aut sunt corporei, aut incorporei. Si cor- 25
porei: eadem erit obiectio quod prius. Si incorporei: contra, ut dicit

3 Quod dicet forte. Dicet enim: Ut hoc dicet forte P   3–4 Quod dicet . . .
subintrat *om. per hom.* V   4 et: anima *add.* P   in quo: quod CP   6 secundum . . . facta: in infinitum esset facta divisio V   6–8 et ita . . . facta *om. per
hom.* CV   8 subintrati: subintratti C; subintrata P   qua: aliqua P   nulla:
non C   9 anime *om.* P   subintrans: subintrantis P   10 spissum: spissitu-
dinem P; longitudinem lati spissi V   ita *om.* P   12 corpus: et *add.* V   13 et¹
om. V   et in continuatione *om.* C   15 illa: ille V   16 quo: ipse *add.* C
16–17 Si continuantur *om.* P   17 est² *om.* V   18 tunc contingit quod: quoniam P
19 alia: altera V   parte *add.* P   sunt: erunt C   20 illo: eo P   puro: poni P;
om. V   vel: et P   23 transmittuntur: transmutantur P   vitales: vegetabiles
P   24 vita: et *add.* P   est *om.* C   25 Contra: Si contra P   25–26 Illi . . .
corporei *om.* P   26 quod: que C   contra: sed C

1 in naturalibus: cf. Arist. *De An.* I. v (409ᵇ3); II. vii (418ᵇ17); *Physic.* IV. i (209ᵃ7);
De Gen. et Corr. I. v (321ᵃ8–9); etc.
18 Arist. in *Physicis*: v. iii (227ᵃ10–16).
19–20 contigua . . . sunt: *Physic.* v. iii (226ᵇ24) iuxta versionem Gerardi Cremonensis.
Versio Graeco-latina habet: 'Tangere autem quorum ultima simul sunt' (cod. Coll.
Corporis Christi Oxon. 111, fol. 65ᵛ).
23–24 Cf. supra, n. 307.
26 quod prius: cf. supra, nn. 306–7.

(this is learnt in the rudiments of natural philosophy), therefore one body enters into another.

313. This is something which perhaps he might say, namely that inasmuch as the soul is corporeal and vegetative it enters into the body in which it is. However, that body into which that soul enters is not completely divided by that entering in, because in this way a division into infinity would be made, and thus that body which is completely divided would perish. Therefore, it follows that there is no complete division at all. Thus, there is there some part of the body entered into in which there is a particle which is not a particle of the soul entering in, and that part of the body entered into has length, width, and depth, and thus in the most of it there will be no growth nor any sensation because of the lack of a vegetative soul and a sensitive soul, and thus that body will be pure body and not vivified. Accordingly, there will only be growth and sensation in contiguity and in continuity of the soul with the body.

314. Moreover, if the sensitive soul and the vegetative soul are corporeal, then the soul and body in which they are are either contiguous or continuous. If they are continuous then there is a continuity of diverse things, which is against what Aristotle says in the *Physics*. If they are contiguous then what happens is that in one part there will be pure body and in the other there will be soul because contiguous things are those whose termini come together. Therefore, there will not be vegetation or sensation in that pure body except in contiguity alone, because in what way would the body sense itself in its middle point, if the soul were externally bordering it?

315. Perhaps someone might say that because vital and breath impulses are transmitted from the soul, and because life is from vital impulses, sensation is from breath impulses.

316. Against: those impulses are either corporeal or incorporeal. If they are corporeal, there will be the same objection as before. If they are incorporeal one can argue against, using what Avicenna says in the

Avicenna in *Metaphysica*, omnis substantia aut est corpus, aut non est corpus. Si non est corpus: aut est pars corporis, aut non. Si est pars corporis: aut est materia, aut forma. Si non est pars corporis: aut est tale quid quod habet ligationem cum corporibus ut moveat illa, et dicitur anima; aut est tale quid quod non habet ligationem cum corpori- 5 bus ut moveat illa, et est substantia separata, et dicitur esse intelligentia. Si igitur spiritus ab anima transmissus per totum corpus sit substantia incorporea, ipse aut erit pars corporis, aut non. Non erit pars corporis, quia ipse nec est materia neque forma; ergo erit vel anima vel intelligentia. Redibit ergo prior obiectio. Hiis ergo obiectionibus et rationibus 10 habetur quod nulla anima est corporea.

XXIII. *Utrum anima sit mortalis vel immortalis*

Sequitur videre utrum anima sit mortalis vel immortalis. Quod sit immortalis sic ostenditur.

317. Omne illud quod in quantum ipsum est seipso est aliquale, se 15 ipso erit tale semper; ut triangulus, in quantum ipse est, habet tres angulos equales duobus rectis: unde triangulus necessario habet tres angulos equales duobus rectis. Cum ergo anima, in quantum ipsa est anima, seipsa vivat, anima non potest non vivere. Ergo anima necessario vivit; ergo anima est immortalis. 20

318. Item. Quicquid corrumpitur, corrumpitur per eius contrarium. Ergo cum anima contrarium non habeat, nec componatur ex contrariis, anima corrumpi non potest; ergo anima est incorruptibilis; ergo anima est immortalis.

319. Item. Ut habetur fere ab omnibus auctoribus de morte loquenti- 25 bus, mors nihil aliud est quam separatio anime a corpore. Sed anima non est corpus, nec animam habet; ergo in ipsa non potest fieri separatio anime a corpore; ergo ipsa mori non potest.

320. Item. Unicuique mutationi necesse est aliquid subici, ut habetur

---

1 est¹, ² *om.* P   4–6 et dicitur . . . illa *om. per hom.* C   7 igitur: ergo V
8 Non erit . . . corporis *om.* V   9 nec: non C   est: erit P   10 prior:
prius C   ergo obiectionibus et *om.* C   13 videre: ut videamus V   15 est¹
om. P   16 ipse *om.* V   17–18 unde . . . rectis *om. per hom.* C   18 ergo:
igitur P   ipsa *om.* P   22 componatur: opponatur C   nec componatur ex
contrariis: non componatur extra contrarium P   23 anima¹ *om.* P   25 fere
om. CV   ab: de C

1–6 Avic. *Metaphys.* II. i (75ʳᵃB). Cf. supra, n. 32.
12 Vid. Avic. *De An.* V. iv (24ᵛᵃA–25ʳᵇC).
21 Cf. Arist. *Physic.* I. v (188ᵇ21–23); *De Caelo*, I. iii (270ᵃ15).
21–24 Cf. Gundiss. *De Immort. An.* (9).
25–26 ab omnibus . . . loquentibus: Aug. *Enarr. in Ps. XLVIII* serm. ii. 2, cit. apud P. Lombard. *Sent.* III dist. xxi. c. i.

Metaphysics: every substance is either a body or is not a body. If it is not a body, then either it is part of a body or not. If it is part of a body, either it is matter or form. If it is not a part of a body, either it is such that it has a link with bodies so that it moves them, and it is called a soul, or it is such a thing that it does not have a link with bodies so that it moves them, and it is a separate substance, and is said to be an Intelligence. If, therefore, the impulse which is sent out from the soul throughout the body is an incorporeal substance, it will either be a part of a body or not. It will not be a part of a body, because it is neither matter nor form; therefore it will either be a soul or an Intelligence. Therefore the previous objection will return. Therefore from these objections and arguments we have it that no soul is corporeal.

XXIII. *Whether the soul is mortal or immortal*

Next we should see whether the soul is mortal or immortal. That it is immortal can be shown as follows.

317. Each thing which, in so far as it is in itself is of some kind, will always remain as such in itself. For example, a triangle, inasmuch as it is, has three angles equal to two right angles, so a triangle necessarily has three angles equal to two right angles. Since, therefore, the soul, inasmuch as it is a soul, in itself lives, it cannot not live. Thus the soul necessarily lives; and so the soul is immortal.

318. Moreover, whatever is corrupted is corrupted by its contrary. Therefore, since the soul does not have a contrary, nor is it made up of contraries, the soul cannot be corrupted. Therefore the soul is incorruptible; and so the soul is immortal.

319. Again, as is held by nearly all authors who speak about death, death is nothing other than the separation of the soul from the body. However, the soul is not a body nor does it have a soul; therefore there cannot occur a separation of the soul from the body in it; therefore it cannot die.

320. Again, it is necessary for something to underlie any change whatsoever, as is held at the end of the first book of the *Physics* of

in fine primi *Physicorum* Aristotelis, quoniam omnis mutatio est in aliquo subiecto. Ergo cum mors sit corruptio, et corruptio sit mutatio,
morti necesse est aliquid subici quod relinquitur actualiter post mortem
existens. Ergo si anima moritur, aliquid de anima relinquitur post
mortem. Illud relictum non erit corpus. Aut ergo illud erit pars cor- 5
poris aut non. Si non est pars corporis, aut est anima aut intelligentia;
quorum neutrum potest esse. Ergo anima mori non potest. Ergo anima
est immortalis.

321. Item. Hec propositio est per se nota: Omne illud quod possibile
est mori anima deserere potest. Sed anima animam deserere non potest. 10
Ergo animam non est possibile mori. Syllogismus est in secunda figura.
Ergo anima est immortalis.

322. Item. Aut ipsa anima habet hoc ipsum quod ipsa est a seipsa,
vel a prima essentia, que summe et maxime est incommutabilis. Si
a seipsa, ergo interire non potest. Si a prima essentia; Contra: essentie 15
nil est contrarium. Ergo essentie, que primitus est et summa, nihil est
contrarium. Sed constitutio speciei existendi et privatio eiusdem sunt
a contrariis. Ergo cum species existendi ipsius anime sit proximo loco
a pura essentia et incommutabili, si privatio et destructio eiusdem contingat esse, ipsa destructio erit ab aliquo quod erit contrarium prime 20
essentie. Sed prime essentie nihil est contrarium. Ergo hoc ipsum quod
est anima, destrui non potest, cum contrarii effectus a contrariis fluant
causis; et ita habetur quod anima est immortalis.

323. Contra. Esse est a perfectione rei; ergo non-esse est ab imperfectione. Sed ut habemus ab Aristotele et ab aliis auctoribus, anima 25
est creata imperfecta et perfectibilis a scientiis et virtutibus. Sed

1 omnis *om.* V   2 cum *om.* C   3 quod: ergo P   5 relictum: derelictum
V   ergo: igitur V   6 est[2] *om.* P   7 neutrum: nullum P   Ergo anima
mori non potest *om.* V   anima[2] *om.* P   9–12 Item. Hec . . . immortalis *om. per
hom.* C   9 possibile: impossibile P   11 non est possibile: impossibile est V
13 ipsa[2]: in ipsum P   14 a: ex C   15 Si a: Sed V   Contra: Similiter
ostenditur quod ipsa est immortalis, sic *add.* V   18 a: in C; ex V   existendi *om.*
V   19 eiusdem: anime *add.* V   20 esse *om.* P   destructio *om.* CP   erit[2]:
est V   21 Sed: Si C   24 Contra: Item CP   25 habemus: habetur
P   ab[2] *om.* C   26 creata: creatura V

1 Arist. *Physic.* I. ix (192[a]29): Versio Graeco-latina habet: 'Si autem fiat, oportet
subici aliquod primum in quo est' (cod. Coll. Corporis Christi Oxon. 111, fol. 13).
Gl. ad loc. '*aliquod* i.e. aliquam materiam *in quo est* i.e. ex quo debet materia generari'.
5–7 Cf. supra, n. 316. Gundiss. *De Immort. An.* (28).
14 Cf. Algazel *Metaphys.* II. iv. 5 (181).
25 ab Aristotele et aliis auctoribus: cf. Gundiss. *De An.* iii (43): 'Omne quod
crescit imperfectum est; nihil enim crescit nisi ad hoc ut perficiatur. Sed anima
crescit quia in sapientia et iustitia proficit; igitur imperfecta est'; Avic. *De An.* I. v
(5[va]E); v. i (22[rb]A–22[va]B). ab Aristotele: forte respicit ad *De An.* III. iv. (430[a]1): 'sicut
in tabula nihil est scriptum actu, quod quidem accidit in intellectu'.
25–26 Cf. supra, n. 304.

Aristotle, because all change is in some subject. Therefore since death is corruption, and corruption is change, it is necessary that something underlies death and is actually left existing after death. Therefore if the soul died, something of the soul would be left after death. That thing which is left will not be a body. Therefore, that thing will either be a part of a body or not. If it is not part of a body it is either a soul or an Intelligence, and it cannot be either. Therefore the soul cannot die. Therefore the soul is immortal.

321. Again, this proposition is self-evident: 'The soul can abandon every thing which can die.' However, the soul cannot abandon a soul. Therefore it is not possible for the soul to die. This is a second-figure syllogism. Therefore the soul is immortal.

322. Again, either the soul itself is constituted as a substance because it exists from itself or from the First Essence, which to the highest and greatest extent is unchangeable. If from itself, then it cannot be destroyed. If from the First Essence, then it can be argued that nothing is contrary to an essence. Therefore nothing is contrary to the essence which is First and Highest. However, the formation of a kind of existence and its privation are due to contraries. Therefore, since the kind of existence of the soul itself is in the place next to the pure and unchanging essence, if the privation and destruction of this happens, this destruction will be from something which is contrary to the First Essence. However, nothing is contrary to the First Essence. Therefore, that which is the soul cannot be destroyed, since the effects of a contrary flow from contrary causes, and so we know that the soul is immortal.

323. Against: Being is from the perfection of a thing; therefore non-being is from imperfection. However, as we know from Aristotle and other authors, the soul is created imperfect but capable of perfection from knowledge and virtue. Yet knowledge is only of true things,

tantummodo verorum est scientia. Ergo quanto aliqua sunt veriora, tanto verior est illorum scientia. Ergo pure veritatis que summe ac primitus est incommutabilis, est pura virtus et summa scientia. Cognitio igitur pure veritatis sine qua nulla virtus est, est summa perfectio anime. Ergo cum veritas et falsitas sint opposita, intuitio pure falsitatis est anime 5 imperfectio. Anima igitur que falsitatem intendit respuendo veritatem summam tali erronea intuitione est imperfecta. Sed ab imperfectione est non esse; sic ergo cum possibile sit animam sequi falsitatem, possibile est animam non esse, et ita possibile est eam mori, quoniam in intuitione falsitatis declinando a bono exequitur viam falsitatis, et sic 10 remanet imperfecta et non ens omnino.

324. Solutio. Ad hoc dicendum quod anima duplex habet esse; unum a sua prima perfectione, scilicet a prima creatione, et illud est immortale et eo privari non potest; secundum eius esse est ab adquisitione scientiarum et virtutum, et hoc esse potest ipsa privari, ut si ipsa puram 15 veritatem respuat et falsitatis viam intendat; sed propter hoc esse non debet ipsa dici mortalis, immo commutabilis et passibilis.

325. Item. In constitutione hominis duo sunt, scilicet corpus et anima. Sed quoad corpus per resolutionem contingit fieri reversionem humane nature in prima corpora elementaria, scilicet in sua principia, 20 post segregationem corporis et anime; pari ratione reliquum componentium, scilicet animam, contingit reverti in suum principium; sed anime principium nihil aliud est quam divina preconceptio que fuit in divina providentia. Sic ergo contingit fieri reversionem anime in suum principium. Sed facta huiusmodi reversione non dicetur tunc anima 25 hominis actu esse, sicut nec post reversionem corporis humani in eius elementa dicetur corpus humanum esse. Cum igitur anima ante reversionem illam fuerit, et post eam non erit, anima est mortalis.

326. Item. In homine duo sunt constituentia eum, quorum unum est corporale, reliquum spirituale. Sed illud quod corporale est contingit 30

---

1 tantummodo *om.* V   2 verior: veriorem V   verior est illorum: earum maior est P   3 virtus et summa *om.* C   4 pure *om.* V   est[2]: et C   5 falsitatis: falsitas *codd.*   6 igitur: enim V   7 imperfecta: inperfectio P   8 cum possibile: inpossibile P   9 animam: animari P   et *om.* P   in *om.* P   11 et *om.* C   15 ipsa[2] *om.* V   18 scilicet *om.* PV   21 post: per C   reliquum: reliqua P   22-23 sed anime principium *om.* C   23 que: quo C   24 reversionem: resolutionem C   25 reversione: resolutione V   28 est *om.* P   30 corporale[2]: spirituale P

14-16 Cf. Gundiss. *De An.* x (101): 'Rationalis ergo anima in tantum vere est in quantum se et Deum intelligit. Sed in tantum a vero esse deficit et quasi moritur, in quantum a contemplatione veritatis avertitur. Et tamen non sic moritur ut esse desinat, sed sic moritur ut sine beata vita semper infeliciter vivat'.

19-21 Vid. p. xx.

therefore the more some things are true, the more true is the knowledge of them. Therefore, pure virtue and the highest knowledge is of the pure truth which is unchanging to the first and highest degree. Therefore, the knowledge of the pure truth (without which there is no virtue) is the highest perfection of the soul. Therefore, since truth and falsity are opposite, the intuition of pure falsity would be the imperfection of the soul. Therefore, the soul which directs itself towards falsity by rejecting the highest truth is imperfect through such an erroneous intuition. However, non-being is from imperfection, so since it is possible for the soul to follow falsity, it is possible for the soul not to be, and thus possible for it to die. This is because in the intuition of falsity it turns away from the good and follows the way of falsity, and thus it remains incomplete and a non-being in every way.

324. Solution: To this it should be stated that the soul has a twofold being, one from its first perfection, namely from the first creation, and that is immortal and it cannot be deprived of it. Its second being is from the acquisition of knowledge and virtue, and it can be deprived of this being, such as if it rejects the pure truth and turns towards the way of falsity. Thus, because of this, this being should not be said to be mortal but rather changeable and capable of being acted upon.

325. Again, there are two components in the make-up of man, namely the body and soul. However, as regards the body, it happens that the return of human nature to the first elemental bodies occurs through dissolution, namely into its principles after the separation of soul and body. For the same reason the remaining component, namely the soul, should revert to its source. However, the origin of the soul is nothing other than the divine plan which was in the divine providence. Therefore, the return of the soul to its source should happen in this way. However, once this return has taken place one would not then say that a human soul exists in act, just as neither would one say that a human body exists after the return of the human body into its elements. Since, therefore, the soul existed before that return, and afterwards it will not be, the soul is mortal.

326. Again, in man there are two things which enter into his make-up, one of which is corporeal and the other spiritual. However, that which is corporeal must return into its first elemental principles. For

reverti in sua principia prima elementaria; pari ratione illud quod spirituale est erit resolvere in sua prima principia spiritualia; sed sua prima principia sunt in divina providentia, quoniam Deus ab eterno previdit animam esse. Sic ergo, ut prius, in illa reversione erit anima mortalis.   5

327. Si forte aliquis ita concederet, secundum hunc modum procedendi contingeret quod, sicut ante creationem mundi omnia erant in divina providentia, similiter contingeret omnia adhuc iterum reverti ad divinam providentiam, et nihil esset actu preter solum Deum.

328. Solutio. Dicendum est quod non contingit reversionem humane 10 nature quoad animam sicut quoad corpus. Revertitur enim corpus humanum in elementa prima, quia prima elementa conveniunt in eius constitutione et salvatur virtus eorum in ipso et materia eorum. Sed ita non est de anima, quia neque divina essentia neque divina providentia est pars anime; quia tunc esset anima omnino incommutabilis et nullo 15 modo demereri posset. Contingit autem fieri reversionem ad Deum per intuitionem divine essentie et eius cognitionem; ab ipso enim processit ipsa anima et ad ipsum est reditura, nisi per eius contemptum fuerit impedita.

XXIV. *Utrum anima sit simplex vel composita*   20

Sequitur ut dicamus: Utrum anima rationalis sit simplex vel composita. Quod sit composita sic videtur posse ostendi.

329. Inter ea que creantur a causa prima in effectu reperiuntur duo genera causatorum, unum corporale et aliud spirituale. Sed ita est in corporalibus quod corporalia habent stabilimentum sue essentie a 25 materia et a forma. Sed spiritualia stabilioris sunt essentie. Ergo cum illa stabilius habeant esse, habent illa stabiliori modo in effectu a materia spirituali et a forma spirituali. Cum igitur anima sit una de spiritualibus

---

1 principia prima: prima corpora P   pari: predicta P   2 prima *om.* C   sed: que V   4 previdit *om.* V   esse *om.* P   6 Si: Et si V   aliquis ita concederet: illud concederetur P   ita: ista V   secundum hunc: et habet P 7 erant: fuerunt P   8 contingeret omnia adhuc: ad hunc contingunt omnia P 10–11 Solutio . . . corpus *om.* C   12 elementa prima: per contrarietatem eorum *add.* V   13 ita *om.* V   14 essentia: scientia V   16 Contingit autem: Et continget C   18 anima *om.* P   et: tamen *add.* V   contemptum: conceptum P 21 Sequitur: Restat PV   dicamus: videamus V   22 ostendi: probari P   24 et *om.* C   spirituale: incorruptibile P   25 corporalibus: corruptibilibus P   corporalia: ea que ibi sunt PV   26 a *om.* PV   27 habeant: habent C   stabiliori: stabiliri C   28 una *om.* V

---

4 ut prius: cf. supra, n. 325.
11–13 Cf. Avic. *De An.* v. iv (24^{vb}A).
20 Cf. Gundiss. *De An.* vii (53–60).

the same reason, that which is spiritual will return into its first spiritual elements. However, its first elements are in divine providence because from eternity God foresees the soul to be. Therefore, as was argued before, in that return the soul will be mortal.

327. If someone perhaps accepts this to be the case, then following this way of proceeding it would turn out that, just as before the creation of the world everything was in the divine providence, similarly it would happen that all things will return again to the divine providence. Nothing, then, will exist in act except for God alone.

328. Solution: It should be stated that as regards the soul the return of human nature does not happen in the same way as the body. The human body returns to the first elements, since the first elements come together in its constitution and their power and matter are preserved in it. However, it is not such where the soul is concerned, because neither the divine essence nor the divine providence is part of the soul, because then the soul would be unchangeable in every way, nor could it lose merit in any way. It happens, however, that a return to God occurs by means of an intuition of the divine essence and the knowledge of it. For the soul proceeded from God and to Him it will return, unless it is impeded by a contempt for God.

XXIV. *Whether the soul is simple or composed*

Next we should discuss whether the rational soul is simple or composed. That it is composed it seems can be shown as follows.

329. Among those things which are created by the First Cause are to be found in effect two types of caused things, one which is corporeal and the other which is spiritual. However, corporeal things are such that they have the foundation of their essence from matter and from form. Spiritual things, however, are of a more stable essence. And so, since they have their being more stably, they have ⟨it⟩ [viz. their being] in effect in a more stable manner as a result of their spiritual matter and spiritual form. Since, therefore, the soul is one of the

creaturis habebit compositionem ex materia et forma, et ita anima rationalis est composita.

330. Preterea. Anima est substantia et non parificatur substantiae; ergo habundat in aliquo quod facit animam non adequari substantie, sicut species habundat suo genere. Ergo cum substantia sit de veritate 5 essentie anime, et sit ibi preter illud quiddam in quo habundat anima a substantia, erit anima composita ex duobus illis, quia, circumscriptis omnibus accidentibus ab anima, adhuc erit invenire illa duo in veritate essentie anime. Est ergo composita ex diversis, ergo non est simplex.

331. Si concedatur; Contra. Anima componitur ex diversis, dicatur 10 unum illorum A, reliquum B. Inde sic: A aut est animatum aut non est animatum. Si est animatum, ergo est habens animam, et anima vege-tativum, et illa anima iterum aliam animam, et sic in infinitum. Si non est animatum, pari ratione nec B est animatum. Sed ex duabus albedini-bus numquam proveniet nigredo. Unum enim contrariorum non causat 15 reliquum. Ergo nec ex duobus inanimatis fiet animatum. Ergo num-quam fiet anima ex illis duobus.

332. Preterea. Unicuique composito respondet sua divisio, quoniam quecumque erit componere erit resolvere. Ergo si anima est composita, ipsa est divisibilis in componentia; ergo est corruptibilis, ergo destrui 20 potest, quod superius improbatum est.

333. Solutio. Ad primam obiectionem. Dicimus quod per illam potius debet ostendi animam esse simplicem quam compositam, quia propter hoc quod simplicis essentie est anima, propter hoc stabilimentum habet sue essentie anima. Prima autem causa simplicissima est; unde ea que 25 immediate exeunt ab ea simplex habent esse: unde anima et intelli-gentie simplex habent esse.

334. Ad aliud. Dicendum est quod compositio illa que est ex genere et substantiali differentia non est nisi dicta compositio ad similitudinem

1 compositionem: esse compositum V   2 rationalis *om.* V   3 parifi-catur: parificatrix C   6 essentie: ipsius C   6–9 et sit . . . anime *om. per hom.* C   7 erit: dicitur V   8 ab anima: anime anima P   10 com-ponitur: est composita P   12 est² *om.* V   13 iterum: habet *add.* V   aliam *om.* C   sic: ita P   15 proveniet: prout P; provenit V   19 resolvere: dividere V   20 est¹ *om.* C   22 primam . . . dicimus: primum dicitur P   23 quia *om.* V   24 propter hoc *om.* V   26 simplex: simpliciter V   29 simi-litudinem: secundum oppinionem et non secundum veritatem. Quae autem recipiuntur secundum divisionem topicam non sunt de veritate essentie rei, imo consequuntur veram rei essentiam *add.* V

3–5 Cf. supra, n. 39.
15–16 Cf. Arist. *Meteor.* IV. vi (383ᵃ7–8).
21 superius improbatum est: cf. cap. XXIII: *Utrum anima sit mortalis vel im-mortalis.*
22 Ad primam obiectionem: cf. supra, n. 329.
28 Ad aliud: cf. supra, n. 330.

spiritual creatures, it has a composition out of matter and form, and thus the rational soul is composed.

330. Moreover, the soul is a substance and yet it is not considered to be just the same as a substance. Therefore, there is something more in it that makes the soul such that it is not taken to be the same as a substance, just as there is more in a species than in its genus. Therefore, since substance is truly of the essence of the soul, and is there besides that something by which the soul is more than a substance, the soul will be composed from those two, since, when all of the accidents have been enclosed by the soul, there will still be found those two in the truth of the essence of the soul. Therefore it is composed out of various things, therefore it is not simple.

331. If this is conceded, it can be objected that the soul is made up of various things. Let one of them be called A and the other B. From which it follows, A either is animated or is not animated. If it is animated, then it has a soul, and this soul is vivified, and that soul again has another soul, and so on to infinity. If it is not animated, for the same reason neither is B animated. However, blackness will never come from two white things, because one of the contraries does not cause the other. Therefore neither does an animated thing arise from two inanimate things. Therefore a soul will never arise from those two.

332. Moreover, to any composed thing corresponds its division because whatever will compose something will dissolve it. Therefore, if the soul is composed, it can be divided into components. Therefore, it is corruptible. Therefore, it can be destroyed, which has been rejected above.

333. Solution: As regards the first objection, we state that by it it will be shown that the soul is simple rather than composed, because of the fact that the soul is a simple essence. It is because of this that the soul has a foundation for its essence. For the First Cause is the most simple; and so those things which immediately come from it have simple being. For this reason, the soul and the Intelligences have a simple being.

334. Regarding the next objection, it should be stated that the composition arising from genus and substantial difference only exists

et proportionem compositionis constantis ex materia et forma, ut potest haberi a Porfirio, et si intellectu tenus sumatur genus anime, et eius differentia divisiva. Non tamen ita fuit in creatione anime quod prius fuerunt illa componentia, scilicet genus anime et eius differentia ante creationem anime. Et etiam, licet in anima esset et materia spiritualis 5 et forma spiritualis, non tamen esset ipsa corruptibilis, cum causa contrarietate careat secundum quam fit corruptio.

xxv. i. *De viribus anime rationalis*

335. Sequitur de viribus anime rationalis. Due autem sunt eius vires, scilicet virtus agendi et virtus sciendi. Virtus autem agendi est illa virtus 10 a qua sunt singule actiones hominis quas ipse eligit pro iudicio et intentione rationis. Huiusmodi autem virtutis triplex est via: Una est secundum quam consideratur comparatio sui ad virtutem vitalem et appetitivam, et est iste modus secundum dispositiones quibus afficitur homo cum accidit ei confusio aut erubescentia, aut risus aut aliquid 15 consimilium. Alius modus consideratur secundum comparationem ipsius ad seipsam in quantum ipsa in se generat usum et intellectum, ut per multa experimenta singularium generatur in anima intellectus principiorum, ex quibus principiis adquiruntur scientie et demonstrationes. Tertius modus consideratur secundum comparationem virtutis 20 active ad virtutem ymaginabilem et estimabilem. Ipsa enim est eis imperans secundum iudicium virtutis contemplative.

336. Virtus autem sciendi est virtus contemplativa qua in homine generantur sapientie per intuitionem eorum que supra ipsum sunt. Est enim humana substantia habens comparationem ad duo, quorum unum 25 est supra eam et reliquum est infra eam; propter debitum autem quod

---

3 divisiva: divisivi C   quod: que V   4 fuerunt: fierent P   6 causa: ipsa C   10 scilicet *om.* V   11 ipse: ipsa P   12 triplex: tripliciter C   via: scientia P   13 comparatio: operatio V   14 afficitur: assignatur P   15 cum: et V   ei *om.* V   aut risus: et visus C   16 comparationem: operationem C   17 ipsa: ipse P   usum: visum V   18 generatur: generantur P   in anima *om.* V   20 comparationem: compositionem C   23 qua: per quam V   25 comparationem: compositionem C   26 est² *om.* V   infra: supra P   propter: per P   autem: enim P

---

2 a Porfirio: *Isagoge* De diff. (18), cum lectionibus variantibus in app. crit. adhibitis.

4–7 Cf. Gundiss. *De Immort. An.* (28): 'Manifestum est quoniam ipsa (anima humana) est pura forma et substantia immateriata et incomposita in se huiusmodi compositione, quae est ex materia et forma. Aut si forte quis hic dicat, quia est ex materia et forma: dicimus tamen formam eius incorruptibilem esse, quia non est ei contrarium per quod corrumpatur'. Avic. *De An.* v. iv (25ʳᵃc).

9–22 Cf. Avic. *De An.* I. v (5ʳᵇE–ᵛᵃ); Gundiss. *De An.* x (84); Algazel *Metaphys.* II. iv. 5 (172–3).

23–p. 92, l. 16 Cf. Avic. ibid. (5ᵛᵃE); Gundiss. ibid. (86–87). Cf. supra, nn. 304, 324.

through a likeness and proportion with the composition which arises out of matter and form, as can be known from Porphyry, and if intellectually the genus of soul is grasped as well as its specific difference. For it was not so in the creation of the soul that previously there were those components, namely the genus of soul and its differences before the creation of the soul. Moreover, even if in the soul there is both spiritual matter and spiritual form, it will not however itself be corruptible, since it is lacking in the cause of contrariety according to which corruption occurs.

xxv. i. *On the powers of the rational soul*

335. What follows concerns the powers of the rational soul. Now there are two powers of the soul, namely the power to act and the power to know. However, the power to act is that power from which the individual actions of a man arise and which he chooses in accordance with the judgement and intention of reason. However, such a power can be taken in three ways. One way is to consider it according to its connection with the vital and appetitive power, and this way is according to the temperament with which a man is affected when he happens to be confused or embarrassed, or laughs or something like these. Another way in which it is considered is according to its relation to itself inasmuch as it produces skill and understanding in itself: for example, by means of the many experiences of individual things an understanding of principles is produced in the soul, principles from which knowledge and demonstrations are acquired. The third way is considered according to the connection of the active power to the imaginative and estimative power, because it directs them according to the judgement of the contemplative power.

336. The power of knowing, however, is a contemplative power by which wisdom is produced in man through the intuition of those things which are above him. For the human substance is such that it has a relation to two realms, one of which one is above it and the other below it. However, because of the obligation which it owes to that

debet ei quod est infra eam habet humana substantia virtutem activam
ad regendum corpus humanum, ex qua virtute activa adquiruntur dispositiones in homine que dicuntur mores esse; et ideo virtus activa
ethice subiacet speculationi, quoniam ab ea perficiuntur mores. Propter
debitum autem quod debet humana substantia ei quod est supra eam, 5
ipsa habet virtutem contemplativam a qua generantur sapientie et perficiuntur virtutes. A vi autem activa est virtutum inchoatio, sed a vi
contemplativa est eius perfectio. Ab hac etiam insunt scientie sine quibus virtutes esse nequeunt. Virtutem enim antecedit scientia et discretio. Unde sunt multi, non considerantes quod scientia habet precedere 10
virtutem, qui dicunt quod virtus est scientia ut prudentia, et est simile:
molestatio precedit iram, non tamen ira est molestatio. Notificans autem
istas duas virtutes de anima, Avicenna dicit in commento quod anima
humana habet quasi duas facies: unam ad id quod est deorsum, et
aliam ad id quod est sursum; ab ea autem que deorsum est generantur 15
mores, et ex ea que est sursum generantur sapientie.

xxv. ii. *De intellectu*

337. Consequenter notandum est quod sub vi contemplativa com-
prehenduntur ratio et intellectus. Intellectus est apprehendere ea que
supra humanam naturam sunt, et naturalia per abstractionem for- 20
marum universalium a materia et ab appendiciis materie. Rationis
autem est conferre intellecta ad invicem et conferendo de eis iudicare.
Distinguitur autem intellectus ab Avicenna quatuor modis, et a multis
aliis auctoribus. Est enim intellectus materialis sive intellectus in potentia, et intellectus formalis sive adeptus, et intellectus in effectu, et 25
intellectus agens.

338. Intellectus materialis est ipsa anima nuda a dispositionibus

---

6 generantur: generatur V   7 virtutes *om.* V   vi: via P   virtutum: virtutis V   8 insunt: homini *add.* V   9 enim: autem V   10 multi: multa V
12 non: nec V   13 duas: diversas C   de anima *om.* V   Avicenna: Abicenna
P; ab Avicenna C   commento: commentato P   15 ab ea autem que: ex eo
quod P   16 et ex ea que: ex eo quod P   est: sunt V   19 ratio *om.*
V   est: autem habet P   20 naturalia: universalia P   abstractionem: altera-
tionem V   21 et . . . materie: apprehendens materie P   25 effectu: effectus V

---

13 Avic. in *Commento*: i.e. *De An.* loc. cit.

19-21 Cf. Avic. *De An.* i. v (5va E); v. v. (25rb); Gundiss. *De An.* x (85, 88).

23 Avic. *De An.* i. v (5va-b F). Vid. supra, n. 59, ubi haec distinctio ascribitur
Aristoteli.

23-24 a multis aliis auctoribus: vid. Alex. Aphrod. *De Intell.* (74); Alkindi *De Intell.*
(9); Alfarabi *De Intell.* (117); Algazel *Metaphys.* ii. iv. 5 (175); Gundiss. *De An.*
x (87-88).

27-p. 93, l. 5 Avic. loc. cit.: Gundiss. ibid. (87).

which is under it, the human substance has an active power in order to direct the human body. From this active power the dispositions are acquired in man which are called morals. Thus the active power underlies ethical speculation because through it morals are perfected. However, because of the obligation which the human substance owes to that which is above it, it has a contemplative power from which wisdom is produced and the virtues are perfected. For the origin of virtues is from the active power, but from the contemplative power is its perfection. Again, knowledge is present from this active power without which the virtues are impossible. For knowledge and discrimination come before virtue. So it is that there are many people who, while not taking into account that knowledge has to come before virtue, say that virtue is knowledge just like prudence, and that it is the same as the following: annoyance comes before anger; anger, however, is not annoyance. While making these two virtues of the soul known, Avicenna says in the commentary that the human soul has, as it were, two faces, one towards that which is downwards and one towards that which is above. Moreover, from that which is below morals are produced, and from that which is above wisdom is produced.

xxv. ii. *On the intellect*

337. It should be noted next that reason and understanding are included under the contemplative power. The task of the intellect is to grasp those things which are above human nature, as well as natural things through the abstraction of the universal forms from matter and from the appendages of matter. The task of reason is to compare those things that have been understood with each other and in so doing to distinguish between them. The intellect is divided by Avicenna and by many other authors in four ways: for there is the material intellect or intellect in potency, the formal or acquired intellect, the intellect in effect, and the agent intellect.

338. The material intellect is the soul stripped of its acquired dis-

adquisitis, et dicitur materialis, quoniam habet se ad simulacrum aptitudinis materie prime ex qua in seipsa non est aliqua forma. Similiter
dicitur esse intellectus in potentia, quoniam esse materie est esse in
potentia, cum eius esse sit imperfectum, potest autem perfici per forme
receptionem.   5

339. Intellectus adeptus est passio generata in anima, que est similitudo rei extra; de quo intellectu habetur ab Aristotele in libro *de Anima*,
quod res intellecta est perfectio anime et de eius essentia. Appellat
autem Aristoteles ibi rem intellectam intellectum adeptum, qui est
ymago rei extra. Et ille intellectus adeptus dicitur esse intellectus 10
formalis, quoniam est habens se ad similitudinem et proportionem
forme; quoniam sicut forma perficit materiam dando ei esse actu, ita
et intellectus adeptus perficit intellectum materialem dando ei esse in
effectu. Est enim anime secunda perfectio ab intellectu adepto.

340. Intellectus autem in effectu est quiddam coniunctum ex intel- 15
lectu materiali et intellectu adepto, et est idem subiecto intellectus
materialis et intellectus in effectu. Diversificantur autem in comparatione: dicitur enim intellectus materialis in quantum habet esse in
potentia nudum a forma adquisita; intellectus autem in effectu dicitur
ille idem intellectus secundum quod ipse actualiter est sub forma que 20
est simulacrum rei extra.

341. Intellectus agens est vis anime apprehensiva rerum universalium
abstrahendo eas ab accidentibus. Intellectus in habitu est intellectus
repositus in anima ad quem si voluntas convertat aciem animi, erit
anima intelligens actu.   25

342. Obicitur autem. In anima constituitur dispositio quedam que
est intellectus formalis et similitudo rei, et illa dispositio non est anime
innata, immo est adquisita. Aliquid ergo est inprimens illam ymaginem
in anima, non ipse intellectus agens; quoniam ipse habet per illam duci

1 simulacrum: similitudinem V   1–2 aptitudinis *om.* V   2 Similiter: Si
materialiter V   3 esse[1] *om.* P   3–4 quoniam . . . in potentia *om. per*
hom. V   6 generata *om.* V   7 ab Aristotele *om.* P   8 et de eius
essentia *om.* V   12 forma *om.* P   esse: rem P   15 quiddam *om.* P
17–18 et intellectus . . . materialis *om. per hom.* P   18 esse: se V   19 adquisita: acquisita V   20 ipse *om.* C   24 quem: quod P   convertat: convertit V   28 ymaginem: mag' C   29 ipse[2] *om.* V   per illam duci: reduci P

7 Arist. in libro *de Anima*: forte respicit ad *De An.* III. iv (430[a]3).
9–10 Dictio 'intellectus adeptus' non occurrit in versione latina Aristotelis.
Fortasse provenit ab Alex. Aphrod. *De Intell.* (77), ubi Aristoteli attribuitur. Vid.
supra, n. 59. Occurrit etiam in Avic. *De An.* Vid. R. de Vaux, *Notes et textes sur*
l'Avicennisme latin, p. 134, not. 2.
10–11 intellectus formalis: cf. *Liber de causis primis et secundis* x (135, 128), ubi cl.
editor dicit: 'Cette épithète est insolite. Je ne l'ai relevée nulle part'.
18–25 Cf. Avic. *De An.* I. v (5vb); Algazel *Metaphys.* II. iv. 5 (175).
21 simulacrum: cf. Algazel ibid. I. iii. 2 (64–65).

positions, and it is called material because its make-up is similar to the disposition of prime matter inasmuch as there is no form in it. Similarly it is said to be the intellect in potency because the being of matter is a being in potency, since its being is incomplete, and so it can be completed by means of the reception of form.

339. The acquired intellect is a modification produced in the soul which is a likeness of an external thing. With regard to this intellect we know from Aristotle in the book *On the Soul* that a thing which has been understood is the perfection of the soul and of its essence. For there Aristotle calls the thing understood the acquired intellect, which is an image of the external thing. Again, that acquired intellect is said to be the formal intellect because it is constituted according to the likeness and proportion of a form. For just as a form completes matter by giving it being in act, in the same way the acquired intellect completes the material intellect by giving it being in effect. Indeed, the second perfection of the soul is from the acquired intellect.

340. However, the intellect in effect is something linked together out of the material intellect and the acquired intellect, and the material intellect and the intellect in effect are the same subject. They are diversified, however, in comparison: the material intellect is said to be inasmuch as it has being in potency stripped of acquired form; the intellect in effect, however, is said to be that same intellect according to which it actually comes from a form which is a likeness of an external thing.

341. The agent intellect is a power of the soul which is perceptive of universal things through abstracting them from accidents. The habitual intellect is an understanding stored in the soul, and if the will turns the eye of the mind towards it, the soul will understand in act.

342. It is objected, however, that a certain disposition is constituted in the soul which is the formal understanding and the likeness of the thing, and that disposition is not innate in the soul, but is rather acquired. Therefore, there is something which impresses that image in the soul, not the agent intellect itself, because by means of it it is

ad rem extra ut eam intelligat. Non generatur in anima a re extra propter distantiam inter animam et rem apprehensam.

343. Si formetur in anima ab affectione in memoria, que est simulacrum rei extra; Contra. Illa ymago que est in memoria est corporalis, cum eius subiectum sit corporale, et anima est incorporalis. Ergo a corporali non potest fieri in ea immutatio. 5

344. Solutio. Ad hoc dicendum est, ut superius ostensum est, quod anima habet convertere se ad corpus quod ipsa regere habet, et ad eius dispositiones, et ad similitudinem ymaginum inventarum in memoria, et inprimitur in anima intellectus formalis mediante primo datore formarum; vel, ut plures auctores videntur velle, est illa forma impressio ab intelligentia ut ministerio eius, et a primo datore formarum ut auctoritate ipsius. Illa autem intelligentia a multis auctoribus dicitur esse angelus, qui minister est anime hominis. Habet enim homo duos angelos, bonum et malum, quorum uterque minister est humane anime. 15

345. Item. Videtur posse ostendi quod plures sint vires anime rationalis sumpte sub vi contemplativa quam ratio et intellectus. Apprehensis enim rebus per intellectum contingit animam uti cogitatione. Sed qua ratione intelligere est a vi intellectiva, cogitare erit a vi cogitativa, et operatio huius virtutis est quandoque circa intellecta et post operationem intellectus, et ita erit vis cogitativa sub vi contemplativa; et ita plures erunt virtutes virtutis contemplative quam ratio et intellectus. 20

346. Preterea. Aliquis in cogitando dispersa simul colligit, ut ipse inter ea illud inveniat quod ipse querit: quoniam cogitare est simul dispersa agitare; et ideo dicit Augustinus in libro *de Trinitate* quod cogitatio ab ipso coactu dicitur. Sed aliud est intelligere, aliud intellecta simul colligere. Ergo aliud est intelligere quam cogitare; et diversi actus a diversis procedunt potentiis: et intelligere est a vi intellectiva, cogitare vero est a vi cogitativa. Ergo aliud est vis intellectiva et aliud est vis cogitativa. 25

30

2 apprehensam: apprehensivam P   3 in anima *om.* P   ab *om.* P   affectione: extra *add.* P   4 corporalis: corruptibilis V   5 est *om.* C   7 est[1]: quod *add.* P   quod *om.* P   8 ipsa: anima V   9 dispositiones: dispositionem P   similitudinem: rerum *add.* V   inventarum: que reponitur C; reponuntur V   10 et inprimitur in: inprimitur etiam P   12 ut[1] *om.* P   ut[2] *om.* P   15 est *om.* P   16 sint: sunt V   19 est *om.* V   23 in *om.* P   26–27 aliud ... simul: et aliud collecta dispersim P   26 intellecta: multa V   27 Ergo *om.* V   28 a[1] *om.* V   29–30 Ergo ... cogitativa *om. per hom.* V

7 Cf. supra, n. 336.
10–11 datore formarum: Avic. *Metaphys.* IX. v (105[va] versus fin.); Algazel *Metaphys.* I. v (125); II. iv. 3 (167); 5 (181).
14 angelus: Algazel ibid. I. v (121); II. iv. 5 (175); Gundiss. *De An.* v (51).
14–15 Cf. P. Lombard. *Sent.* II. dist. xi. c. i.
25 Augustinus ... *Trinitate*: rectius *Confess.* x. xi. 18.

itself led to the external thing so that it can understand it. It is not produced in the soul by the external thing because of the distance between the soul and the thing perceived.

343. What if it were formed in the soul by a modification in memory, which is a likeness of an external thing? Against this it can be argued that the image which is in memory is corporeal, since its subject is corporeal. The soul, however, is incorporeal, therefore a change cannot take place in it from a corporeal thing.

344. Solution: With regard to this it should be said, as has been shown above, that the soul has to turn itself towards the body which it has to direct, and towards its dispositions, and towards the likeness of images found in memory. The formal intellect is impressed on the soul by means of the First Giver of Forms; or, as many authors seem to wish, that form is an impression from an Intelligence as its servant, and from the First Giver of Forms as by its authority. That Intelligence is said by many authors to be an angel, who is a servant of the soul of man. For a man has two angels, one good and one bad, each of whom is an attendant of the human soul.

345. Again, it seems that it can be shown that there are more powers of the rational soul to be included under the contemplative power, other than reason and understanding. For when things are perceived by means of the intellect it happens that the soul uses thought. However, for the same reason that to understand is from the intellective power, so to think will be from the cogitative power. Again, the activity of this power is sometimes directed towards those things which are understood and comes after the operation of the intellect. Thus the cogitative power will come under the contemplative power; and so there will be more powers of the contemplative power than reason and understanding.

346. Moreover, when someone thinks, they bring together scattered things at once so that they find what they were looking for among them, because to think is to consider scattered things at the same time. And so Augustine says in the book *On The Trinity* that 'thinking' [*cogitatio*] is said from 'to be brought together' [*coactus*]. However, it is one thing to understand and another to link things which have been understood together at the same time. Therefore, to understand is other than to think, and diverse acts proceed from diverse potencies. To understand is from the intellective power; to think, however, is from the cogitative power. Therefore, the intellective power is one thing and the cogitative power is another.

347. Solutio. Dicimus quod una et eadem vis est intellectiva et vis cogitativa; sed diversimode sunt in comparatione ad diversos actus relate. In comparatione enim ad actum cogitandi dicitur esse vis cogitativa, et in comparatione ad actum intelligendi dicitur esse vis intellectiva.   5

348. Item. Ab Aristotele habetur in *Physicis* quod ars spernit materiam: per unam enim et eamdem artem contingit operari in diversis materiebus, ut contingit aurifabrum operari in auro, in argento, in plumbo, in cupro, nec propter diversitatem materiarum variatur essentia artis. Pari ratione cum contingat animam per vim appre- 10 hensivam tum apprehendere res singulares et sensibiles tum res universales et intelligibiles non erunt propter diversitatem apprehensorum alia vis apprehensiva illa et alia illa, immo una et eadem vis in essentia. Illa autem vis qua apprehenditur sensibile est sensus, illa autem qua apprehenditur universale est intellectus; ergo una et eadem vis in es- 15 sentia est intellectus et sensus.

349. Item. In qua proportione se habet ymaginatio ad ymaginabile in eadem proportione se habet sensus ad sensibile; ergo permutatim in qua proportione se habet sensibile ad ymaginabile in eadem proportione se habet sensus ad ymaginationem. Sed omne sensibile est 20 ymaginabile; quoniam illud idem quod est sensibile est ymaginabile. Ergo idem est sensus quod ymaginatio. Ergo una et eadem vis est sensus et ymaginatio.

350. Consimili modo potest ostendi quod omnes vires anime apprehensive sunt una et eadem vis in essentia. Quod bene concedimus, 25 dicentes quod apprehensive vires ipsius anime humane non sunt diverse in essentia, immo in accidente tantum. Nulla enim est diversitas inter eas nisi in modo apprehendendi. Sensus enim apprehendit rem presentem; ymaginatio rem absentem et prius sensatam; intellectus autem abstrahit rem ab accidentibus, ut patere potest manifestius rationibus 30 supradictis.

351. Item. Superius dictum est quod in anima rationali due sunt

---

2 diversimode: diverse C   3 relate *om.* P   4 in comparatione: ad comparationem P   ad *om.* C   8 in², ³ *om.* C   9 in *om.* C   materiarum: materierum V; non *add.* P   11 res² *om.* P   13 illa et alia illa: ista quam illa V   vis¹: est *add.* V   14 qua: que C   14–15 sensibile . . . apprehenditur *om. per hom.* C   15 universale: insensibile P   18 se habet *om.* C   20 se habet *om.* C   20–21 omne . . . ymaginabile: idem est sensibile et ymaginabile V   21 quoniam . . . ymaginabile *om.* CV   22 sensus *om.* C   24–25 apprehensive *om.* P   25 Quod: Et C   26–27 dicentes . . . tantum *om.* V   32 Item: ut *add.* P

---

6 in *Physicis*: Arist. *Physic.* II. i (193ª14–21).
32 Cf. supra, nn. 335–6.

347. Solution: We state that the intellective power and the cogitative power are one and the same, but they are diversified in comparison with diverse acts in respect of each other. For the cogitative power is said in comparison with the act of cognition and the intellective power is said in comparison with the act of understanding.

348. Again, we know from Aristotle in the *Physics* that art holds matter to be of little importance, for with one and the same art it happens that someone works upon various materials. So it happens that a goldsmith works with gold, silver, lead, and copper, nor does the essence of the art vary because of the variety of materials. For the same reason, since it happens that the soul perceives both individual and sensible things as well as universal and intelligible things by means of the perceptive power, because of this diversity of things perceived there will not be another power apprehending one and the other; rather there will be one and the same power in essence. However, that power by which the sensible is grasped is sensation, and that power by which the universal is grasped is the intellect; therefore the intellect and sensation are one and the same power in essence.

349. Again, in the proportion according to which the imagination is constituted in respect of something which can be imagined, sensation is constituted in the same proportion in respect of something which can be sensed. Therefore, changing things around, in the proportion by which the sensible is constituted in respect of what can be imagined, sensation is constituted in respect of imagination in the same proportion. However, every sensible thing can be imagined, because that same thing which is sensible can be imagined. Therefore sensation is the same as imagination. Therefore sensation and imagination are one and the same power.

350. In a similar way it can be shown that all perceptive powers of the soul are one and the same power in essence. Which we readily concede, saying that the perceptive powers of the human soul itself are not diverse in essence, but only accidentally. For there is no diversity among them except in the mode of apprehending, because sensation perceives a thing which is present and the imagination a thing which is absent and which was previously sensed. The intellect, however, abstracts the thing from the accidents, as may be seen more clearly through the arguments presented above.

351. Again, it was said above that in the rational soul there are two

vires, vis activa et vis contemplativa, et a vi contemplativa sunt sapientie
et a vi activa perficiuntur mores. Quoniam a vi activa est regimentum
corporis humani secundum iudicium virtutis contemplative. Sed sicut
in corpore humano est regimentum, ita et in universitate totius orbis
est regimentum regens discordantia, ut contrarietates corruptibilium; 5
qua ratione ergo ponitur esse anima in corpore humano ut ipsum regat,
debemus ponere animam mundi esse que regat ipsum mundum.

352. Item. Nutrimentum proportionaliter transmittitur ad omnes
partes corporis nutriendas, et hec transmissio non est casualis, sed fit
cum discretione. Divisio enim alimenti, cum ipsum conferatur pro- 10
portionaliter corpori secundum quod partes corporis egent eo, fit cum
electione puri et fit cum expulsione impuri; omnis autem discretio et
electio rei est cum apprehensione rei electe. Alimentum ergo quod
transmittitur ad partes corporis propter restaurationem deperditorum
apprehenditur; sed omnis animi apprehensio est ab anima; anima ergo 15
apprehendit et intuetur illud alimentum quod intromittitur, et anima
non est res que nutritur; est ergo alia, et hec servit nature; quecumque
autem illa sit, illa est anima mundi, que etiam transmittit spiritus per
nervos et musculos corporis.

353. Item. Hoc corpus humanum in quo est hec anima non est ita 20
stabilis essentie sicut universitas mundana, quia facilius corrumpitur
corpus hominis quam totus mundus. Anima autem incorruptibilis est;
ergo maiorem habet affinitatem anima cum universo, scilicet, cum
mundo, quam cum corpore humano. Potius ergo ponendum est mundum habere animam, cum stabilius sit eius regimen, quam corpus 25
humanum habere animam.

354. Dicet forte aliquis quod regimentum totius mundi est a prima
causa et ministerio et auctoritate, et quod non est aliquod causatum
regens universum mundum. Pari ratione debet dici de corpore humano

---

1 et² *om.* P   2 et *om.* P   Quoniam: Et P   4 regimentum: et *add.* V
5 contrarietates corruptibilium: contrarietatis P; contraria et corruptibilia V   7 que
regat: qui regit C   mundum *om.* P   8–19 Item . . . corporis *om.* P   9 casualis: causalis C   12 impuri: puri C   autem: et V   15 sed *om.* C   animi
om. V   16 illud: nutrimentum vel *add.* V   intromittitur: nutritur C   et: hec
add. V   17 res: rei V   servit: deservit V   18 autem *om.* C   illa²: ipsa
V   mundi: vivendi V   20 Hoc *om.* P   hec anima non: anima illa P
21 essentie: esse C   quia: illud *add.* P   22 corpus . . . mundus *om.* P
23–24 ergo . . . humano *om.* P   23 anima *om.* V   24 ergo *om.* C   25 cum
. . . regimen *om.* C   sit: esset V   regimen: regimentum V   27 Dicet . . .
aliquis: Dicatur P   mundi *om.* C   a *om.* C   28 causatum: creatum V
29 universum *om.* V   mundum *om.* P

---

7 animam mundi: Calcidius *In Platonis Timaeum* clxxvii (206). Vid. T. Gregory,
'L'*anima mundi* nella filosofia del xii secolo', *Giorn. critico di filos. ital.* 30 (1951),
494–508.

powers, an active power and a contemplative power, and that wisdom is from the contemplative power, and that morals are perfected by the active power. This is because the governance of the human body according to the judgement of the contemplative power is from the active power. However, just as there is governance in the human body, so in the totality of the entire globe there is governance ruling discord, such as the oppositions between corruptible things. Therefore for this reason, given that the soul is posited to be in the body so as to govern it, we have to posit that a World Soul exists which governs the world itself.

352. Again, nourishment is transmitted proportionately to all parts of the body that require nourishing, and this transmission is not by chance but happens with discernment. For the division of food, since it is bestowed proportionately on the body according as the parts of the body need it, happens with the selection of the pure and with the expulsion of the impure. However, all discernment and choice of a thing is together with the perception of the thing chosen. Therefore, the food which is transmitted to the parts of the body in order to restore what has been lost is perceived. However, all perception of the mind is from the soul, therefore the soul perceives and intuits the food which is let in. However, the soul is not the thing which is nourished; it is therefore something else and it preserves nature. However, whatever that is, it is the World Soul, which also sends out impulses through the nerves and muscles of the body.

353. Again, this human body in which the soul is is not so stable in its essence as is the visible universe, since the human body is more easily corrupted than the entire world. The soul, however, is incorporeal; therefore the soul has a greater affinity with the universe, namely with the world, than with the human body. Therefore we should posit that the world has a soul more than the human body since its governance is more stable.

354. Perhaps someone might say that the governance of the entire world is from the First Cause and from its support and authority, and that there is nothing caused which governs the visible universe. For the same reason it should be said concerning the human body that it is

quod ipsum non regitur a causato nec ab alio quam a divina essentia, quod falsum est.

355. Preterea, forte dicet quis quod in corpore humano ubique est continuatio partium, et propter illam continuationem vivificatur corpus omnino ab anima. Sed ita non est de mundo, quoniam in omnibus suis 5 partibus non est continuatio; unde ab una anima regi non potest.

356. Contra. Ossa hominis et medulla non continuantur, nec cerebrum cum osse; et tamen in ipsis est vegetatio ab una et eadem anima; quoniam in eis est augmentum et nutrimentum. Pari ratione ab una et eadem anima potest procedere regimentum per universitatem orbis, 10 nec erit ibi impedimentum quin ipsa sit propter decontinuationem in partibus mundi.

357. Item. A Boetio habemus in libro *Consolationum*, ut videtur, quod mundus habeat animam ubi ipse invocat primam causam dicens:

> 'Tu triplicis mediam nature cuncta moventem   15
> Connectens animam per consona membra resolvis'.

Per hoc enim quod dicit: 'animam moventem cuncta', videtur innuere mundum habere animam que mundum moveat regendo ipsum. Postea vero dicit Boetius continue de anima:

> 'In semet reditura meat, mentemque profundam   20
> Circuit, et simili convertit imagine celum'.

Per hoc autem innuit celum moveri ab anima. Sed qua ratione celum movetur et regitur ab anima, pari ratione debet poni quod mundus et regitur ab anima et movetur ab anima, vel quare non. Si concedatur, mundum habere animam;   25

358. Contra. Corpus huius hominis, demonstrato corpore, regitur ab anima et vivificatur, nec est alia anima qua vivificatur pes huius

---

1 causato: creato V   essentia: et nihil aliud est anima quam Deus *add.* V   2 quod falsum est *om.* P   3 dicet: dicit V   quis *om.* C   ubique *om.* V   4 corpus *om.* P   6 unde: et ideo V   8 cum: et V   tamen: cum C; tunc V 9-10 quoniam . . . anima *om. per hom.* CP   10 potest *om.* V   11 nec: non V   13 habemus: habetur P   ut videtur *om.* P   14 habeat: habet P 15 triplicis: triplici V   16 connectens: connectans P   connectens . . . resolvis *om.* V   17 enim *om.* PV   videtur innuere: modo invenire P   18 que . . . ipsum *om.* P   19 vero *om.* P   continue *om.* P   23 et regitur *om.* C   pari ratione *om.* C   24 ab anima¹ *om.* C   25 mundum . . . animam *om.* P 26 corpore: hominis *add.* V   27 et *om.* P   nec . . . vivificatur *om.* V   qua: quam C   huius *om.* P

3-4 Cf. Algazel *Metaphys.* I. i. 2 (13-14).
13 Boetio . . . in libro *Consolationum*: III metr. ix. 13-14.
19 Ibid. 16-17.
22 celum moveri ab anima: cf. supra, n. 7.

not ruled by something caused nor by anything other than the divine essence, which is false.

355. Moreover, perhaps someone might say that everywhere in the human body there is a continuity of parts, and because of that continuity the body is in every way vivified by the soul. However, it is not thus regarding the world, because there is not a continuity in all of its parts, thus it cannot be governed by one soul.

356. Against: The bones and marrow of a man are not continuous, nor is the brain with the cranium, and yet there is life in these from one and the same soul, since there is growth and nourishing in them. For the same reason, from one and the same soul governance can proceed through the whole of the globe, nor will there be an impediment there but rather it will exist as such because of the discontinuity in the parts of the world.

357. Moreover, as we know from Boethius in the book of *Consolations*, it would seem that the world has a soul, where he calls upon the First Cause, saying:

'At the midpoint of threefold nature you attach a soul and release it throughout harmonious parts, moving all things.'

By that which he says, 'a soul . . . moving all things', it seems that he suggests that the world has a soul that moves the world while governing it. However, afterwards Boethius continues saying with regard to the soul:

'Soul goes forth and returns to itself, circles the lower mind, and turns the heavens into an image of the same.'

Thus, in this way he implies that the heavens are moved by soul. However, because the heavens are moved and governed by soul, for the same reason it should be posited that the world is governed by the soul and moved by the soul, or if not, why not? If it is conceded that the world has a soul:

358. Against: The body of this man, once the body has been designated, is governed and vivified by the soul. Nor is there one soul

hominis et alia qua vivificatur cerebrum et caput; immo ab una sola
anima est vivificatio totius corporis. Pari ratione, cum mundus habeat
animam, totus mundus et quelibet eius pars animabitur et regetur ab
eadem anima. Sed corpus huius hominis et corpus illius hominis sunt
partes mundi; ergo vivificatur ab eadem anima. Ergo non est ponere 5
aliam esse animam huius hominis et aliam esse illius hominis, quia qua
ratione haberet aliquis homo duas animas, haberet duo corpora, vel
quare non?

Item. Si mundus habeat animam sic obicitur.

359. Illa anima aut est vegetabilis tantum, ut anima arboris; aut est 10
vegetabilis et sensibilis, ut anima brutorum animalium; aut vegetabilis,
et sensibilis, et rationalis, ut anima hominis. Si rationalis, ergo mundus
est corpus animatum, sensibile, rationale; ergo mundus est homo, quod
falsum est. Si vegetabilis et sensibilis; ergo mundus est animal; ergo
homines et boves, et asini, et similia, non sunt animalia, immo partes 15
unius animalis: sicut non dicimus caput alicuius hominis esse animal,
nec caput asini esse animal, immo partes animalis.

360. Solutio. Dicimus mundum non habere animam. Sed Plato et
alii auctores videntur velle mundum habere animam: unde Augustinus
loquens de opinionibus Platonis neque manifestat se consentire vel dis- 20
sentire ei quod dicit Plato, scilicet mundum habere animam. Non
credimus enim aliam mundi esse animam quam Spiritum Sanctum
regentem et vivificantem universum.

361. Ut haberi potest ex predictis, a natura est preparatio in corpore
organico ut ipsum sit convenientius ad animam rationalem recipiendam 25

1 qua: a qua C; que V   1–2 una . . . corporis: uno anima vivificatur corpus
totius P   sola anima *om.* V   3 animam: regitur *add.* P   animabitur et regetur:
et vivificatur P   4 hominis *om.* V   6 aliam: animam *add.* V   7–8 vel quare
non *om.* P   9 sic obicitur *om.* P   10 Illa: ergo *add.* P; autem *add.* V   anima
arboris: in arboribus et P   est *om.* P   11 anima . . . animalium: in brutis P
12 rationalis: voluntaria P   anima *om.* C   13 corpus animatum: animal P
13–14 mundus . . . animal *om. per hom.* P   15 et boves *om.* P   et asini *om.*
C   similia: consimilia V   16 unius *om.* PV   16–17 sicut . . . animalis
om. V   16 alicuius: unius C   17 immo . . . animalis *om.* P   18 Dicimus . . . animam *om.* V   18–19 Sed . . . unde *om.* CV   19 Augustinus:
tamen *add.* C; cum *add.* P   20 manifestat se: videtur P   se: eum V   21 ei:
ea P   animam: animal P   Non: Nihil P   22 mundi *om.* V   24 Ut:
autem *add.* P   25 organico: cogitato C   convenientius: conveniens P   recipiendam: accipiendam C

18–19 Plato et alii auctores: cf. supra, n. 351.
19–21 Augustinus: *De Consensu Evangelistarum* I. xxiii. 35: 'Utrum autem universa
ista corporalis moles quae mundus appellatur, habeat quamdam animam vel quasi
animam suam, id est rationalem vitam, qua ita regatur sicut unumquodque animal,
magna atque abdita quaestio est: nec affirmari debet ista opinio, nisi comperta quod
vera sit; nec refelli, nisi comperta quod falsa sit'.
21–23 Cf. locos Guill. de Conchis et Petri Abaelardi cit. apud Guillaume de Conches
Glosae super Platonem, p. 145 et not. c.

through which the foot of this man is vivified and another by which
the brain and head are vivified, rather from one soul alone is the en-
livening of all of the body. For the same reason, since the world has a
soul, the whole world and each of its parts will be animated and ruled
by the one soul. However, the body of this man and the body of that
man are parts of the world, therefore they are vivified by the same
soul. Thus, one should not posit that the soul of this man is one soul
and the soul of that man is another, because by the same argument if
a man had two souls, he would have two bodies, or if not, why not?

Moreover, if the world had a soul, it would be objected as follows:

359. That soul is either vegetative only, such as the soul of a tree,
or is vegetative and sensitive, such as the soul of brute animals; or it
is vegetative, sensitive, and rational, such as the soul of man. If it is
rational, then the world is an animated, sensible, and rational body.
Therefore the world is a man, which is false. If it is vegetative and
sensitive, then the world is an animal. Therefore men and cattle, and
asses and others like these, are not animals, but rather parts of one
animal, just as we do not say that the head of some man is an animal,
nor the head of an ass is an animal, but rather they are parts of an
animal.

360. Solution: We state that the world does not have a soul. However, Plato and other authors seem to have wanted the world to have
a soul. Thus, when Augustine spoke about the opinions of Plato, he
does not reveal whether he agrees or disagrees with that which Plato
says, namely that the world has a soul. However, we do not believe
there to be any other soul of the world than the Holy Spirit, who rules
and gives life to all.

361. As can be gathered from what has been said, there is from
nature a preparation in an organized body so that it will be better
disposed towards receiving the rational soul than towards receiving

quam ad aliud recipiendum; sufficienti autem preparatione corporis et
appropriatione existente ut anima ei infundatur, a primo datore formarum ei infunditur anima, et ita nature ministerio precedente subsequens est a prima causa perfectio, scilicet anima que est perfectio
corporis organici viventis potentialiter.   5

xxv. iii. *Inquirit utrum nova anima infundatur corpori vel antiqua*

Sed videtur quod potius corpori sufficienter apto ad recipiendum
animam rationalem prius debeat ei infundi anima antiqua quam anima
nova creari et ei infundi; et hac ratione.

362. Facilius est existens in effectu accipere et illud dare quam 10
novum facere et illud accipere et illud dare. Ergo, cum aptatum fuerit
corpus a natura ut ipsum per animam vivificetur, facilius est ei animam
iam factam infundere quam novam facere et ei eam infundere; et ita
illa anima que fuit unius hominis potest esse alterius hominis.

363. Preterea. Anima existens in corpore meretur et demeretur. 15
Bonum est autem animam salvari, malum autem animam dampnari: et
bonum est appetendum, malum autem respuendum. Bonum est ergo
infundere corpori organico potentialiter viventi animam antiquam mole
peccatorum oppressam, ut ipsa existens in corpore mereatur purgando
se a peccatis suis et summum bonum sibi adquirat; malum autem est 20
ipsam remanere a corpore separatam et non mereri, nec perfecte purgari,
sed ad mortis penam perduci. Melius est ergo animam prius factam
corpori infundere, ut in eo possit mereri per scientiarum et virtutum
inquisitionem, quam novam animam corpori infundere. Sed Creatoris
est ita bene agere quod melius non potest fieri; ergo dat potius corpori- 25
bus animas antiquas, ut ipse et mereantur et purgent se a vitiis, ut
eternam vitam habeant, quam novas eis conferat animas, et ne anime
veteres, que possent per coniunctionem ipsarum cum corporibus
mereri, dampnentur. Et ita ut prius.

---

1 sufficienti: sustineri autem existere P   autem *om.* C   1–2 et appropriatione
existente *om.* P   2 ei: ea P   2–3 formarum ei *om.* P   3 infunditur:
ea *add.* P   3–4 subsequens: sequens C   4 scilicet . . . perfectio *om.* V
6 Inquirit . . . antiqua *in marg.* C; *om.* PV   7 potius . . . apto: corpus habente
in effectu P   apto: aptato V   8 prius debeat ei *om.* CV   anima *om.* P
9 creari: fieri P   ei *om.* C   et² *om.* P   11 novum: novam P   accipere et
illud *om.* CV   aptatum: aptum C   12 corpus: organicum *add.* P   13 eam
om. C; eum *add.* P   13–14 et ita illa anima: et in illa P   14 potest: iterum
add. V   hominis *om.* P   15 Preterea: item P   16 Bonum est autem: et
bonum est P   16 et: sed P   17 autem *om.* P   21 et non mereri *om.* V
22 mortis *om.* P   perduci: deduci P   factam: formatam C   23–24 ut in eo
. . . inquisitionem *om.* CV   24 animam corpori infundere *om.* C   25 ita: tam
P   fieri: agere V   26 purgent: purgant C   27 eis: ei C   ne *om.* P
28 possent: possunt C   28–29 corporibus mereri: corpore salvari P

XXV. ii. ON THE INTELLECT

anything else. For when there has been a sufficient preparation of the body and an existing adaptation so that the soul may be infused in it, the soul is infused in it by the first Giver of Forms, and thus through the preceding assistance of nature the perfection from the First Cause comes afterwards, namely the soul which is the perfection of a potentially alive organized body.

xxv. iii. *The author examines whether a new or an old soul is infused into the body*

It seems that it would be better for a body which is already capable of receiving a rational soul that an old soul should be infused in it before creating a new soul and infusing it in it; and for the following reason.

362. It is easier for something which exists in effect to receive and to give something than to make something new and for it to receive it and to give it. Therefore, since the body has been made suitable by nature to be vivified by the soul, it is easier to infuse into it a soul which has already been made than to make a new soul and to infuse it in the body; and thus that soul which was one man's can be another man's soul.

363. Moreover, a soul existing in a body acquires merit and loses it. For it is good for the soul to be saved; it is bad, however, for the soul to be damned. Again the good is to be desired, but evil rejected. Therefore it is a good thing to infuse an old soul burdened with a weight of sins into an organized body potentially alive so that when it exists in the body it will acquire merit through purging itself from its sins and acquiring the highest good for itself. It is an evil, however, for it to remain separated from the body and not to acquire merit, nor to be perfectly cleansed but to be led to the punishment of death. Therefore, it is better for a soul which has been made beforehand to be infused in a body, so that it can acquire merit in it through the acquisition of knowledge and virtue, rather than to infuse a new soul into the body. However, the activity of the Creator is so good that a better cannot happen; therefore he gives old souls to bodies, so that these will acquire virtue and purge themselves from vices so that they will have eternal life, rather than bestow new souls upon them, lest old souls, which can acquire merit through being joined with bodies, should be damned. And thus as before.

364. Solutio. Ad primum dicendum est quod non est facilius ipsi creatori animam prius creatam infundere quam novam creare et illam infundere; sicut si lignum aliquod habilitetur per caliditatem et siccitatem sufficienter ut ipsum igniatur, nova confertur ei igneitas et non transfertur in ipsum vetus igneitas prius existens in alio corpore. 5 Creando autem animam Deus eam infundit corpori organico potentialiter viventi. Si autem non sit in eo organum cum aptitudine vitam recipiendi non infundit ei animam. Non enim dat animas nisi secundum precedens nature ministerium et competentem preparationem existentem in corpore. Deficiente vero debito organo corporis recedit anima ab ipso ex 10 quo per ipsum introducebatur; et si iterum per naturam redeat eiusdem corporis organum, in idem corpus redibit eadem anima que in eo prius fuit, et naturalis secundum hoc erit mortuorum resurrectio; sed non redibit naturaliter eiusdem corporis organum, et ideo non erit resurrectio mortuorum naturalis, sed miraculosa. 15

365. Ad aliud dicendum quod ratio et intellectus sunt vires ipsius anime per quas potest ipsa anima discernere bonum a malo et servare voluntatem suam in rectitudine et a bono non declinare. Si ergo ipsa bonum cognoscens malo adhereat, cum ad ipsum non sit coacta, immo sponte bonum contempnens malo se submittat, non datur ipsa post 20 separationem eius a corpore suo iterum alii corpori ne ipsa declinet a malo iterum, sicut prius declinavit. Melius est ergo ei novam animam infundere mundam a labe, preterquam ab originali peccato, ut ipsa, utens ratione et intellectu, rationis arbitrio rectitudinem sue voluntatis conservet propter ipsam rectitudinem. Tunc autem conservatur voluntas 25 in rectitudine quando solum vult homo illud quod Deus vult ipsum velle.

366. Item. Queritur, utrum anima exuta a corpore possit intellectu uti actualiter.

Quod non, sic videtur. Cum anima intelligit, ut habetur ab Augustino

1 est[1], [2] om. P   2 creatam: factam P   4 confertur: confert P   4-5 et non... igneitas om. P   6 eam om. V   8 infundit: infundet C   9-10 existentem in corpore... organo: existentem. Corpore vero deficiente a debito organo C   10 vero: autem P   corporis om. C   11 introducebatur: anima add. V   per naturam redeat: per naturam redeat, per naturam redeat add. C   12 in idem: ibidem V   13-15 sed non ... miraculosa om. P   16 dicendum: est add. V   17 anime om. V   anima om. V   18 in rectitudine: et rectudinem P   ergo ipsa: ipsum igitur P   19 malo adhereat: malum faciat P   immo: sed C   20 se submittat: adhereat P   21 eius om. P   suo om. V   ipsa: ipsum P   declinet: declinat P   22 malo: bono P   ergo ei: igitur P   23 peccato om. P   24 arbitrio: arboribus P   25-26 propter ... velle om. P   27 Item om. C   possit: posset V   29 sic om. P   Cum: quando P

1 Ad primum: cf. supra, n. 362.
16 Ad aliud: cf. supra, n. 363.
29-p. 101, l. 1 ab Augustino ... auctoribus: Aug. Tract. in Iohan. Evang. xv. 19; Arist. De An. I. i (403ª8-10), III. viii (432ª3-10); Gundiss. De An. x (101).

364. Solution: To the first it should be stated that it is not easier for the Creator to infuse an already created soul than to create a new one and infuse it. It is the same as when some wood is sufficiently prepared by means of heat and dryness so that it ignites: new fieriness is conferred upon it and an old fieriness already existing in another body is not transferred to it. For in creating a soul God infuses it into an organized body which is potentially living. If there is not an organ in it with the ability to receive life, he does not infuse a soul into it. For he only bestows souls following on a previous assistance of nature and an appropriate preparation existing in the body. However, if the required organ of the body is lacking, the soul withdraws from that because of which it was introduced. Moreover, if by nature the soul were to return to the organ of the same body, the same soul will return to the same body in which it was previously, and in this way the resurrection of the dead would be natural. However, the soul will not return naturally to the organ of the same body, and so the resurrection of the dead will not be natural but miraculous.

365. Regarding the next objection, we state that reason and understanding are powers of the soul by means of which the soul itself can distinguish good from evil and preserve its will in uprightness and not turn away from good. If, therefore, while knowing the good it chooses to follow evil, although it is not forced to go with it, but freely spurning good submits itself to evil, it is not given to it after its separation from its body to go to another body again lest it turns the good[12] down again, just as it did previously. Therefore, it is better to infuse a new soul into the body, clean from stain, apart from that of original sin, so that it may, through using reason and understanding, preserve the uprightness of its will with the choice of reason for the sake of that uprightness. For then the will will be preserved in uprightness only when a man wishes that which God wants him to wish.

366. Again, it is asked whether the soul when divested of the body can actually use the intellect.

That it cannot appears as follows. When the soul understands, as

---

[12] Reading 'bono' from P (as suggested by Werner) rather than 'malo' in the edition.

et ab Aristotele et ab aliis auctoribus, ipsa intelligit per ymaginem in ea constitutam representantem ei rem illam, cuius similitudo est illa ymago, et anima convertens aciem animi ad illam ymaginem coniungit et copulat eam cum re cuius est illa ymago, et sic intelligit actualiter. Sed anima cum est a corpore exuta nulli est coniuncta, nec habet 5 ligationem cum corpore. Ergo mediante corpore aliquo non formatur in anima aliqua passio que sit rei similitudo, et omnis passio que formatur in aliquo formari habet per contactum vel per aliquam ligationem anime cum corpore quando in eo ipsa est. Sed anima exuta a corpore non habet ligationem cum corpore ut ipsa illud moveat, nec accidit 10 ipsi contactus cum aliquo. Ergo in anima exuta a corpore non formatur alicuius rei ymago. Sed si anima intelligat, per ymaginem intelligit; sed a corpore exuta non habet in se ymaginem. Ergo a corpore exuta non habeatur intellectum.

367. Contra. Ut habetur ab Augustino et Ieronimo et aliis auctoribus, 15 anima in corpore existens mole carnis oppressa ita obfuscatur quod uti intellectu non potest nisi sensus ministerio precedente; unde dicit Aristoteles in principio *Posteriorum*: 'Omnis doctrina et omnis disciplina intellectiva fit ex preexistenti cognitione', scilicet sensitiva. Anima autem exuta a corpore libera est. Ergo potius potest tunc uti intellectu 20 quam prius; sed quando est in corpore intelligere potest; ergo et extra.

368. Preterea. A pluribus habemus auctoribus quod post separationem anime a corpore, quandoque contingit unam animam certificare aliam animam et de ignoto dare certificationem ei et cognitionem. Sic ergo contingit unam animam maiorem habere scientiam alia anima.   25

369. Contra. Anima exuta a corpore nullo est impedita quin ipsa quidlibet intelligat: est enim habens se ad similitudinem oculi existentis in centro et videntis totam speram. Ergo nulla anima alia est sapientior.

1 ymaginem: ymaginationem P   2 representantem: que presentat P   est: alia *add.* P   5 cum est *om.* CV   coniuncta: iuncta C   7 rei: ei V   8 contactum: contractum P   vel: et V   9–10 anime . . . ligationem *om. per hom.* P   11 ipsi contactus: ipsius esse contactum P   aliquo: alio P   12 ymaginem: ymaginationem C   13 in se *om.* P   13–16 non habeatur intellectum. . . . quod *om.* CV   16 obfuscatur: elata^{ur} P   18 principio: primo V   omnis² *om.* V   19 intellectiva *om.* P   sensitiva: Contra *add.* C   20 autem *om.* C   21 sed: et V; *om.* C   ergo et extra *om.* P   22 Preterea: Item P   post: prius V   24 aliam animam *om.* P   de: ab P   certificationem *om.* P   et² *om.* P   26 a: de V   nullo: a nullo V   impedita: impedimenta P   28 videntis: videns V   anima *om.* P

15 ab Augustino . . . auctoribus: Aug. *De Trin.* VIII. ii. 3; locum Ieronomi non invenimus; Avic. *De An.* V. v (25^{va}B).
18–19 in principio *Posteriorum*: Arist. *Anal. Post.* I. i (71^{a}1–2) iuxta versionem Iacobi Venetici (5).
19–21 Cf. Avic. *De An.* V. v (25^{va}B); v. vi (26^{va}E).
22 Cf. Gregorius M. *Dial.* IV. 33.

we know from Augustine, Aristotle, and other authors, it understands by means of an image which is set up in it which represents that thing to it, and the image is a likeness of it. The soul, when it turns the eye of the mind towards that image, links and joins it with the thing which the image is of, and thus it is actually understood. However, when the soul has been divested of the body it is not joined to anything, nor does it have a tie to a body. Therefore, no modification which is the likeness of a thing is formed in the soul by means of any body, and every modification which is formed in any body has to be formed through some connection between the soul and the body when the soul is in the body. However, when the soul has been divested of the body, it does not have a tie with the body so that the soul could move the body, nor does it happen that the soul has contact with any body. Therefore, no image is formed of anything in the soul when it has been stripped of the body. However, if the soul understands, it understands by means of an image, yet when the soul has been stripped of a body it does not have an image in itself. Therefore, when it has been stripped of a body it does not have understanding.

367. Against: As we know from Augustine, Jerome, and other authors, the soul as existing in the body is weighed down by the burden of the flesh and is so darkened that it cannot use the intellect except with the previous assistance of sensation. For this reason Aristotle says in the *Posterior Analytics*: 'All teaching and every intellectual discipline arises out of pre-existent knowledge', namely sense knowledge. However, when the soul has been stripped of the body it is free. Therefore, it can use the intellect then more than before. However, when it is in the body it can understand, therefore it can do so also when outside.

368. Moreover, we know from many authors that after the separation of the soul from the body it sometimes happens that one soul informs another soul and gives assurance and knowledge to it concerning what is unknown. Thus it happens that one soul has more knowledge than another soul.

369. Against: Once the soul has been stripped of the body, there is no impediment to why it should not understand something, for it is constituted like an eye which is in the centre and which sees all of the sphere. Therefore, no soul is wiser than another.

370. Item. Potest similiter hic queri, utrum anima philosophi cuiusdam sit sapientior separata a corpore quam anima alicuius idiote post separationem ipsius a corpore.

Quod non sit sapientior videtur per proximam rationem precedentem.

371. Quod sit sapientior videtur per hoc quod scientie quas firmavit 5 ipse philosophus in anima sua per ymagines rerum in ea repositas, remanent in anima post remotionem ipsius a corpore, quia nullam est assignare causam destructionis illarum scientiarum firmatarum in anima, cum tam anima quam scientia contrarietate careat, et anima aptior est ad retinendum scientias quando est extra corpus quam quando 10 est in corpore; quia quando est in corpore convertit ipsa aciem animi et intentiones suas ad regimen corporis et eius dispositiones et ad multa que accidunt corpori a causis extrinsecis.

372. Solutio. Dicimus quod anima separata a corpore intellectu uti potest, et una intelligit plus alia et est sapientior alia; quia quanto anima 15 magis particeps est illuminationis procedentis a pura veritate, que summe lucida et incommutabilis est, illa purificatione habet intellectum perspicaciorem, et maioris est sapientie quam illa que mole peccatorum est oppressa, minus existens particeps nature illuminationis. Sicut enim lux facit ad operationes sensus, ita et illuminatio pure veritatis facit ad 20 operationem intellectus. Unde bene concedimus quod anima cuiusdam idiote, quando est separata a corpore, est sapientior quam anima philosophi, si anima illa que fuerit idiote fit magis particeps pure veritatis et lucis ab ea irradiantis quam anima philosophi sit particeps; et propter hoc contingit quod anima minus sapiens scientiam recipit a 25 sapientiori intuendo eam et quicquid est in ea. Probabile tamen est quod anima philosophi plus sciat de quibusdam quam anima idiote, licet ipsa sit particeps pure veritatis, et de illis plus scit ipsa que recipit per receptionem discipline in corpore que non sunt de divina contemplatione.

1 Item *om.* C   hic *om.* C   2–3 sapientior . . . corpore: cum ipsa fuerit exuta sit sapientior quam anima alicuius idiomate cum ipsa fuerit separata P   4 proximam . . . precedentem: precedentem rationem proximam C   6 ymagines: ymaginem P   7 remotionem *om.* P   quia: quod C   8 illarum *om.* P   9 careat: caret P   10 est: sit P   11 quia . . . corpore *om.* V   et: ad V   12 ad: supra P   et² *om.* P   14 Solutio: ad hoc *add.* V   15 una: aliqua C   17 illa purificatione: per illam purificationem V   purificatione: puriorem P 18 perspicaciorem *om.* P   et: etiam *add.* C   19 minus existens: et minus est P 20 ita et illuminatio: in eluminatione P   21 bene *om.* C   23 illa *om.* PV   fuerit: fuit C   fit: fuerit C   24 quam: pro P   particeps: pure veritatis *add.* V   26–29 intuendo eam et quicquid . . . contemplatione *om.* P   28–29 et de . . . contemplatione: ut de illis que non sunt de divina contemplatione plus scit anima

5–11 Cf. Gundiss. *De An.* x (98).
15–19 Cf. Gundiss. *De Immortalitate Animae* (19); *Liber de causis primis et secundis* x (130).
19–21 Cf. Avic. *De An.* v. v (25^{rb}A).

370. Again, it can likewise be asked at this point whether the soul of any philosopher is wiser when separated from the body than the soul of any idiot after the separation of his soul from the body.

That it is not wiser appears to be the case through the argument immediately preceding.

371. It seems that it is wiser inasmuch as the knowledge which the philosopher has secured in his soul, through the images of things stored in it, will remain in the soul after its removal from the body. This is because no cause can be assigned to the destruction of that knowledge which has been secured in the soul, since both the soul and knowledge are lacking in contrariety, and the soul is more apt to retain knowledge when it is outside the body than when it is in the body. For when it is in the body the soul directs the eye of the mind and its intentions towards the direction of the body and its dispositions and the many other things which happen to the body from extrinsic causes.

372. Solution: We state that the soul when separated from the body can make use of the intellect, and that one soul understands more than another and is wiser than another. This is because in so far as a soul participates more in the illumination proceeding from the Pure Truth, which is the Highest Brightness and Unchanging, by that purification it has a keener understanding, and it has a greater wisdom than that soul which is burdened by the weight of sin, and exists as a lesser participant in the illumination of its nature. For just as light helps the functioning of sensation, so also the illumination of pure truth helps the activity of the intellect. Thus we willingly concede that the soul of some idiot, when it is separated from the body, is wiser than the soul of a philosopher, if that soul which was the idiot's becomes a greater participant in the Pure Truth and of the light irradiating from it than the soul of the philosopher. Again, because of this it happens that the soul which is less wise receives knowledge from the wiser soul in gazing upon it and whatever is in it. However, it is probable that the soul of the philosopher knows more about some things than the soul of an idiot, even if the latter is a participant in Pure Truth, and of those things, the soul of the philosopher knows more about what it has gained through learning in the body than from divine contemplation.

373. Sed queritur. Quomodo una anima possit recipere scientiam ab alia, cum utraque ipsarum careat et auribus et ore et instrumentis significandi aliquid quod ipsa intelligit? Nos enim id quod intelligimus significamus aliis per voces, vel per aliqua alia signa, ut claustrales per nutus et signa corporalia suos intellectus aliis ex- 5 primunt.

374. Similiter queritur. Quomodo anima seipsam possit intelligere, cum nihil in seipsum agat vel patiatur a seipso, ut eodem oculo quo quis videt, eundem oculum videre con contingit; et eodem baculo quo quis percutit, non contingit eundem baculum percutere.   10

375. Ad primum. Dicendum est quod sicut compositum se habet ad compositum, ita simplex se habet ad simplex; unde sicut compositum recipit inmutationem a composito, ita simplex inmutatur a simplici; et sicut nos oculo composito manente anima videmus rem compositam, ita anima separata a corpore seipsam videt et intuetur, simplex cum ipsa 15 sit simplex; unde intuetur aliam animam, et eam intelligendo, percipit in ea inmutationes in ea repositas, que sunt similitudines rerum extra, et per illarum similitudinum perceptionem venit in cognitionem rei prius ignote, et secundum quod plus et plus illuminatur a causa prima, secundum hoc clarius eas intuetur, et plus scientie sibi adquirit; et 20 quia sic venit in scientiam rei ignote, per hoc quod convertit aciem animi sui ad aliam animam puriorem et magis illuminatam, dicitur quod una anima recipit scientiam ab alia; et per hoc quod ipsa percipit in illa alia anima ymaginem sui ipsius anime intelligentis, intelligit ipsa anima seipsam per sui ymaginem in alia anima perceptam. Et 25 est simile apud sensum: oculus in speculo percipit ymaginem sui, et per illam ymaginem seipsum videt. Et sic patet solutio secunde questionis.

philosophi quam anima idiote, que, scilicet anima philosophi, apud se retinet ex disciplinis receptis in quam adminiculo corporis quando ei fuit coniuncta V

2 ipsarum *om.* P   auribus: aure P   2–3 instrumentis significandi: instrumenti signis P   3 enim: autem C   id: idem V   4 significamus: signamus P   vel: et P   aliqua alia *om.* P   5 corporalia: corporum V   5–6 corporalia . . . exprimunt *om.* P   7 anima *om.* V   8–9 quo quis videt *om.* P   9 contingit: aliquem *add.* P; nisi per reflexionem radiorum *add.* V   11 sicut *om.* V   12 ad simplex: ad simplicem V   14 manente: iuvante C   anima *om.* P   15 a corpore *om.* P   videt et *om.* C   15–16 cum . . . simplex *om.* P   17 inmutationes in ea *om.* C   repositas: receptas V   rerum: esse P   18 perceptionem: receptionem P   19 a causa prima *om.* P   20 clarius: plus et plus P   21 sic *om.* V   22 sui *om.* PV   aliam *om.* P   et magis illuminatam: ea P   23 quod *om.* CP   24 alia *om.* P   ymaginem sui: magne P   25 anima *om.* P   in *om.* C   perceptam: receptam P   26 apud: quod P   ymaginem: ymaginationem P

7–9 Cf. Aug. *De Trin.* IX. iii. 3.   11 Ad primum: cf. supra, n. 373.

373. However, the question arises, in what way can one soul receive knowledge from another soul, since both are missing ears and a mouth and the organs to communicate something that the soul understands? For we communicate what we understand to others by means of words, or by means of some other signs—as, for example, cloistered monks who express their ideas to others by means of nods and other bodily signs.

374. Similarly it is asked, in what way can the soul understand itself, since there is nothing in it which acts or is acted upon by itself? For example, with the same eye with which someone sees, it is not the case that the eye itself is seen, and with the same stick with which someone strikes, it does not happen that the same stick is struck.

375. Regarding the first question, it should be stated that just as a composite thing is constituted in respect of a composite thing, in the same way a simple thing is to a simple thing. Thus just as a composite thing receives an impression from another composite, in the same way a simple thing is changed by a simple thing. Again, as when the soul is present we see a composed thing with a composed eye, so the soul separated from the body sees and intuits itself, a simple thing, since it is simple. Thus it intuits another soul, and in understanding it, it perceives the impressions which are stored in it, which are the likenesses of external things, and through the perception of those likenesses it comes to the knowledge of a thing which previously was unknown. Moreover, according as it is more and more illuminated by the First Cause, in this way it intuits them more clearly and it acquires more knowledge. Thus since it comes to the knowledge of an unknown thing in this way, inasmuch as it turns the eye of its mind towards a soul which is purer and more illuminated, it is said that one soul receives knowledge from another.

375bis.[13] Moreover, inasmuch as the soul perceives an image in that other soul of the soul itself which is understanding, the soul itself understands itself by means of the image of itself perceived in the other soul. Again, it is similar in the case of sensation: the eye perceives an image of itself in the mirror and by means of that image it sees itself. Thus the solution to the second question is clear.

---

[13] This is the reply to the objection in no. 374.

XXV. iv. *In qua parte corporis sit anima*

Restat ut videamus, cum anima sit in corpore et non in qualibet parte corporis essentialiter, ut prius ostensum est, in qua parte corporis humani sit anima essentialiter.

376. Quod sit in corde sic videtur posse ostendi. Ut habemus a pluri- 5 bus auctoribus, ex constrictione cordis provenit dolor, ex dilatatione provenit gaudium. Sed in anima est dolor propter cordis offendiculum, quia ipsa tenetur regere. Similiter gaudium est in anima. Indicium ergo est animam esse in corde, quia ex constrictione cordis provenit dolor et ex dilatatione gaudium.   10

377. Preterea. Ab Aristotele habemus quod ira est ascensus sanguinis circa cor; unde cum ascendit sanguis circa cor, tunc est ira. Sed plura sunt membra corporis vegetati ab anima quam cor; potius videtur ergo animam esse in corde quam in alio membro.

378. Contra. Anima est in corpore humano ad regimentum corporis 15 humani; cor autem est unum principalium membrorum hominis: unde si contingat per aliquod accidens fieri timorem in homine fugit sanguis ad cor propter ipsum consolandum, ne ipsum deficiat propter nimiam timiditatem. Regit ergo sanguis ipsum corpus in hoc casu. Est ergo hoc indicium quod anima debens regere corpus humanum sic sit essentialiter 20 in sanguine quia ipse sanguis est nutrimentum totius corporis, et ita est ipse eius consolatio.

379. Quod autem anima sit in cerebro potest haberi per hoc quod cuiuslibet sensus perfectio est in cerebro, et cuiuslibet alterius apprehensionis, ut eius que fit per ymaginationem, vel per estimationem, vel 25 per memoriam. Cum ergo apprehensio sit ab anima, et in cerebro per inmutationes ibi receptas fiant apprehensiones, indicium est quod potius est anima in cerebro quam in alia parte corporis humani. Quod bene

---

2 ut videamus: videre P   3 corporis *om.* P   prius *om.* P   5 sit *om.* P
7 offendiculum: ostendiculum C   8 quia: quod C   9 quia: quod C   cordis: eius P   11 Preterea: Item P   12 circa cor *om.* P   ascendit: accedit P
12–13 Sed . . . cor *om.* P   13 potius . . . ergo: et ita potius videtur P   14 alio:
aliquo C   15 Contra: Item P; *om.* V   regimentum: regimen P   16 autem:
enim V   principalium: principium C   hominis: hominum P   18 deficiat:
desicceat P   20 quod: quia V   21 sanguine: et ideo *add.* C; et ideo etiam
add. V   quia: eo quod P   21–22 et ita . . . consolatio *om.* CV   23 autem
om. V   25 ymaginationem: ymaginem P   26 ergo: igitur P   sit ab
anima et: anima sit P   cerebro: et *add.* P   27 ibi: sibi C   receptas: repositas V

5–6 a pluribus auctoribus: Arist. *De An.* I. iv (408ᵇ5–8); Constantinus *Viat.* III.
xiv (cliiiiᵛ): 'Cor enim eorum cum timeat constringitur, et calor suus naturaliter
coadunatur et non dilatatur . . . In letitia dilatatur cor subito et aperitur'.
11 Aristotele: *De An.* I. i (403ᵃ31).

XXV. iv. *In what part of the body is the soul?*

It remains for us to see, since the soul is in the body and not essentially in some part of the body, as has already been shown, in what part of the body the soul is essentially.

376. It appears that it can be shown as follows that it is in the heart. As we know from various authors, pain comes from the constriction of the heart and joy from the dilation of the heart. However, there is pain in the soul owing to a cause of offence to the heart, because the soul is expected to be in charge. In a similar fashion there is joy in the soul. Therefore, it seems that the soul is most likely in the heart, because pain arises out of the constriction of the heart and joy arises out of its dilation.

377. Moreover, from Aristotle we know that anger is due to a rising up of blood in the area around the heart. Thus when blood rises up around the cardiac region, then there is anger. Yet there are more organs of an enlivened body from the soul than from the heart. Therefore it seems more likely that the soul is in the heart than in any other organ.

378. Against: the soul is in the human body in order to direct the human body. The heart, however, is one of the principal organs of man, thus if it happens through some circumstance that fear arises in a man, blood rushes to the heart in order to strengthen it, lest it fail because of excessive timidity. Therefore, blood directs the soul in this case. Therefore, this is evidence that the soul, which should direct the human body, is thus essentially in the blood because blood itself is the nourishment of all of the body, and so is itself the consolation of the body.

379. However, that the soul is in the brain can be known inasmuch as the completion of any sense is in the brain, as well as any other kind of perception, such as that which occurs by means of imagination, or by means of estimation, or through memory. Therefore, since perception is from the soul, and perceptions occur in the brain by means of the impressions received there, this is evidence that the soul is more likely to be in the brain than in any other part of the human body.

concedimus dicentes quod non est in corde. Multi tamen auctores videntur velle animam secundum se totam esse in qualibet parte corporis, sicut Deus ubique est secundum se totum.

380. Ad predictas autem rationes sic resistimus. Dicendo quod ex constrictione cordis provenit dolor, quia cor est unum principalium 5 membrorum; et ideo quia anima appetit esse in corpore, appetit anima ut cor existat in debita dispositione; unde si adveniat constrictio molestatur ipsa anima, quia iam recedit cor a debito organo: et ideo ex constrictione cordis provenit dolor in anima ne cor deficiat per constrictionem. Ex dilatatione vero cordis provenit gaudium in anima, quia 10 cor est naturaliter calidum: unde suum debitum esse est esse in caliditate. Caliditatis autem est dissolvere et dilatare; et ideo ex dilatatione cordis provenit gaudium in anima. Propter hoc quod cor debet regi ab anima est in organica dispositione et debita sibi et regulari quando sic dilatatum est cor. Interdum tamen potest esse cordis dilatatio vehe- 15 menti gaudio existente, quod cor ultra modum dilatabitur ita quod homines debite memorie et debite scientie viam amittant. Ex ascensu autem sanguinis circa cor provenit ira, quia cor naturaliter est calidum, et sanguis naturaliter est calidus, et ita calidum adicitur calido, et tanto magis habundat ibi caliditas ex calore dissoluto et rarefacto, et ita est 20 ibi timor et levitas: quoniam ex dilatatione timor, et ex rarefactione levitas. Sed quando cor est inflatum et agile, tunc habet homo spem habendi victoriam de aliqua molestia sibi illata. Sed ira est cum spe habendi victoriam, et ita ex ascensu sanguinis circa cor provenit ira. In

1-3 Multi ... totum *om.* P   6 appetit[2] anima *om.* P   7 existat: existens P
8-10 et ideo ... constrictionem *om.* P   9 per: propter V   10 vero: autem C
11 naturaliter: de natura V; *om.* P   debitum: proprium P   13 cordis *om.*
P   provenit: prout P   Propter hoc: Et propter hoc etiam V   cor: quod *add.* P
14 organica: propria P   et: existente P   14-17 quando ... amittant *om.* P
17 ascensu: accessu P   18 naturaliter: in se P   19 et sanguis ... calidum
om. per hom. V   20 caliditas: calidum C   21 timor: tumor C   et[2] *om.*
P   rarefactione: calefactione V   24 victoriam: memoriam P   ascensu: accessu P

---

1 Multi . . . auctores: vid. locos citatos ab H. Ostler, *Die Psychol. des Hugo von St. Viktor* (Beiträge VI. i, 1906), p. 62 not. 4, et praecipue Guill. de S. Theod. *De Natura Corporis et Animae* II (719D): 'sicut Deus in mundo, sic quodammodo ipsa (anima) sit in corpore suo, ubique scilicet . . . tota . . . in singulis partibus'.
15-17 Cf. Constantin. *Pantegn.* Theor. IV. vii (xvi[va]): 'Tota sua excluditur substantia cum nimium gaudium subito dissolvit et exire eum (calorem naturalem) facit. Unde cito quis moritur'; Anon. *Quaest. Salernitan.*: 'Queritur quare quislibet levius moritur ex nimio gaudio quam ex tristitia' (cod. Bodl. Auct. F. 3. 10, fol. 139).
18-23 Cf. Constantin. ibid. IV. viii (xvi[vb]): 'Ira enim est fervor cordis per quem calor naturalis subito extra prorumpit cum anima ob illatas iniurias excogitatam vindictam explere desiderat'.
23 victoriam: cf. Avic. *De An.* IV. iv (20[rb]A): 'Que vero vult vincere et id quod putatur nocivum repellere est irascibilis . . . ex ira habetur intensio voluntatis ad victoriam'; cf. supra, nn. 55, 71.

This is something that we readily concede, and we state that the soul is not in the heart. However, many authors seem to hold that the soul itself is complete in every part of the body, just as God is in Himself complete everywhere.

380. However, regarding the aforementioned arguments, we oppose them as follows, by saying that out of the constriction of the heart comes pain, because the heart is one of the principal organs. Therefore since the soul desires to be in the body, the soul desires that the heart exists in the correct manner. Thus, if a constriction comes about the soul itself is disturbed, because the heart has already fallen away from its proper functioning. Thus from the constriction of the heart suffering arises in the soul lest the heart should fail because of its constriction. However, joy arises from the dilation of the heart in the soul because the heart is naturally hot, and so its correct state of being is to be hot. The effect of heat, however, is to loosen and dilate, and thus from the dilation of the heart joy arises in the soul. Thus, since the heart must be directed by the soul, it is in its organic disposition and in its regular and required manner when it is dilated in this way. Sometimes, however, there can be a dilation of the heart when there is an intense joy, so that the heart is dilated excessively, so much so that men stray from the way of what they should remember and know. Anger occurs from the rising up of blood to the area around the heart because the heart is naturally hot, and blood is naturally hot, and thus heat is added to heat. And so great does the heat abound there from the diffused and rarefied heart[14] that there is fear and foolishness because fear arises out of dilation and foolishness out of rarefaction. However, when the heart is full and active, then a man has the hope of winning victory over any disturbance inflicted upon it. However, anger exists together with the hope of having victory, and thus out of the rising up of blood to the area around the heart anger

[14] Reading 'corde' with Werner rather than 'calore'.

huiusmodi autem casu contingit quod sanguis ascendit circa cor, quia cum cor molestetur ab aliquo accidente fugit sanguis ad ipsum consolandum; et sic, ut preassignatum est, cum spe habendi victoriam nascitur ira.

381. Non oportet autem quod anima sit in sanguine propter hoc quod, 5 cum anima sit vis nutritiva corporis et eiusdem conservativa, quantum in se est, transmittit ipsa per sui illuminationem et vivificationem ipsum sanguinem per totum corpus ubi opportunum est eum esse, nisi aliquid eam impediat, cum eius intentio convertatur in corporis conservationem, ut per existentiam eius in corpore adquirat ipsa anima sibi pure veritatis 10 participationem.

XXVI. i. *De libero arbitrio*

Restat autem ut videamus differentiam inter rationem et libertatem arbitrii.

382. Quoniam videtur idem esse rationem et liberum arbitrium pro- 15 pter hoc quod, cum intellectus aliqua apprehendit, ea ratio postea comprehendit et de eis iudicat, et quod ei melius videtur, illud eligit; et ob hoc est libertas in anima. Est enim potestas in anima eligendi bona et mala secundum examinationem a ratione procedentem. Sed quid aliud est hec ratio quam illa eadem potestas, cum ab eadem potestate 20 sint iudicium et electio?

383. Preterea. Ut inferius ostendetur, libertas arbitrii est potestas conservandi rectitudinem voluntatis propter ipsam rectitudinem. Sed conservatio voluntatis in rectitudine per se et proprie est a ratione, et ita eadem potestas anime est liberum arbitrium et ratio.   25

384. Contra. Arbitrium videtur esse liberum ob hoc quod in ipso est eligere prout vult vel bonum vel malum; electio autem mali a ratione non est, immo potius per rationis absentiam vel eius impotentiam et debilitatem. Non est ergo ratio libertas arbitrii.

1 autem *om.* P   ascendit: accidit P   2 molestetur: molestatur P   2–3 ad ipsum consolandum: ut ipsum consoletur P   6 cum: ab *add.* C   7 transmittit ipsa: transmutat ipsum P   ipsa *om.* V   7–8 ipsum sanguinem *om.* P   8 ubi: in quo C   9 cum: eum P   10 existentiam: coniunctionem P   in: cum P   ipsa anima *om.* P   13 Restat autem: Sequitur P; autem *om.* V   16 hoc *om.* V   apprehendit: apprehenderit C   17 et[1]: quod P   illud *om.* V   18 est[1]: ibi *add.* P   in anima[1]: quia P   enim: ibi P   in anima[2] *om.* P   18–19 bona et mala *om.* P   19 procedentem: procedente P   20 illa: hec P   21 sint: sit P   22 ostendetur: habetur P   23 conservandi: servandi P   24–27 et ita . . . a ratione *om. per hom.* C   25 liberum arbitrium: libertas arbitrii V   26 est *om.* P   28 potius *om.* P   29 debilitatem: debitum P

13–p. 111, l. 6 Impress. ap. O. Lottin, *Psychologie et morale au XIIe et XIIIe siècles,* Gembloux, 1949, III. ii, pp. 610–14.
22 Cf. infra, n. 396.

arises. In a case of this kind it happens that blood rises up to the area around the heart, because when the heart is disturbed by any incident blood flows to strengthen it, and thus, as has already been pointed out, together with the hope of obtaining victory anger is born.

381. It is not, however, necessary that the soul is in the blood because of the fact that, since the soul is a nutritive power of the body and preserves it, inasmuch as it is in itself, it transmits blood throughout all of the body, by means of its illumination and giving of life, to where it is opportune for it to be, unless something impedes it. This is because its intention is directed towards the preservation of the body, so that by means of its existence in the body the soul itself acquires for itself a share in pure truth.

XXVI. i. *Concerning free will*

It remains, however, for us to examine the difference between reason and the freedom of the will.

382. However, it seems that reason and free will are the same, because of the fact that when the intellect grasps some things, reason afterwards understands them and judges them, and chooses what seems better to it, and because of that there is freedom in the soul. For there is a power in the soul to choose good things and bad things according to the investigation which comes from reason. But what else is this reason than that very same power, since judgement and choice are from the same power?

383. Moreover, as will be shown below, the freedom of the will is a power to preserve the uprightness of the will for the sake of that uprightness. However, the preservation of the will in uprightness in itself and properly is from reason, and thus reason and free will are the same power of the soul.

384. Against: The will appears to be free because there is choosing in it, namely it chooses either good or evil. The choice of evil, however, is not from reason; on the contrary, it is more due to the absence of reason, or its weakness. Therefore reason is not free will.

385. Solutio. Dicimus quod unum et idem in essentia est ratio et libertas arbitrii: quoniam eadem est hec vis et illa, sed reputantur esse diverse propter diversas comparationes eius ad diversa. Dicitur enim illa vis esse libertas arbitrii ob hoc quod rationalis creatura secundum eam potens est nòn peccare et non succumbere servituti peccati; et 5 ideo potens est secundum eam rationalis creatura adherere rectitudini et respuere eius oppositum. Illa eadem vis dicitur esse ratio ob hoc quod ipsa ratiocinatur arbitrando quid sit bonum et quid sit malum et ut, percepto bono, conservet se in illo et respuat eius oppositum. Patet ergo solummodo esse differentiam in accidente inter rationem et liber- 10 tatem arbitrii; quoniam in eo est ratio quod est potestas arbitrandi per ratiocinationes quid sit bonum et quid malum, quid verum et quid falsum; et in hoc est libertas quod, dato iudicio, potens est sequi verum et conservare se in bono. Unde sicut unum et idem est pater et filius, sed diversa sunt in accidente; sicut unus et idem homo miles et 15 dux, sed in diversa comparatione secundum diversa accidentia, ita ratio et libertas arbitrii una et eadem est vis, sed differunt in accidentibus. Unde, sicut, si proprie loquamur, non concederetur hec: pater est filius, nec ista: miles est dux; ita nec ista concederetur: ratio est libertas arbitrii, si propria sumatur locutio et per se. 20

Item. Queritur utrum potestas declinandi a bono sit pars libertatis arbitrii.

386. Quod sit eius pars videtur. Quia liberior est ille qui sibi relinquitur ut ipse de duabus viis illam eligat quam voluerit, quam iste qui coactus est procedere secundum alteram illarum tantum. Due autem 25 sunt vie: una in bono et alia in malo; cuius ergo arbitrio relinquitur eligere utrum illarum voluerit liber est. Si autem non posset procedere nisi per alteram illarum tantum, non esset adeo liber sicut qui potest

---

1 Solutio: Ad hoc P   2 quoniam: quia P   3 eius ad diversa om. P
4 arbitrii om. P   rationalis creatura: ratio vel creatura aliqua C   5 eam: hanc
P   non¹ om. C   6 ideo: ita P   est om. C   secundum eam om. V   7 vis
om. V   hoc: eo P   quod om. V   8 ipsa om. C   quid: quod P (bis)   ut
om. P   9 in illo et respuat: respuendo P   Patet: Est P   10 esse differentiam: differentia P   inter: et V   10-11 libertatem arbitrii: liberum arbitrium P
11-12 ratio . . . ratiocinationes: est ibi potestas per rationes arbitrii P   12 quid:
quod P (quater)   13 et om. P   potens est: potest P   14-15 unum . . .
accidente om. P   15-16 sicut . . . accidentia om. C   15 homo: quandoque
est add. V   16 sed . . . accidentia: et diversitas est tunc ibi in accidente V
17 arbitrii una et om. P   accidentibus: accidente P   18-20 Unde . . . per se
om. P   21 libertatis: liberi C   23 eius om. P   24 quam voluerit: que
sibi placuit P   iste om. C   25 alteram illarum tantum: altitudinem P   26 et
om. P   ergo: autem V   27 illarum: illorum P   posset: possit P   28 per
alteram illarum tantum: tantum secundum alterum illorum P   28-p. 108, l. 2 sicut
. . . Ita om. P

4-7 Cf. Anselm. De Lib. Arb. i (208, 18-21).

385. Solution: We state that reason and free will are one and the same in essence, because both one and the other are the same power, but they are regarded as diverse because of their various relations to diverse things. Free will is said to be that power because of the fact that due to it a rational creature is able not to sin and not to succumb to the servitude of sin. Therefore a rational creature according to it is able to adhere to uprightness and to reject its opposite. That same power is said to be reason because it reasons when deciding what good is and what evil is, and so, having perceived the good, it preserves itself in that and rejects its opposite. Therefore, it is clear that there is only an accidental difference between reason and free will. This is because in it there is reason, which is a power of judging by means of reasoning what good is and what evil is, what is true and what is false. Again, freedom is in it because, given the judgement, it is able to follow the true and to preserve itself in the good. Thus, just as a father and a son are one and the same, but are diverse accidentally, just as a soldier and a commander are one and the same man, but are different in relation to different accidents, so reason and free will are one and the same power, but they differ accidentally. So it follows that, strictly speaking, this would not be conceded: 'The father is the son', nor this: 'The soldier is the commander'. Nor will this be conceded: 'reason is free will', if this is understood strictly speaking and as such.

Again, it is asked whether the power to turn away from the good is a part of the freedom of the will.

386. It seems that it is a part of it, because he is freer who is left to himself so that he may choose between two paths the one which he wants, than he who is forced to proceed according to one of those ways only. For there are two ways, one towards good and one towards evil; therefore he whose will is left to choose which of the two ways he wishes is free. If, however, he cannot go forward except by one of the two ways only, he would not then be free like someone who could go

procedere per utramlibet illarum. Sed, si malo adhereat, declinat a bono
et habet potestatem declinandi a bono. Ita ergo potestas declinandi
a bono est de libertate arbitrii, cum ipsa libertas sit potestas procedendi
in bono et potestas declinandi a bono.

387. Preterea ad idem. Primus homo non peccasset, si non liberum 5
arbitrium habuisset; sed peccatum est a potestate peccandi. Videtur
ergo quod potestas peccandi sit liberum arbitrium, aut pars libertatis
arbitrii.

388. Si concedatur; Contra. Ponatur quod hic sint duo homines, et
uterque sit liber. Unus tamen sit tante libertatis quod nullo modo in 10
servitutem redigi possit; alius autem habeat ita se in sua libertate quod
eam amittere queat et in servitutem redigi. Uter istorum est liberior?
Constans est quod ille est liberior qui servituti nullo modo succumbere
potest. Potestas ergo succumbendi servituti non facit ad libertatem. Sed
potestas peccandi est potestas succumbendi servituti. Ergo potestas 15
peccandi non est pars liberi arbitrii.

389. Item. Imbecillitas rei non facit ad esse rei, immo magis ad non
esse. Libertas autem ad esse facit. Ergo imbecillitas non facit ad libertatem; sed potestas declinandi a bono est ab imbecillitate; ergo potestas
peccandi non facit ad libertatem.   20

390. Item. Potestas peccandi rationali creature addita minuit eius
libertatem, quoniam eius verum esse minuit, quod est liberum, eo quod
potens est non errare; et ita potestas peccandi minuit libertatem, et
separata et remota auget libertatem. Ergo libertas arbitrii non consistit
in posse peccare.   25

391. Item. Si aliquis peccat, ipse servit peccato; ergo si ipse potest
peccare, ipse potest servire peccato; et si potest servire peccato, ei
potest dominari peccatum. Ergo a primo: si potest peccare, ei potest
dominari peccatum. Sed in hoc quod ei potest dominari peccatum,
diminuitur eius libertas, et in hoc similiter quod potest servire 30
peccato. Ergo similiter in hoc quod potest peccare diminuitur

2 Ita *om.* V   3 libertas: liber C   4 in: a P   5 Preterea: Item P
6–7 Videtur ergo quod: Sic ergo P   7 ergo *om.* V   7–8 sit . . . arbitrii: est
de libero arbitrio P   11 servitutem: servitute P   11–12 alius . . . redigi *om.* P
12 Uter: Uterque PV   est *om.* V   13 Constans est: Constat P   succumbere
potest: incumbet P   14 ad *om.* P   15–16 potestas peccandi *om.* P   17 magis
om. P   19 ab imbecillitate: becillitas P   19–20 potestas peccandi: illa potestas P
22 quoniam: quia P   24 et remota *om.* P   24–25 libertas . . . peccare: non
est pars libertatis P   26 aliquis: quis P   ipse² *om.* PV   27 ipse *om.* P   potest²
om. V   30 diminuitur: dampnatur P   31 peccato *om.* P   similiter:
et P   hoc: sit *add.* P

9–15 Anselm. *De Lib. Arb.* i (208, 13–17).
21–25 Cf. ibid. (208, 26–209, 5).
26–31 Cf. ibid. ii (209–10).

by either of the two ways. However, if he cleaves to evil, he turns away from good and so has the power to turn away from good. Therefore the power of turning away from good comes from the freedom of the will, since this freedom is a power to proceed towards the good as well as a power to turn away from the good.

387. Moreover, concerning the same: The first man would not have sinned if he did not have free will. Sin, however, comes from the power to sin. Therefore it seems that the power to sin is free will, or part of the freedom of the will.

388. If this is conceded, the following is objected against it: Given two men, and each of them is free—one man, however, has such freedom that he cannot in any way be led back into servitude; the other man is so constituted, however, in his freedom that he can lose his freedom and be led back into servitude. Which of them is the freer? It is certain that that man is freer who can in no way succumb to servitude. Therefore the power to succumb to servitude is of no use to freedom. But the power to sin is a power to succumb to servitude. Therefore, the power to sin is not a part of free will.

389. Again, the weakness of something is of no use to the being of a thing, rather more to the non-being of a thing. For freedom benefits being. Therefore, weakness does not benefit freedom; but the power to turn away from good is due to weakness, therefore the power to sin does not benefit freedom.

390. Again, when the power to sin has been added to a rational creature it diminishes its freedom, because its true being, which is free because it is able not to err, is diminished. Thus the power to sin lessens freedom, and when separated and removed increases freedom. Therefore, the freedom of the will does not consist in the ability to sin.

391. Again, if someone sins, he is a slave to sin. Therefore, if he can sin, he can become a servant to sin; and if he can become a servant to sin, sin can rule over him. Therefore, from the first, if he can sin, sin can rule over him. However, inasmuch as sin can dominate over him, his freedom is diminished, and in this similarly he can become a slave of sin. Therefore similarly inasmuch as he can sin the freedom of a

libertas rationalis creature. Potestas ergo peccandi non est pars liberi arbitrii.

392. Solutio. Ut ostensum est, bene concedimus quod potestas peccandi neque est liberum arbitrium, neque pars libertatis arbitrii. In hoc enim non est arbitrium liberum quod potest habere se indifferenter 5 secundum electionem ad bonum et ad malum. Si amoris intentione eligat bonum, bonum est; si autem malum, malum est. Sed in hoc est liberum quod potest non peccare et nulla re potest cogi ad peccandum; nulla enim necessitate peccat, sed ex sola voluntate, eo, scilicet quod vult quod non debet et quod non expedit. Et sciendum quod potentia pec- 10 candi revera non est potentia, sed potius impotentia; ex debilitate enim mentis contingit aliquem velle quod non debet. Et tunc vult aliquis quod non debet, quando vult quod Deus non vult eum velle.

393. Dicitur autem a pluribus auctoribus quod peccatum est per liberum arbitrium: non ideo autem quod est liberum, sed ideo quia 15 male conservatur libertas arbitrii, hoc est potestas non peccandi.

394. Per hoc ergo patet responsio ad primam obiectionem, que fit per similitudinis collationem. Non enim dicimus in hoc consistere liberum arbitrium quod nullo modo possit peccare, sed in hoc quod potens est non peccare et nulla re cogi potest ad peccandum. 20

395. Et nota quod hec argumentatio non valet:—iste potest servire peccato; ergo peccatum potest ei dominari—licet videatur ibi esse locus a correlativis, et ibi est fallacia accidentis. Quia potestas que est hominis postea redditur peccato. Non enim est in peccati potestate quod peccatum dominetur homini, sed in hominis potestate est et, ut verius 25 dicam, in eius impotentia est. Hec autem locutio: peccatum potest dominari, hic improprie sumitur quando aliquis eam concedit, eo quod sic redditur peccato potestas que eius non est. Et potest haberi simile ab Aristotele in *Topicis*, ubi ipse dicit quod male assignat proprium

---

1 liberi: libertatis V   3–4 Solutio . . . arbitrii *om.* V   3 bene *om.*
P   potestas: potentia P   4 neque[1]: non P   libertatis: liberi P   5 non:
et ideo C   5–6 indifferenter secundum electionem: in electione sui P; secundum electionem indifferentem et V   6 Si: Et si V   amoris: accidentis P   8 liberum:
arbitrium *add.* P   9 nulla enim: et ex nulla P   peccat: peccati C   ex *om.* C
10–12 et quod . . . non debet *om. per hom.* C   10 sciendum: est *add.* V
11 potius: est *add.* V   12 Et: Quod P   14 a: in P   15 ideo autem
quod: quia P   ideo[2] *om.* P   17 ergo *om.* P   18 dicimus: debemus P
19 modo *om.* V   20 potest *om.* P   21 nota: notandum V   22 ibi *om.* C
23 correlativis: relativis P   que *om.* V   24 enim *om.* P   25 dominetur:
dicitur esse P   27 hic: huic V; *om.* C   concedit: concedentem P   28 sic
om. PV   haberi: habere P   29 ipse *om.* P   assignat: assignatur P

7–10 Cf. ibid. (210, 2–10).
14 a pluribus auctoribus: cf. P. Lombard. *Sent.* II. dist. xxiv. c. xii.
21–28 Cf. Anselm. *De Lib. Arb.* ii (210, 11–19).
29 in *Topicis*: Arist. *Topic.* v. viii (138^{b}31–37).

rational creature is diminished. Therefore the ability to sin is not a part of free will.

392. Solution: As has been shown, we readily concede that the power to sin is neither free will nor a part of the freedom of the will. For free will is not such that it can be indifferent regarding the choice of good or evil. If it chooses good with the intention of love, it is good. If, however, it chooses evil, it is evil. However, it is free inasmuch as it can not sin and it cannot be compelled by anything to sin, for through no necessity does it sin, but only by the will, namely in so far as it wishes what it should not and what is not advantageous. And it should be known that the ability to sin is in truth not an ability but rather an inability, for it is out of weakness of mind that it happens that someone chooses what he ought not to. Moreover, when someone wishes what God does not want him to wish, then he wants what he should not wish.

393. However, it is stated by many authors that sin exists because of free will, not because it is free but rather because the freedom of the will is badly preserved, which is the ability not to sin.

394. In the following way, therefore, the reply to the first objection is clear, and it comes about through the drawing up of a likeness. For we do not state that free will consists in this, that it can in no way sin, but in so far as it is able not to sin and it cannot be forced to sin by anything.

395. Again, note that this argumentation is not valid: 'this man can become subservient to sin; therefore sin can rule over him'; namely, we can see that this is an argument from correlatives, and there is a fallacy of the accident there. This is because the power which belongs to a man is afterwards surrendered to sin. For it is not in the power of sin to rule over a man but it is in the power of man, and, as I truly say, it is in his weakness. For this statement: 'sin can rule' is taken improperly here when someone concedes it, because in this way power is handed over to sin which it does not have. And the same can be known from Aristotle in the *Topics*, where he says that someone incorrectly

aeris qui dicit: spirabile esse proprium aeris; quia potestas spirandi non
est in aere, sed est animalis spirantis proprietas.

396. Item. Ex premissis haberi potest quod libertas arbitrii est
potestas servandi rectitudinem voluntatis. Et in hoc est error quod aliquis habens illam potestatem ea abutitur, rectitudinem voluntatis non 5
observans.

397. Sic autem obicitur. Non-ens servari non potest; quia qualiter
posset aliquis servare rem non existentem? Sed, homine existente in
peccato non est ei rectitudo voluntatis. Ergo non est existens sue voluntatis rectitudo. Ergo rectitudinem voluntatis servare non potest. Ergo 10
liberum arbitrium non habet.

398. Solutio. Potestatum quedam est semper cum suo actu, ut
potestas solis qua sol movetur; quedam non est semper cum suo actu,
ut potestas videndi non est semper cum suo actu videndi, ut quando res
existens in tenebris videri non potest, et tamen aliquis habet potentiam 15
videndi eam. Similiter qui in peccatis est, in tenebris est, voluntatem
in rectitudine non servans, habet tamen potestatem eam conservandi; et
quia non servat, nec cogitur non servare, peccat ideo.

399. Item. Videtur forte alicui quod liberum arbitrium male sit
rationali creature collatum; quia si liberum arbitrium non habuisset, 20
non peccasset.

400. Ad hoc dicendum quod iste qui dicit liberum arbitrium causam
peccati esse consimilis est opinionis Ade, qui dixit Domino sese pec-
casse per hoc quod ei tradidit Evam, volens refundere peccatum in
Dominum. Similiter est de homine habente libertatem arbitrii, qui ideo 25
peccat, quia ipse eam non servat recte; non autem peccat quia libertatem habet, sed quia libertatem arbitrii a Deo ei collatam in rectitudine
non servat. Et est liberum arbitrium collatum rationali creature ad

---

1 potestas spirandi: illa potestas V   2 est² . . . proprietas: ipsius animalis P
3 Item *om.* CV   4 servandi: conservandi C   4–5 aliquis: quis P   6 observans: observat CV   7 Sic autem obicitur: Obicitur autem P   quia: et
V   7–8 quia . . . existentem *om.* P   9 ei: in eo C   rectitudo: in rectitudine V
9–10 Ergo . . . rectitudo *om.* C   12 Potestatum: Potentialiter V   semper *om.*
CV   14 ut *om.* V   14–15 res existens: aliquis est P   15 videri: videre P
15–16 videri . . . in tenebris *om. per hom.* C   15 aliquis *om.* P   16 eam
om. P   17 in rectitudine: rectitudinis P   servans: peccat ideo *add.* V   conservandi: servandi P   18 peccat ideo: non peccat P   19 forte *om.* P
20 collatum: consignatum P   non *om.* C   20–21 non habuisset . . . peccasset:
non haberet, non peccaret P   22 iste *om.* CV   23 Ade: cum Ada C; cum
Adam V   qui *om.* V   24 refundere: confundere P   25 habente: habens
P   libertatem arbitrii: liberum arbitrium P   26 eam: eum P   recte: et ideo
peccat *add.* P   27 arbitrii *om.* P   in rectitudine *om.* P

3–4 Cf. Anselm. *De Lib. Arb.* iii (212, 19–20).
12–18 Cf. Anselm. ibid. iii (213, 5–25); iv (214, 2–8); xii (224, 13–22).
28–p. 111, l. 1 Cf. ibid. iii (211, 5–8).

assigns a property to air who says: 'to be breathable is a property of air', because the power to breath is not in the air but is a property of the animal who breathes.

396. Again, from what has been put forward it can be ascertained that the freedom of the will is a power to preserve the uprightness of the will. Moreover, there is an error in this because someone who has that power misuses it in not observing the uprightness of the will.

397. However, it is objected as follows: Non-being cannot be preserved; because in what way can someone preserve a non-existing thing? However, when a man exists in sin, there is no uprightness of the will in him. Therefore the uprightness of his will does not exist. Therefore he does not have free will.

398. Solution: Some powers are always accompanied by their realization, such as the power of the sun by which the sun is moved; some are not always with their realization—for example, the power to see is not always accompanied by the act of seeing, such as when a thing which exists in the darkness cannot be seen, and yet someone has the ability to see it. Similarly someone who is in sin is in darkness, and does not preserve the will in uprightness, and yet has the power to preserve it; and because he does not preserve it, nor is he forced not to preserve it, so he sins.

399. Again, perhaps it seems to someone that it is bad for free will to be joined to a rational creature, because if it did not have free will it would not sin.

400. Regarding this it should be stated that he who says that free will is the cause of sin is of a similar opinion to Adam, who said to the Lord that He himself had sinned inasmuch as the Lord had given Eve to him, wanting to fling the sin back to the Lord. It is the same with regard to a man who has freedom of the will, who sins in this way, because he does not preserve it correctly. Indeed, he does not sin because he has freedom but because he does not preserve the freedom of the will which God has joined to him in uprightness. Moreover, free will has been joined to a rational creature to choose what is advantage-

volendum quod expedit et quod debet. Si autem nullo modo posset
peccare, tunc non mereretur, si coactionis necessitate non peccaret; non
etiam demereretur; et ita in tali statu esse suum sumpsit a creatore
rationalis creatura, ut hanc libertatem haberet, scilicet quod posset non
peccare et nulla re cogeretur ad peccandum, sed voluntati sue relinque- 5
retur quoad boni executionem vel mali electionem.

XXVI. ii. *Utrum liberum arbitrium sit in Deo et in angelis*

Consequenter queritur, utrum liberum arbitrium sit in Deo et in
angelis.

401. Quod non sit sic videtur. In libro *Sententiarum* habetur quod 10
liberum arbitrium est facultas rationis et intellectus qua, gratia assistente,
eligitur bonum, et qua, gratia desistente, eligitur malum. In Deo autem
non est facultas eligendi malum; quoniam in eo non est potestas peccandi; ergo in ipso non est liberum arbitrium. Similiter in angelis qui
confirmati sunt non est facultas rationis et intellectus qua eligatur 15
malum, gratia deficiente; electio enim mali est ex defectu; ergo in ipsis
non est libertas arbitrii.

402. Quod sit in Deo liberum arbitrium primo loco ostenditur per
beatum Anselmum, qui dicit: 'Absurdum est dicere liberum arbitrium
non est in Deo nec in angelis; quoniam Deus liber est ad arbitrandum 20
quicquid vult'. Similiter et angeli libertatem habent rationis ad arbitrandum de bono et malo, ut eligant bonum et respuant malum:
quoniam si eligant malum, hoc non esset de libertate rationis sed de
defectu, et defectus rationis potius diminuit libertatem quam eam
augeat. A Boetio habemus et ab Augustino consimile, quod magis viget 25
libertas arbitrii in angelis quam in hominibus: quoniam angeli per-
fectiores et veriores sunt in rationis arbitrio quam homines: et ideo
maioris sunt libertatis in arbitrio, quoniam ipsi sunt modo confirmati

---

1 et quod debet *om.* P   2 mereretur: vel *add.* P   coactionis necessitate non
om. P   3 etiam *om.* P   suum: situm P   6 electionem: extentionem V
8 sit *om.* V   11–15 gratia . . . intellectus qua *om. per hom.* P   11 assistente:
existente V   12 et *om.* V   16 gratia deficiente: gratia differentie V
20 est¹: esse V   21 angeli: in angelis P   22 ut: unde P   bonum *om.* C
23 esset: erit P; est V   23–24 sed de defectu *om.* C; rationis *add.* V   24 et
defectus *om.* P   rationis *om.* V   25 habemus: habetur P

10 in libro *Sententiarum*: II. dist. xxiv. c. iii, sed mire discrepat a diffinitione
P. Lombardi: 'liberum vero arbitrium est facultas rationis et voluntatis'.
19–21 Cf. Anselm. *De Lib. Arb.* i (207, 11–13): 'Libertatem arbitrii non puto esse
potentiam peccandi et non peccandi. Quippe si haec esset definitio, nec deus nec
angeli qui peccare nequeunt liberum haberent arbitrium; quod nefas est dicere'.
25 Boetio: *De Consol. Phil.* v pr. 2. Augustino: cf. *De Civ. Dei* XXII. xxx. 3. cit. apud
P. Lombard. *Sent.* II. dist. xxv. cc. iii–iv.

ous and what he ought to. If, however, he could not sin in any way, then he would not obtain merit if because of a necessity of compulsion he did not sin, nor indeed would he lose merit. Thus in a state such as this the rational creature takes its being from the Creator, so that it has this freedom, namely that it can not sin and cannot be compelled to sin by anything, but it is left to his will to do good or choose evil.

XXVI. ii. *Whether there is free will in God and in the angels*

Next it is asked whether there is free will in God and in the angels.

401. That there is not seems as follows: In the book of the *Sentences* it is held that free will is a faculty of reason and understanding by which, with the help of grace, the good is chosen, and by which, when grace ceases, evil is chosen. For there is no faculty of choosing evil in God because there is no power to sin in him, therefore there is no free will in him. Similarly in the angels who have been confirmed in grace there is not a faculty of reason and understanding by which evil is chosen while grace is lacking, for the choice of evil is out of lack of grace, and therefore there is not freedom of the will in them.

402. That there is free will in God is shown in the first place by blessed Anselm, who says 'It is absurd to state that there is not free will in God and in the angels, because God is free to choose whatever he wishes.' Similarly the angels also have the freedom of reason to choose between good and evil, so that they choose good and reject evil. Thus, if they were to choose evil, this would not be out of the freedom of reason but from its lack, and a lack of reason more diminishes freedom than increases it. From Boethius we know, and from Augustine likewise, that the freedom of the will is more vigorous in angels than in men, because angels are more perfect and truer in the judgement of reason than men, and thus have a greater freedom in judgement, because they are now confirmed in freedom. Men, however, are, up

in libertate. Homines autem adhuc sunt potentes succumbere servituti, quoniam peccato; et ideo liberiores sunt angeli quam homines.

403. Solutio. Dicimus quod liberum arbitrium est in Deo et in angelis.

404. Ad hoc quod obicitur in contrarium, scilicet quod liberum arbitrium est facultas rationis et intellectus etc., dicendum est quod illa 5 descriptio non est data per se secundum essentiam liberi arbitrii, sed est data secundum accidens; et hoc potest haberi per hoc quod ibi dicitur, 'qua eligat malum, gratia deficiente'. Quod enim gratia deficiat, hoc est per accidens, quia ex imbecillitate et imperfectione rationis est, quod ipsa eligit malum; et cum voluntate eligendi malum est ibi absen- 10 tia gratie, et non est substractio gratie causa efficiens electionis mali, vel voluntatis eligendi malum: electio enim mali est a mala voluntate, et eius non est aliqua causa efficiens, sed potius causa deficiens, ut habetur ab Augustino in XII libro *de Civitate Dei*.

405. Si quis obiciat, hominem esse potentiorem Deo et angelo, eo 15 quod homo potens est peccare, et Deus non possit peccare. Dicendum est, ut patet ex predictis, quod potestas peccandi non est potentia, immo potius inpotentia; sicut homo mortuus non est homo, sed potius non-homo, eo modo potestas peccandi est imperfecta potestas, quoniam ex infirmitate est peccatum et non est ex fortitudine. 20

406. Item. Queritur, cuius sit per se et proprie illa libertas que denominat arbitrium, a qua dicitur arbitrium esse liberum.

407. Non est ipsius arbitrii; quoniam prius est in natura libertas quam sit arbitrium. Quoniam antequam aliquis arbitretur, in eo est libertas. Si quis dicat: arbitrium sumitur ibi pro potestate arbitrandi; 25 patens est hoc non esse verum, per hoc quod ipsa potestas arbitrandi est vis rationalis, et arbitrium ab illa vi procedit; et illa vis est in rationali; quoniam arbitrium non est in effectu.

408. Si libertas arbitrii sit hominis per se et proprie, patet hoc esse falsum per hoc quod, ut ostensum est, libertas arbitrii est in Deo et in 30 angelis.

3 Solutio. Dicimus: Ad hoc dicendum P; Dicendum est V   4 quod¹ *om.* V
6 essentiam: viam P   7 est data *om.* CV   7–8 dicitur: dicit P   8 deficiente: difficiente P   10 eligit: eligat P   12 mala *om.* V   13 aliqua *om.* CV   sed *om.* V   potius: immo V   causa² *om.* V   13–14 ut . . . Dei *om.* P
16 potens *om.* P   Deus: ipsi P   17 peccandi *om.* V   18 immo: sed P   sed: immo V   22 a qua dicitur arbitrium *om.* P   23 est¹ *om.* V   prius: prior V   24 arbitretur: arbitratur V   24–25 antequam . . . arbitrium *om.* P
25 dicat: quod *add.* V   26 est¹ *om.* P   27 rationalis: rationabilis P   in rationali: effectu rationalis P   29 arbitrii *om.* C   30 quod *om.* C

14 Aug. *De Civ. Dei* XII. vii.
17 ut patet ex predictis: cf. supra, n. 392.
21–p. 114, l. 23 Impress. apud Lottin, *Psychologie et morale*, III. ii, pp. 615–17.
30 ut ostensum est: cf. supra, nn. 402–3.

to now, able to succumb to servitude because of sin. Thus angels are more free than men.

403. Solution: We state that free will is in God and in angels.

404. Regarding that which was objected against, namely that free will is a faculty of reason and understanding, etc., it should be stated that that description is not given as such according to the essence of free will, but is given in an accidental manner. Again, this can be known through what is said there: 'by which he chooses evil, lacking grace'. For grace to be lacking is to be by accident, because it is due to the weakness and imperfection of reason that it chooses evil. Moreover, with the will to choose evil there is an absence of grace there, and the taking away of grace is not the efficient cause of the choice of evil, nor of the will to choose evil. Indeed, the choice of evil is from an evil will, and there is not any efficient cause of it but rather a deficient cause, as we know from Augustine in the twelfth book of *The City of God*.

405. If someone were to object that a man is more powerful than God and an angel because a man is able to sin, and God cannot sin, it should be said, as is clear from what has already been stated, that the ability to sin is not a power; rather, it is more a lack of power. Just as a dead man is not a man but rather a non-man, in the same way the ability to sin is an incomplete power, because sin is due to weakness and is not due to strength.

406. Again, it is asked, whose is that freedom in itself and properly which indicates the will, and from which the will is said to be free?

407. It is not of the will itself, because freedom is prior in nature to the will, because before someone decides, there is freedom in him. If someone were to say: 'the will is taken there as the power to choose', it is clear that this is not true, inasmuch as this power to choose is a rational power, and the will proceeds from that power; again, that power is in the rational power because the will is not in the effect.

408. If the freedom of the will belongs to a man properly and in itself, it is clear that this is false inasmuch as, as has been shown, freedom of will is in God and in the angels.

409. Si sit ipsius rationis; Contra. Ratio est forma; libertas est forma; ergo libertas per se et proprie non est in ratione. Si dicatur ei inesse, hoc erit per accidens, et non secundum eius propriam intentionem.

410. Ad hoc dicendum est quod rationalis essentie per se et proprie est libertas arbitrii, et illud subiectum non excedit libertatem arbitrii 5 nec ab ea exceditur, quoniam libertas arbitrii est in omni, et dicitur in quo est ratio, et convertitur. Sed tamen hec libertas quandoque impedita est a suo effectu, ita quod non potest exequi suum actum, ut in demone. Est enim in eo libertas arbitrii, sed non effectus eius, quoniam in ipso est potestas conservandi rectitudinem voluntatis, tamen rectitudinem 10 illam non conservat propter defectum illuminationis procedentis a prima causa. Sicut aliquis existens in tenebris potestatem habet videndi, non tamen videt propter lucis defectum.

411. Per premissa igitur haberi potest quod in brutis animalibus non est liberum arbitrium, cum ipsa rationis examinatione careant. Est 15 tamen in eis vis concupiscibilis et vis irascibilis, cum in eis sit anima sensibilis. Est autem in eis vis concupiscibilis qua appetunt eis expedientia, vel que apparent esse expedientia. Vis irascibilis in eis est qua respuunt que eis vere vel apparenter sunt nociva; et propter huiusmodi appetitum, qui est cum electione, et propter respuitionem nocivi, que 20 est cum fuga, videtur in eis esse discretio inter expediens et nocivum; et ita in eis videtur esse liberum arbitrium, cum eis relinquatur unum vel aliud eligere.

412. Item. Experimento videtur idem haberi posse per hoc quod ovis eligit suum agnum inter alios agnos, et per hoc quod videtur discernere 25 eum ab aliis. Consimile patet in canibus et apibus. Apes enim eligunt flores sibi competentiores et sibi nocivos respuunt.

413. Solutio harum obiectionum patet per hoc quod superius dictum est in tractatu de viribus anime sensibilis. A vi enim estimativa est huiusmodi electio in brutis animalibus, et non a ratione. Tamen non est 30 dicendum quod in eis est discretio, immo distinctio, cum discretio fit a ratione. Bruta enim distinguunt rem a re per accidentia rerum, et

---

2 per: in C   Si: Et si P   dicatur: dicam P   inesse: inest P   4 est *om.* P
5-6 et illud . . . omni *om.* C   7 Sed: Est P   8 est *om.* P   effectu: affectu
P   ita . . . actum *om.* P   actum: effectum V   9 ipso: ipsa P   11-12 procedentis . . . causa *om.* P   14 igitur *om.* CV   16-17 cum . . . concupiscibilis *om.* V   17 appetunt: expetunt V   19 respuunt: ea *add.* V   que eis
om. C   vel: que *add.* V   20 appetitum: appetitionem P; appetunt V   respuitionem; reparationem P   25 quod *om.* P   26 eum: unum V   27 competentiores: competentes P   29 est² *om.* P   30-32 Tamen . . . a ratione *om.*
per hom. CV

---

28 superius: cf. supra, nn. 261-2.

409. If it is of the reason itself, then it can be argued against this that reason is a form, freedom is a form, therefore freedom properly and in itself is not in reason. If it is said to be in reason, this will be in an accidental manner, and not according to the proper understanding of it.

410. Regarding this it should be stated that the freedom of the will is properly and in itself of a rational essence, and that subject does not go beyond the freedom of the will nor is it exceeded by it, because the freedom of the will is in everything, and is said to be in that in which reason is and vice versa. And yet this freedom is sometimes impeded by its effect so that it cannot carry out its act, such as in a demon. For there is freedom of the will in him but not its effect, because in him there is the power to preserve the uprightness of the will, but he does not preserve that uprightness because of the lack of the illumination proceeding from the First Cause. Just as someone who is in the dark has the ability to see but does not see because of the lack of light.

411. Therefore, from what has been put forward, it can be held that there is no free will in brute animals since they lack that investigation of reason. However, there is in them the concupiscible power and the irascible power since the sensitive soul is in them. For there is in them the concupiscible power by which they desire what is useful to them, or what appears to be useful to them. The irascible power in them is that by which they reject those things which apparently or truly are harmful to them. Moreover, because of desire of this kind which is accompanied by choice, and because of the rejection of the harmful which is accompanied by flight, it appears that there is in them a discrimination between the useful and the harmful. Thus it appears that there is free will in them, since it is left to them to choose one or the other.

412. Again, it seems that we can know the same through experience in that a sheep picks out its lamb from among other lambs, and inasmuch as it appears to discern it from others. The same is clear in dogs and bees. For bees choose the flowers which are more suitable to them and reject the ones which are harmful.

413. The solution to these objections is clear through what has been said above and in the discussion on the powers of the sensitive soul. For choice of this kind in brute animals is from the estimative power and not from reason. It should not, however, be stated that there is discernment in them, but rather distinguishing, since discernment arises from reason. For brute animals distinguish one thing from another through the accidents of things and they do not discern by the

non discernunt. Aliquis tamen loquens improprie diceret aliquod brutum animal discernere unum ab alio, ut canem et simiam.

414. Item. Videbitur alicui quod libertas arbitrii est iste tres vires coniuncte, scilicet vis concupiscibilis, vis irascibilis et vis rationalis; vel quod liberum arbitrium sit in anima secundum illas tres, ita quod non 5 tantum secundum aliquam illarum; ob hoc quod a vi concupiscibili est appetitus boni, vel apparentis boni; a vi irascibili mali detestatio; a vi rationali discretio et iudicium.

415. Ad hoc dicendum est quod ad esse liberi arbitrii non exigitur vis concupiscibilis, vel vis irascibilis, sed solummodo sufficit vis rationalis. 10 Quoniam rationalis creatura intellectu apprehendit, et ratione inter intellecta discernit et arbitratur. In homine enim est inchoatio tendens ad discretionem a vi concupiscibili, secundum quam aliquis homo appetit aliquid; vel a vi irascibili, secundum quam homo aliquid respuit. Sed, licet aliquis homo aliquid appetat, non tamen tenetur illud eligere. 15 Similiter, licet aliquis aliquid respuat per vim irascibilem, non tenetur illud fugere, nisi prius convertat rationem suam ad id quod ipse appetit, et ad id quod ipse respuit, et secundum arbitrium rationis eligat rem appetitam vel fugiat. Quoniam ratio, ut dicit Aristoteles in libro de Proprio, imperat vi concupiscibili et vi irascibili, nisi cum perversa 20 fuerit anima hominis. Non enim secundum vim concupiscibilem vel irascibilem tenetur homo aliquid exequi, nisi rationis arbitrium subvenerit ad regimentum utriusque illarum virtutum.

1 diceret: dicit PV   aliquod om. V   2 simiam: similia P   3 Videbitur: Videtur P   iste: ille P   4 et om. CV   5 liberum arbitrium om. P   in anima om. P   6 aliquam illarum: unum illatum P   7 vel apparentis boni om. C   8 discretio: dilectio P   et: vel V   9 est om. CP   10 concupiscibilis: rationalis P   vel: sed P   solummodo: solum P   11 rationalis: ratiocinatur V   12 intellecta: intellectum P   enim: autem CV   tendens om. P   15 tenetur om. V   16 respuat; respuit C; respiciat P   per vim irascibilem: a vi irascibili P   17 nisi: ubi V   18 respuit: respicit P   18–19 eligat rem appetitam: eligit appetitum P   19 fugiat: fugat C   Aristoteles: Af. P   19–20 in libro de Proprio om. P   20 cum om. V   23 Explicit iste liber: sit scriptor crimine liber add. V

3 alicui: cf. Stephanus Langton Quaestiones, apud Lottin, Psychologie et morale, 1, p. 60, et Alex. Nequam Speculum Speculationum, ibid. III. ii, pp. 608–9.
19–20 in libro de Proprio: i.e. Arist. Topica, lib. v. Respicit ad v. i (129ª10–15): 'Rationabilis proprium ad concupiscibile et irascibile quod hoc quidem imperat, illa autem ministrant. Nam neque rationabile semper imperat sed quandoque et imperatur, neque concupiscibile et irascibile semper imperantur sed et imperant quandoque, cum fuerit perversa anima hominis' (cod. Coll. Trin. Oxon. 47, fol. 32).

intellect. Yet someone speaking incorrectly will say that some brute animal will discern by the intellect one thing from another, such as a dog and a monkey.

414. Again, it might seem to someone that the freedom of the will is in these three powers joined together, namely the concupiscible power, the irascible power, and the rational power. Or it might seem that free will is in the soul according to those three powers, so that it is not there according to one of them only. Thus it is the case that from the concupiscible power is the desire for the good, or what appears to be good, from the irascible power is the hatred of evil, and from the rational power is discernment and judgement.

415. Regarding this, it should be stated that for free will to exist, the concupiscible power is not required nor is the irascible power, rather the rational power on its own suffices. Since a rational creature perceives by means of understanding and by reason, it discerns and judges between those things that have been understood. For in a human being there is the beginning of a tendency towards discernment from the concupiscible power, inasmuch as someone desires something, or from the irascible power, inasmuch as a man rejects something. Yet although someone desires something, he is not, however, bound to choose. Similarly, although someone rejects something by means of the irascible power, he is not bound to flee unless he first turns his reason towards that which he desires and towards that which he rejects, and according to the judgement of reason he chooses the thing desired or he rejects it. Because reason, as Aristotle says in the *Topics*, rules by the concupiscible power and the irascible power, except when the soul of a man is perverse. So it is that a man is not bound to carry out anything according to the concupiscible or the irascible power, unless the judgement of reason intervenes to direct each of these powers.[15]

[15] The text breaks off at this point without the promised ch. xxvii 'On divine providence'.

APPENDIX

Diagrams and Explanations

Note. The diagrams contained in this appendix have been copied from those given in John Blund, *Traktat über die Seele*, ed. and trans. Dorothée Werner, Freiburg i.Br., 2005, Anhang, pp. 331–42; they are reproduced here by kind permission of Verlag Herder. The text of the explanations has been freely adapted from the German translation of Werner's appendix made for me by Conleth Loonan.

As Werner comments (p. 331), the drawings and explanations given serve to illustrate Blund's geometric and optical arguments, even if he is not always correct.

Nos. 91, 93, 102–107

In no. 91 Blund attempts to solve the supposed paradox that one can generally see something that is *larger* than the diameter that passes through the midpoint of one's own eye. He first of all cites an opinion which states that this should not be possible. It follows from this opinion that no object which appears in front of us can appear to be *smaller* than the diameter of one's own eye.[1]

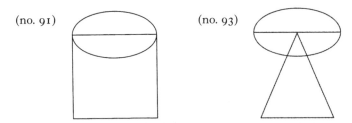

(no. 91)   (no. 93)

Blund's argument is that our field of vision has the shape of a parallelogram which is formed between the diameter through the midpoint of the eye and the diameter of the object lying opposite. Thus, nothing can be seen which is either *smaller* or *larger* than the diameter of the eye because every object which is seen must be located between the two extended sides of the parallelogram, which must always have the same distance from each other. It also follows that nothing can be seen which is *larger* than the diameter of one's own eye.

In no. 93 Blund offers his solution to the paradox by arguing that our field of vision does not take the form of a parallelogram but is rather in the shape

---

[1] Werner comments (p. 331) that it is unclear why Blund asks if something can be seen that is larger than the diameter that intersects the eye, and then without any further explanation states that no object which is in front of us can appear smaller than the diameter of one's own eye. A possible explanation might be that Blund wishes to show that his solution applies to both cases (larger *and* smaller) in the sections which follow.

of a triangle (owing to the convergence of two lines), whose base is formed by the diameter of the observed object. Thus an explanation is also offered as to why things can be seen which are *larger* than one's own eye.

In no. 102 Blund next attempts to solve the following contradiction: something (in the example given, an angle) which was *larger* will later be seen as *smaller*. How is this apparent contradiction arrived at? That which is seen within a greater angle *appears* larger (one of Euclid's theorems) and that which is seen within a smaller angle appears *smaller*. In addition, it will be accepted as a further theorem (it is not always clear whether Blund means a geometric or a non-geometric theorem) that the further the observed object is from the eye, the *smaller* it appears.

On the strength of this another supposition will also be accepted, which contradicts the first set: the *smaller* the circumference, the *greater* the angle of incidence (*angulus contingentiae*).

The argument may be stated as follows:

Assumption 1

(1) That which is seen within a *greater* angle *appears* larger.
(2) Every object appears to be *smaller* or *larger*, according to distance of the viewer from the object.

Assumption 2

(1) The *smaller* the circumference, the *greater* the angle of incidence.
(2) Every object appears to be *smaller* or *larger*, according to distance of the viewer from the object (=Assumption 1(2)).

Assumption 3

However, if we take Assumptions 2(1) and 1(2) together, it follows from 2(1) that when something is moved away from its original position, the angle of incidence becomes *greater*, yet this contradicts 1(2), which states that *everything* must become *smaller* when the viewer moves further away.

In no. 103 Blund solves this contradiction as follows: something that is further away will *not* be seen as *larger*. Thus as the angle in which the circle is seen reduces because of its increasing distance from the eye, so too will the angle in which the angle of incidence is seen, become *smaller*. Thus, it applies to the angle in which one sees the angle of incidence.

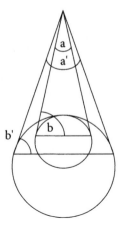

The angles of incidence are b and b′, and a and a′ are the angles within which the angles of incidence are seen.

In this way Blund can solve the contradiction that something seems larger the further away it is seen. However, what exactly the angle should be within which one sees the angle of incidence remains problematic. Blund's argument simply consists in the fact that one cannot speak of one and the same angle when a circle is in different positions.

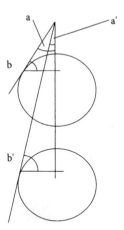

As the diagram makes clear, at an almost constant angle of the angle of incidence within which it is observed, in the case of an object which alters its position, at any given time it is larger or smaller.

In no. 104 the presupposition of no. 102, that something seems *larger* the closer one is to it, is questioned. In no. 105 it is objected that when one draws a semicircle and positions an eye (a) at the midpoint of the circumference so that it sees the entire diameter, and another eye (b) which is nearer to the end of the circumference and sees exactly the same diameter, then to eye (b), which is closer to the diameter, the diameter must seem larger since the theorem has already been demonstrated that the closer one comes to something the larger it appears. Hence it would also follow that the angle of eye (b) is greater than that of the other eye (a), following what was said above (no. 102), that when something is seen as larger, the greater the angle. Hence it follows that (b) must be larger.

In no. 106 Blund refers to a theorem of Euclid, according to which all angles of triangles drawn from a point in the circumference of a semicircle are right angles and are therefore equal in size. Thus, the diameter cannot seem larger to one eye rather than to another. In his *solutio* in no. 107 Blund agrees with this theorem of Euclid, and he refers to the fact that one eye is not closer to the *entire* diameter than the other. Blund has also succeeded in showing that nos. 105 and 106 contradict each other, and also that the conclusion (in no. 105) was false.

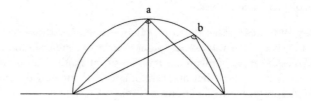

Nos 137–144

No. 137: types of shadow

$\bigcirc$ = bright body

● = illuminated body

(a)–(c)
(a) infinitely large, cylindrical shadow

(b) finite, conical shadow

(c) finite, basket-shaped shadow

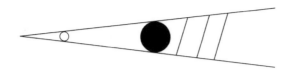

(d) Comparison between shadows of type (b): shadows where the first are larger than in the second

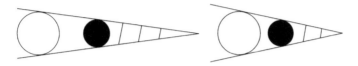

(e) Comparison between shadows of type (c): shadows with the first are larger than in the second

(f) Comparison between shadows of type (a): shadows remain equal independently of the distance between both bodies

Nos. 138–140: the three ways in which the arguments run

1. (No. 138)

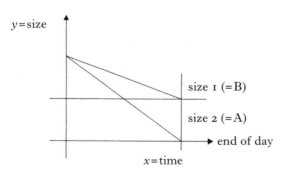

A and B contract continually in the same shape. A becomes smaller, however, more sharply in comparison with B, and that is why, at the end of the day, A is smaller than B.

2. (No. 139) The same as no. 138 only inverted.

3. (No. 140)

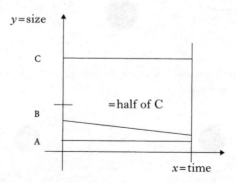

A and B are smaller than C. Thus as A and B contract at the end of the day they are still smaller than C.

No. 141: against no. 138

(*a*)

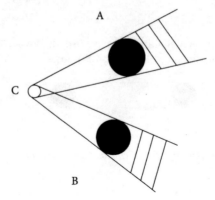

Initial position: Both shadows—A and B—have a basket-shaped, same-sized, infinite shadow, so A and B are equal in size.

(*b*)

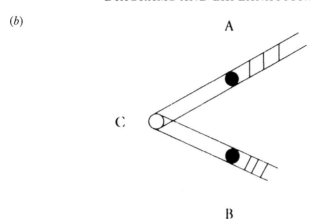

A and B contract, becoming exactly the same size as C, and thereby acquire a cylindrical shadow. [Thus there is an alteration from shadow shape (*c*) to shadow-shape (*a*), including the additional presuppositions from no. 137(*f*).]

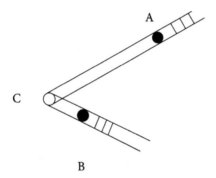

Although A moves further away from C, the shadow shape does not itself change from A, more or less like (*d*) and (*e*) in no. 137, and so remains cylindrical.

No. 142: against no. 139

(*a*)

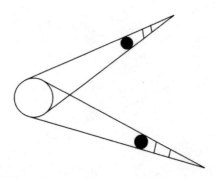

At their starting-point A and B are equal in size but smaller than C. A and B are at the same distance from C and have a finite, cylindrical shadow.

(*b*)

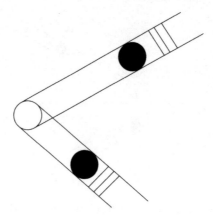

At their finishing-point A and B have become exactly the same size as C, but A is further away from C than B. And again, although A has become further away from C, the shadows of the two are nevertheless of the same size. Both have cylindrical shadows of infinite size.

No. 143: objection to no. 140

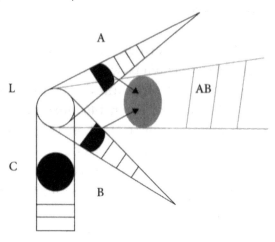

FIRST IMAGE (*black shading*)

A and B are here two dark hemispheres; both at any given time are smaller than the bright body (L), but each in itself is larger than half of L. A and B have conical, finite shadows. C is a dark sphere, which is exactly the same size as L and has a cylindrical, infinite shadow.

SECOND IMAGE (*grey shading*)

A and B in the course of the day move towards one another, and contract in themselves through the entire day so as ultimately to form a shape AB which is still larger than L. AB has at the end of the day a basket-shaped, infinite shadow.

No. 144

In no. 144 Blund now points out an inconsistency which he spots as arising out of Figures 1 and 2 in no. 143: A and B both have a conical and finite shadow which is smaller than the shadow of C. At the end of the day, as they both contract the cone AB results, and their combined shadow is larger than that of body C. Blund's counter-example allows him to attack the other arguments as containing a fallacy inasmuch as the final 'growth' of the shadow (as represented in no. 143) is not a successive and same-shaped increase: rather, there is an instantaneous change.[2]

---

[2] Werner (p. 342) points out that Blund's presentation of the problem contains some difficulties which leave the chapter as a whole unclear. For example, the group in no. 140 cannot to be compared with that in no. 143 even if Blund formulates no. 143 as an exception to no. 140. In no. 140 A and B together are never greater than C, while in no. 143 at the end of the day A and B (hence AB) are greater than C. Moreover, Blund's classifications are not always very exact. In no. 140, when he is talking about 'size' or quantity, it is not clear whether this size refers to the further movement of the body itself, to the shadow of the body, or to the type of shadow. Again, that there are different magnitudes of infinity according to Blund's explanations (for example, in no. 143 that of C and that of AB would be larger than that of C) presents no problem for him in the text.

INDEX AUCTORUM

Editiones adhibitae tantum memoratae sunt ubi ex re visum est. Locos asterisco notatos in praefatione (pp. xviii–xx) lector inveniet. Numeri sectiones textus respiciunt

SIGLA

Beiträge   *Beiträge zur Gesch. der Philosophie und Theologie des Mittelalters*
C.S.E.L.   *Corpus scriptorum ecclesiasticorum Latinorum*
P.G.   Migne, *Patrologia Graeca*
P.L.   Migne, *Patrologia Latina*

ADELARDUS BATHONIENSIS
De Eodem et Diverso
(ed. H. Willner, *Beiträge*, IV. 1, 1903)
32: 9

Quaestiones Naturales
(ed. M. Müller, *Beiträge*, XXXI. 2, 1934)
xii (15): 111
xxi (25): 146
lxxi–lxxiv (63–65): 10
lxxii (64): 9
lxxiv–lxxvi (65–69): 13
lxxvi (68): 3

ALANUS DE INSULIS
Contra Haereticos
(*P.L.* 210, cols. 305–430)
I. xxvii: 297, 299

ALEXANDER APHRODISIENSIS
De Intellectu et Intellecto
(ed. G. Théry, *Autour du décret de 1210.*
II. *Alex. d'Aphrodise*, Bibl. Thomiste
VII, 1926)
74 seq.: 59, 337
77: 339

ALEXANDER NEQUAM
(ed. T. Wright, Rolls Series, 1863)
De Laudibus Divinae Sapientiae
D. ix (490): 111
De Naturis Rerum
I. ix (45): 9
I. xx (66): 169, c. XIII
II. clii (233): 34
II. cliii (234–6): ix, 91, 102–3, 108, 110, 111
Speculum Speculationum
viii, 414

ALFARABI
De Intellectu et Intellecto
(ed. E. Gilson, *Archives d'hist. doctr. et litt. du moyen âge*, 4 (1929), 115–26)
117: 337
Opus incertum: 96

ALFREDUS ANGLICUS
De Motu Cordis
(ed. C. Baeumker, *Beiträge*, XXIII. 1–2, 1923)
Prol. (1–4): 16
VII (28–29): 26
X (37–45): 307
XI (53): 27

ALGAZEL
Metaphysica
(ed. J. T. Muckle, Toronto, 1933)
I. i. 2 (13–14): 355
I. iii. 2 (64–65): 340
I. iv. 1 (92): 4
I. iv. 3 (104–6): 5, 9, 13
(112–15): 13
I. v. (121, 125): 344
II. iv. 1 (162–3): 46
II. iv. 3 (165): 217
(167): 102, 344
(168): 101
II. iv. 5 (172–3): 274, 335
(175): 337, 340, 344
(181): 322, 344
II. v. 3 (185): 304

ALKINDI
De Intellectu
(ed. A. Nagy, *Beiträge*, II. 5, 1897)
9: 337

ANONYMI
Liber de causis primis et secundis
(ed. R. de Vaux, *Notes et textes sur l'Avicennisme latin*, Bibl. Thomiste XX, 1934)
X (128, 135): 339
(130): 372

Quaestiones Phisicales
(ed. B. Lawn, *The Salernitan Questions* (Oxford, 1963), pp. 160–77)
37 (p. 172): 111

Quaestiones Salernitanae: 111–17*

INDEX NOMINUM
ET VERBORUM POTIORUM

Numeri sectiones textus respiciunt

augmentum 46–49, 51–54, 144, 356.
Augustinus 34, 290, 360, 366–7, 402; in libro de Civitate Dei 404; in libro de Libero Arbitrio 237; in libro de Trinitate 346.
Avicenna 13, 18, 34, 37, 73, 85, 101, 208, 240, 254, 337; in Commento 246, 336; in Commento de Anima 89, 145, 237, 252; in Commento super librum de Anima 64, 193; in Commento Metaphysice 32; in Metaphysica 316. v. Commentator.

basis 93, 97, 106.
Boethius 402; de Consolatione 7, in libro Consolationum 357; in secunda ed. super Porphirium 87.
bonitas: pura 13.
bonum: summum 13, 286, 363.

cado in cum accus. 30, 85, 87, 193, 233, 253–5, 257, 266.
calatoydes (-dos) 137, 144.
carnalitas 52.
caruncula 193–4, 196, 203.
causa 25, 33, 90, 95, 99–100, 110, 125, 165; efficiens 404; extrinseca 371; prima 13, 282–4, 286, 329, 333, 354, 357, 361, 375, 410; nihil sine -a quod est a creatore rebus est insitum 251; contrariarum -arum contrarii sunt effectus 287.
causatum 329, 354; -a sunt suarum causarum vestigia 273; in -is causantibus alia ex se 293.
causo 161, 218, 271, 293, 331.
celebro: -ata digestione 231.
cellula: cerebri 305; memorialis 303; ymaginativa 244.
cerebrum 193, 196, 221, 237–8, 240, 250, 254, 262, 305, 356, 358, 379.
certificatio 368.
certifico 368.
chilindreidos 137, 141, 144.
circumferentia 105, 308.
claritates oculorum 111.
claustralis 373.
coaccidens 235.
coequevus 39.
cogitativus v. vis.
cognitio 266–8, 286, 323, 328, 367–8, 375; sensitiva 2.
collatio 76; per similitudinis -nem 394.
coloratus 89, 96, 120.
coloro 123.
combinatio 195.
commensuratio 165–6.
commensuro 101, 165.
Commentator (i.e. Avicenna) 123, 254.
comparatio 21, 82, 123, 162, 166, 335–6, 340, 347, 385.

complexio 46, 54, 194, 221; -io temperata 212, 215, 218.
complexionatus 54, 226–7.
componens subst. 325, 332, 334.
compositio 252–3, 258–9, 261, 334.
compositum 226–7, 230, 259, 375.
comprehensio 30.
concavitas 115, 146, 177–8, 250, 254, 262.
concavus 89, 177.
concomitans 39, 43.
concretio 16, 21, 71, 201.
concretive 51, 60, 71, 201.
concupiscibilis v. vis.
confricatio 201.
conoidos (-des) 137, 144.
conservatio 46, 221, 225, 383.
conservativus 381.
consideratio 17–20.
consistentia 166, 170.
constituens 326.
constrictio 145–6, 161, 171, 173, 190, 225, 376.
contemplativus: v. virtus, vis.
contiguatio 314; in -ne et in continuatione 313.
contiguo (continguo) 223, 302, 311, 314.
contingentia 102–3.
contrarietas 33, 293, 334, 351, 371.
contrarior vb. 287.
contristatio 202.
cor 196, 376–8, 380.
corda (i.e. chorda) 55, 171–3, 175.
corporalitas 46.
corpus 2, 28–29, 32; animatum 9, 35–36, 38, 297, 301, 311, 359; animatum et sensibile 311; coloratum 89; concavum 177; elementare 325; inanimatum 297; lucidum 58–59; luminosum 137; mobile 9, 18; odoriferum 199–201; organicum 14–16, 19, 21, 41, 170, 361, 363–4; porosum 115–16; purum 313–14; sensatum 312; sphericum 116, 141–3; tenebrosum 137; tersum 89; vegetatum 39, 302, 310–12, 377.
correlativum 395.
corruptivus 50.
creatio 17, 324, 327, 334.
creator 251, 286, 363–4, 400.
cristallinus 89, 92, 100, 241.

dator formarum 344, 361.
decontinuatio 356.
decrementum 141, 144.
deforis 55, 83.
deintus 55, 61.
delectamentum 55.
deperditio 54.
deperditus 46, 352.
descriptio (discriptio) 209, 214, 404.

detectio 122, 149.
determinatio 260.
Deus 282, 284, 326–8, 364–5, 379, 400–3, 405, 408.
dextrorsum 112.
diameter 91, 105–7, 143.
digestio: -ne facta 54; celebrata 231.
dilatatio 376, 380.
diminutio 49, 51, 103, 111.
diminutivus 49.
disciplina 2, 67, 367, 372.
dispositio 1, 4, 10, 161, 166, 335–6, 338, 342, 344, 371, 380.
distendo 217, 221, 235, 306, 309–10, 312.
distensio 175, 309.
distentio (distensio P) 300.
diversifico 340.
divisio 174, 258–9, 261, 313, 332.
divisivus 39, 334.
doctrina 2, 367.
dolor 81, 376, 380.
draco 96.
duratio 293.
dux 385.

eccho c. XIII.
effectus 51, 110, 286, 410; immediati 291–2; remoti 291; in -u 25, 37, 59, 118, 210, 213, 273, 329, 339–40, 362, 407; ymagines constitute in -um (-u C an recte?) 281; exit (-iens) in -um 51, 59; fluit in -um 80, 95; venientes in -um 89; per -um 125; unus et idem -us non est a diversis 75; contrariarum causarum contrarii sunt -us 287, cf. 322.
elementaris 226, 325–6.
elementatus 227–8.
elementum 27, 195–6, 325; -a prima 212, 226, 328; pura 229; sensibilia 228.
elongatio 103, 107.
elongo 102.
ens 85, 87.
equedistans 91, 109, 137, 142.
equedistantia 91; equa distantia 93.
esse subst. 8, 15, 17–18, 38, 52, 118, 129, 161, 170, 172, 206, 211–12, 215, 218, 271, 293, 329, 380, 389, 415; concretum 21; duplex 324; generale 41; naturale 17; possibile 13; simplex 333; speciale 41; verum 170, 390; e. est a perfectione rei 323.
essendi 38.
essentia 22, 39, 329–30, 348, 353, 404; divina 328, 354; prima 322; propria 31; pura 322; rationalis 410; simplex 333; de -a 57, 339; ex -a 123; in -a 62, 82, 348, 350, 385; secundum -am 301, 304, 308–9, 404.
essentialiter 375, 378.

estimabilis 335.
estimatio c. XIX, 262, 265, 379.
estimativus v. vis.
ethicus 336.
Euclides: in Geometria in tertio E. 106.
Eva 400.
existentia 381.
experimentum 335; per sensus -um 137; -o habemus (-etur) 155, 238; -o videtur 412; quod -o patet (percipimus) esse falsum 169, 278, cf. 101; -o potest haberi hoc esse falsum 179.
extremum 121–2, 202.

facio ad 172, 372, 389.
facultas 18, 401, 404; geometrica 21.
fallacia accidentis 87, 124, 395.
firmamentum 6, 8, 10, 12, 300, 303.
flexibilis 215.
fluo 13, 49, 51, 95, 109, 212, 222, 237–8, 244, 246, 293, 322; in effectum 80, 95, 218.
fluxilis 109–10.
forma 15, 21, 89, 92, 94, 117, 138, 152, 169, 206, 234, 245, 339, 409; cogitata 294; nova 48; propria 52, 54; sensata 204; sensibilis 196, 299; separata 234; spiritualis 329, 334; substantialis 47; universalis 295, 337; visibilis 115; -e intelligibiles 244; -a/materia 32, 96, 232, 284, 316, 329, 338; -a dat esse 15; nulla -a est res per se existens separata a substantia 15. v. dator.
formalis v. intellectus.
fortitudo 72, 79–80, 405.
fundus aque 108.

gaudium 81, 286.
generativus v. vis.
genero 95–96, 99–100, 110, 122–5, 155, 185, 245, 293, 335–6.
genus 121.
geometricus 21; -ice 137.
gleba 8.
gratia 401, 404.
grossus 293.

habeo cum vb. infinitivo 4, 27, 157, 210, 245, 265, 308, 344.
habilitas 110.
habilito 364.
hic et ibi 301.
hinc et inde 58, 125, 160.
homogenius v. omogenius.
humor 24, 89, 92, 100, 166, 241.
hypoteticus 270.
idemptitas 125.
idiota 370, 372.
Ieronimus 367.

SUBJECT INDEX

JOHN BLUND

TREATISE ON THE SOUL